KW-266-312

Environmental Challenges in the Mediterranean 2000-2050

edited by

Antonio Marquina

Department of International Studies,
Universidad Complutense, Madrid, Spain

Kluwer Academic Publishers

Dordrecht / Boston / London

Published in cooperation with NATO Scientific Affairs Division

Proceedings of the NATO Advanced Research Workshop on
Environmental Challenges in the Mediterranean 2000-2050
Madrid, Spain
2–5 October 2002

A C.I.P. Catalogue record for this book is available from the Library of Congress.

ISBN 1-4020-1948-3 (HB)
ISBN 1-4020-1949-1 (PB)

Published by Kluwer Academic Publishers,
P.O. Box 17, 3300 AA Dordrecht, The Netherlands.

Sold and distributed in North, Central and South America
by Kluwer Academic Publishers,
101 Philip Drive, Norwell, MA 02061, U.S.A.

In all other countries, sold and distributed
by Kluwer Academic Publishers,
P.O. Box 322, 3300 AH Dordrecht, The Netherlands.

Printed on acid-free paper

All Rights Reserved
© 2004 Kluwer Academic Publishers
No part of this work may be reproduced, stored in a retrieval system, or transmitted in any
form or by any means, electronic, mechanical, photocopying, microfilming, recording or
otherwise, without written permission from the Publisher, with the exception of any
material supplied specifically for the purpose of being entered and executed on a computer system, for exclusive use by the purchaser of the work.

Printed in the Netherlands.

Environmental Challenges in the Mediterranean 2000-2050

WITHDRAWN
FROM
UNIVERSITY OF PLYMOUTH
LIBRARY SERVICES

90 0588159 5

This book is to be returned on
or before the date stamped below

18 JUN 2004

UNIVERSITY OF PLYMOUTH

PLYMOUTH LIBRARY

Tel: (01752) 232323
This book is subject to recall if required by another reader
Books may be renewed by phone
CHARGES WILL BE MADE FOR OVERDUE BOOKS

NATO Science Series

A Series presenting the results of scientific meetings supported under the NATO Science Programme.

The Series is published by IOS Press, Amsterdam, and Kluwer Academic Publishers in conjunction with the NATO Scientific Affairs Division

Sub-Series

I. Life and Behavioural Sciences	IOS Press
II. Mathematics, Physics and Chemistry	Kluwer Academic Publishers
III. Computer and Systems Science	IOS Press
IV. Earth and Environmental Sciences	Kluwer Academic Publishers
V. Science and Technology Policy	IOS Press

The NATO Science Series continues the series of books published formerly as the NATO ASI Series.

The NATO Science Programme offers support for collaboration in civil science between scientists of countries of the Euro-Atlantic Partnership Council. The types of scientific meeting generally supported are "Advanced Study Institutes" and "Advanced Research Workshops", although other types of meeting are supported from time to time. The NATO Science Series collects together the results of these meetings. The meetings are co-organized bij scientists from NATO countries and scientists from NATO's Partner countries – countries of the CIS and Central and Eastern Europe.

Advanced Study Institutes are high-level tutorial courses offering in-depth study of latest advances in a field.
Advanced Research Workshops are expert meetings aimed at critical assessment of a field, and identification of directions for future action.

As a consequence of the restructuring of the NATO Science Programme in 1999, the NATO Science Series has been re-organised and there are currently five sub-series as noted above. Please consult the following web sites for information on previous volumes published in the Series, as well as details of earlier sub-series.

http://www.nato.int/science
http://www.wkap.nl
http://www.iospress.nl
http://www.wtv-books.de/nato-pco.htm

Series IV: Earth and Environmental Sciences – Vol. 37

UNIVERSITY OF PLYMOUTH

Item No. 9005881595

Date 1 2 MAY 2004

Class No. 333.7091822 ENV
Com. No.

PLYMOUTH LIBRARY

Table of Contents

Foreword

This book collects most of the contributions prepared for the NATO Advanced Research Workshop (ARW) on Environmental Challenges in the Mediterranean 2000-2050. The ARW took place in Madrid from 2 to 5 of October 2002.

Scholars and researchers from North and South of the Mediterranean and other countries from NATO gathered in Madrid for discussing several important factors that could affect Mediterranean security in the next 50 years. In this regard the first chapter presents the different debates regarding environmental security that took place in the last decade. This chapter is an introductory chapter and takes into account the possible audience in the Southern Mediterranean and the wishes of some contributors. For this reason the chapter deals *in extenso* with the question of human security and the different approaches.

The ARW dealt with issues that may affect the supply of fundamental goods for human living. To this end the complex question of climate change, its impact in the Mediterranean, the predicted increase in temperatures and predictions about rainfall are analysed. As part of this approach a chapter about solar variability and climate change is also added.

There are uncertainties regarding the increase (decrease) of seasonal precipitation and temperature variability in the next decades in the Mediterranean. Probably some predictions on climate change for the next decade have become a reality now in the Mediterranean. But there is a consensus that significant changes will take place in this regard and that they will have serious implications for human health, economic activities and food production, biodiversity, and human comfort and will require substantial adaptation and mitigation strategies to be managed successfully.

In regard to desertification, it is defined, according to the United Nations Convention to Combat Desertification. Several authors explain from different angles the importance of this question in the Mediterranean. The adaptive strategies put in place in the Maghreb; the magnitude, trends and projections in Egypt and Lybia and the positive prospects of human efforts in adaptation; the practices of combating desertification in the Negev, the new technologies used and the alternatives proposed for development of arid and hyper-arid areas; the situation and policies to be implemented in Turkey, Syria, Lebanon and Iraq; the insufficiency of some policies of restoration to reverse the relict desertification in Southern Europe and the implications and transformations induced by changes in the European Common Agricultural Policies and international market forces. All this research help to improve our knowledge about the extent, the causes and the factors triggering desertification processes and the relevance of adaptive policies and strategies already implemented and the need for more ambitious strategies and policies.

With respect to water, there is a consensus among the different authors that water demand will increase in the Mediterranean and taking into consideration factors such as climate change, population growth, increasing urbanisation and pollution of water, the prospects are dim. The countries bordering the Mediterranean are experiencing and will experience serious scarcities and will need to reform their water sector. The book presents some plans and policies implemented or to be implemented in several Mediterranean countries to reform the sector. But as regards the policies of conservation, use of ground water aquifers, desalination, waste water reclamation and transformation of agricultural sector, the political will and ingenuity differ depending on the country. Reforms are slow in the Maghreb. A medium term plan has been put in place in Egypt. A very sophisticated approach has been carried out in Israel. In Jordan, ambitious projects have to be implemented. In Southern Europe the long term structural water shortages should remain concentrated in a few regions. For instance, in Spain, the Ebro water transfer is just in the process of implementation for the alleviation of the situation in the Spanish Levant.

New inputs for understanding the adaptive policies in progress can be find in the chapters dealing with population growth in the Southern Mediterranean and in particular in the chapter dealing with the population decline in the European Union. The author concludes that social rather than demographic changes are needed to adapt to the expected population trends. The second chapter emphasizes the need for co-operation for helping countries with a high rate of population growth to complete the third stage of demographic transition. These pressures and the process of increasing urbanisation have important implications on land-use, greenhouse emissions and food security and consumption. The chapters dealing with food trade, food security and EU agricultural policy show the dependence of the Mediterranean countries, and in particular the non-EU Mediterranean countries, on world markets for importing considerable parts of their food needs. This dependence will persist due to population growth and environmental changes that affect and may affect severely the agriculture of the Mediterranean region. The gradual liberalisation of the Euro-Mediterranean agricultural trade will be an important step in the adaptive responses. But access is not a panacea and the removal of subsidies on the part of the major OCDE countries will affect the food prices. On the other hand, the gradual adjustments in the EU Common Agricultural Policy will be centred on territorial, human factors and the contribution of agriculture to preserving the environment and the rural landscape.

Two examples of increasing urbanisation are included: The cases of Cairo and Istanbul. Several important changes are taking place. Greater Cairo has ceased to attract a large proportion of migratory population in Egypt. The trend points to a transition from a compact city to a vast diffused spatial development. The quality of life has greatly degenerated and immigration rates have declined and it is questionable whether future growth projections will materialise. Regarding Istanbul the environmental problems are difficult to manage but Turkey´s future membership in the EU implies the progressive adaptation of the EU rules on environment. This will also imply an adaptation process that will improve the present situation. The prospects for the stabilisation of the population by the year 2020 have to be added to this.

The book ends with a chapter on the pollution of the Mediterranean and the prospects for implementation of the protocol for the Protection of the Mediterranean Sea Against Pollution from Land-Based Sources and the Strategic Action Programme.

The chapters of this book constitute an important input for further research on adaptation, mitigation, conservation and risk reduction policies to be implemented to increase resilience to environmental change in the Mediterranean. There are significant uncertainties to be clarified, but the partial conclusions of this book do not permit bold interpretations and predictions, although the situation is worrying. Significant adaptation, mitigation, conservation and risk reduction strategies are taking place and new strategies can be implemented. The cost, however, will probably become a crucial barrier for the adoption of these strategies. It is a question of evaluating the cost of inaction or weak policies for national, international and human security. Conflict prevention, focusing on a middle term time scale, is crucial in this regard.

Finally I have to emphasise that this ARW received generous support by NATO Scientific Committee. Later, El Real Instituto Elcano (Madrid) provided the funds needed to fill the gap in the cost of the projected activities, including the book editing, thus substituting the promises and commitments of the General Secretary of the Spanish Ministry of Environment that were never fulfilled. I have to give my thanks to Rafael Bardají (deputy director) and Emilio Lamo de Espinosa (director) for their generous support.

I also give my thanks to all the participants of the ARW and the contributors to this book for their support, given the economic uncertainties we had to suffer during more than twelve months for the reasons already explained. To the different reviewers of the chapters. And to Gracia Abad and Alberto Priego, UNISCI members, for their support and dedication to the organisation of the ARW and the editing of this book.

Antonio Marquina
Director of UNISCI

Section I

Environmental Security

Chapter 1

Environmental Security and Human Security

A. Marquina
UNISCI-Universidad Complutense
Facultad de CC. Políticas, Campus de Somosaguas. 28223 Madrid.

This chapter will try to explain the process of elaboration of the concept of environmental security and the difficulties in its definition. Later the link with the concept of human security is presented, explaining the low priority of environmental issues in the different approaches and definitions of the concept of human security. The chapter concludes explaining the process of securitization and the difficulties of avoiding al least some statist perspective of the concept of environmental security, taking into consideration the policies of mitigation, adaptation and risk reduction that have to be adopted to diminish the vulnerability of human collectives and individuals and increase resilience of these referent subjects

1. Introduction

The definition of environmental security is a debatable issue. There are many books and studies dealing with security and environment. The approaches are different. It is not a question of repeating the different phases of the research on environmental security started in the late 1970's with the works of Lester Brown. It is important to emphasise the different approaches and meanings of security depending on the author. The words used are often unacceptable or produce confusion. What is the meaning of security? Does security mean "freedom from danger, fear, want and deprivation"? [58]. Is the state the referent subject of environmental security or are human beings? Is the environment a "threat" to national security or human security? Is it a risk? What are the vulnerabilities that have to be reduced? What are the agents providing security? These are several questions that in the last decade gave place to important debates among researchers. The problem is that the debates on environmental security were conducted by specialists coming from different disciplines, not only from the international security field. Thus, the use of the different terms is sometimes not very accurate. Let me start explaining what I mean with the different terms I will use in this chapter.

A. Marquina (ed.), Environmental Challenges in the Mediterranean 2000-2050, 5–25.
© 2004 *Kluwer Academic Publishers. Printed in the Netherlands.*

2. The Concept of Security

The concept of security is a concept difficult to define. The concept of security is historically connected to nation states and armed forces. From this point of view a definition is not difficult at all. It meant protection against a military attack. The world system is composed of sovereign states. Barry Buzan [11] has shown the difficulties of the definition, stressing that the concept becomes more difficult to define than other concepts given the overlap existing between the concept of security and the concept of power, the lack of interest in security issues by different authors that were critical of political realism, the low priority given to the clarification of conceptual questions and the utility that politicians find in the ambiguity of the concept of security. It serves to justify different alternatives and policies, even wishes and passions.

Important disparities can be found in definitions made during the Cold War. For instance, Walter Lippmann considered that " a nation is secure to the extent to which it is not in danger of having to sacrifice core values if it wishes to avoid war, and is able, if challenged, to maintain them by victory in such a war" [11]. Stephen M. Walt stated that the main focus of security studies is the phenomenon of war and, as a consequence, security studies may be defined as the study of the threat, use, and control of military force [80]. Arnold Wolfers tries to link two essential aspects: perceptions and reality. In this sense he explains that "security in an objective sense, measures the absence of threats to acquired values, in a subjective sense, the absence of fear that such values must be attacked" [81]. Barry Buzan, after discussing the inadequacy of different definitions, considers that the central aspect is the discussion about the pursuit of freedom from threat. In the context of the international system, " security is about the ability of states and societies to maintain their independent identity and their functional integrity" [11].

In these state-centred definitions, conceptualizations focused on war and military force, or core values to be maintained and defended against threats, can be found.

Let me start with the questions connected with the definition [56]:

1.-The referent subjects: The human collectives—states and non-states, in particular ethnic and national groups—the individual as a person and citizen, members of these communities, and global humanity. There is a significant disagreement on the subjects supposedly being secured.

2.-The core values to be secured and protected. The referent subjects above mentioned have different values to be secured: sovereignty, integrity, autonomy, physical security, human rights, well-being, the maintenance of the biosphere. It implies an expansion of the field beyond the military affairs and the vital interests of the states. However, security studies can not become the study of everything bad [19].

3.-Threats and risks that endanger core values. An issue becomes part of the security agenda depending on the threats and risks that can affect existentially the values that have to be securitised. The precise nature of these threats and risks is a matter of disagreement among the different authors. After the cold war non-military threats and risks have become more important. What is still problematic is the clarification of the linkages existing among them.

4.-Vulnerability and resilience of referent subjects to risks and threats.

5.-The actors and mechanisms of response.

The threats and risks together with the vulnerability of the subjects of reference that can be considered existentially threatened and the limits of the possible responses, permit an understanding and delimitation of the concept of security. The traditional reference subject has been the state and its securitization was not problematic. The difficulty appears when the referent subject is the individual or global human beings, because security is an inter-relational concept. Rivalry and interactions permit the creation of a sense of identity that has to be defended with regard to other human collectives [12]

The importance of the issues affecting global humanity is increasing, in particular environmental risks and hazards. This has been facilitated by the globalisation process.

The individual as referent subject of security was accepted by liberal thinkers, but the problem is how to operationalise the singularity of every person. Security can only be achieved in a collective enterprise [67].

Given the fact that the security of the different subjects can be affected by different factors in different sectors: military, political, economic, social and environmental, the concept is broad and has been expanded from the traditional political-military security issues.

From this perspective the concept of security has to include the low probability of damage to the survival of the referent subjects mentioned above in the different sectors.

2.1. THREATS, RISKS AND VULNERABILITIES

Threats are always intentional. We cannot define something as a threat if there is no will to do damage. Intention is fundamental.

Everybody admits that greenhouse gas emissions are the result of global activities. This is an unintentional consequence of human activities in the search of economic development and well-being. There is not a single state responsible for these emissions, even taking into account the high rate of emissions by some developed states. It can be said that these actions are the result of irresponsible behaviour and are not a threat unless there has been an agreement to penalise these emissions or that it can be discovered that there is an intention to cause damage when producing these greenhouse emissions.

However, risk does not have this meaning. It means a contingency or the proximity of damage. It is linked to the concept of possibility, proximity or danger of causing damage. Martin Soroos considers that "threat is present when circumstances exist that have the potential for significant adverse impacts on humans" [70]. But, traditionally, the question of voluntarism is crucial. I do not see how inadvertent actions or a climate change can be considered a threat. There is a clear tendency to consider nature and environmental issues as a hostile force. Formulating environment as an external or internal threat permits the construction of environmental security with the same logic as state security. Security is something inter-relational. If nature is reified as a threatening entity in Hobbesian terms, then environmental security is a mechanical representation of state security. This deduction is clearly wrong and the least that can be said is that conceptualisation of environmental security is largely underdeveloped [16].

Another problematic question is the possibility of protection against all significant threats to core values.

Vulnerabilities are related to threats and risks. A vulnerability exits when the subjects to be secured are exposed to potentially harmful developments and lack the means to effectively limit or cope with the damage that may occur from them. They can try to reduce the vulnerabilities or reduce the threat/risk. Let me say that it is an illusion to think that states and human beings have the means to cope with all possible threats and risks, and that their core values can become more or less invulnerable. Human beings and human social and political constructions are limited. Human beings are not gods. Achieving total security is an illusion. This kind of eschatological metaphor can be found more or less inadvertently hidden in many presentations and definitions. It is important to underline the philosophical, anthropological and even theological presumptions that accompany many attempts to redefine the concept of security linked to the environment or the concept of environmental security linked to human security.

3. Environmental Security

The call for the expansion of the concept of security linked to environmental issues started in the late seventies. Lester Brown presented a new vision of national security emphasizing climate change, deforestation, soil erosion and food scarcity as new challenges to national security. Several years later, in 1983, Richard Ullman [77] included in the category of threats to national security an action or a sequence of events that could degrade the quality of life for the inhabitants of a state. Environment was considered a backdrop to human activities, and environmental security was not centred on the safeguarding of strategic natural resources. Later, in 1989, J.T Mathews [60] presented her vision of environment and security. The concept of security needed to be extended to encompass resource availability questions as well as more broadly understood environmental issues. From another angle, Peter Gleick [25] argued in that same year that resources and environmental issues were likely to lead to violent conflicts.

But the term environmental security was also perceived as contradictory. It may be invoked to defend the *status quo* of the present world ecological order, blocking change and thus producing insecurity [9].

The interest and expansion of the research increased little by little. Several steps in the clarification of the concept can be mentioned:

1.-Discussion on security and environment. In this first stage, mentioned above, the literature can be qualified as very general and anecdotal. The state still was the referent subject of security.

2.-Research on links between environment and security   [4,31,32,43]. Environmental scarcity in some circumstances affects security by fuelling violent conflicts. The causes of conflict were not explored. By replacing the focus on security by a focus on conflict, the difficulties in the definition of the concept environmental security were only partially avoided.

3.-Delimitation of environmental issues that can be considered security issues. Environment was increasingly considered the human-made context for our lives [16].

There is much scientific evidence to convince us that human-generated environmental degradation is increasing and transnational [58], though claims about the overall condition of the planet and its effect on the human species is still very much under discussion [46]

4.-Accommodation of environmental security and human security.

5.-Research on links between global consumption, environmental change and security [16].

3.1. ENVIRONMENT AND NATIONAL SECURITY. CRITICAL ARGUMENTS

Regarding this concept, it is important to note that the initial reasoning started from a pre-existing theoretical framework of the conceptualisation of national security. Understanding environmental security issues as national security issues was not very problematic. There are several reasons to explain this. If the economic base of national security is eroded by environmental change, the possibilities of defence against external aggression diminish. Environmental degradation has a very negative impact on national interests given its impact on the economy, health, quality of life and exacerbation of inequalities between people [7]. However, given the fact that environmental problems cross national borders the connection between environmental problems and national security is problematic. Other critical arguments [7,15,16,18,19,27,42,58,68] stressed the following points:

1.- Consideration of environmental change as a security issue invites the participation of the military establishment, but environmental scarcity rarely leads to interstate war. And few authors outside the military/security establishment suggest that a military response is required to meet environmental problems. Thus military tools are not generally of much value.

2.-The extension of the concept of security such as Myers makes in the book Ultimate Security [61] encompasses the problematic interaction between social and ecological systems. At this level security ceases to be a term of distinction and becomes a synonym for an environmental world-view.

3- Traditionally, security has been focused on threats existing within and between political-social systems. And as long as security is considered within the framework of a realist perspective of maintaining and expanding power within the state, the inclusion of environmental issues is counterproductive. It will link environment with threats and conflicts.

Threats from organised violence and the "threats" from environmental degradation and scarcity are inherently different in scope, extent and character. The assumption that environmental scarcity will necessarily lead to war is very questionable. Thus linking environmental issues to national security is a mistake. It reproduces outdated habits of mind.

4.- The sense of "us against them" that is typical of national security does not fit well in the environmental context and its core values. On the other hand, national security and the "we versus they" thinking is antithetical to the interdependent nature of many environmental problems. Environmental challenges require a co-operative approach. Finally, declaring environmental change a threat to national security may produce undesirable side-effects.

5.- Only certain environmental problems can be considered as threatening national interests. These are ozone depletion and climate change.

6.- Linking environment and security blurs the fact that world armies are a leading cause of environmental degradation.

7.- The concept represents nothing more than a way for the North to cloak its interventionist intentions in the South using environmental rhetoric and crisis management. But the sources of the global environmental problems tend to emanate from the most developed countries.

8.- The concept confuses subjective individual feelings of security and insecurity with the inherently relative and competitive concept of security as it applies to state security.

9- The time to solve environmental problems is longer than the public's emotive time-span.

10.- The concept leads to exclusivity as opposed to the universality that is inherent in the concept of environment.

In order to be more pragmatic and efficient Richard Matthew [56] proposed limiting environmental security to four areas:

1.-Greening the military, avoiding environmental degradation.

2.-Using security assets to support environmental policies without compromising traditional roles.

3.-Developing strategies for tracking and responding to those areas where the environmental factors are likely to intensify conflict or pose a threat to national interests.

4.-Incorporating environmental expertise into conflict resolution capabilities.

Later he expanded the conceptualisation explaining three perspectives on linking environment and security:

1.- The statist view

2.- The humanist view

3.- The ecologist view.

The statist view is concerned with the protection of territory, sovereignty, citizens and culture and in this regard there are clear state preoccupations for the environment :

1.- The protection of access to environmental goods beyond its borders.

2.- The protection from negative externalities such as crossborder air pollution or sudden population flows.

3.- The prediction as to where and when environmental change will create a situation in which they may have to use force. In this regard the state has to monitor and address instabilities, tensions or crises that may be caused or triggered by environmental factors.

4.- The greening of the military.

5.- Impacts of warfare on environment. Promoting dialogue, building confidence and transferring technology.

6.-The use of security assets, such as spy satellites to support environmental initiatives.

7.- The provision of disaster and humanitarian assistance [58].

There are clear possibilities of damage to national values and interests coming from environmental factors. This is the line of the researchers trying to clarify the links between environmental degradation and scarcity and conflicts. But there are several insufficiencies in this statist approach, taking into consideration the role of states as actors providing security. They implement policies of conservation, adaptation, mitigation and also risk reduction, not only protection and prediction. Policies to be implemented by the state or group of states will try to increase the resilience to risks and challenges coming from environmental degradation, scarcity and stress. Thus environmental security does not only deal with protecting people from external threats. It also takes into account issues and phenomena that cross the borders of states, the dynamics of change they induce, and policies of conservation, mitigation, adaptation and risk reduction that are applied.

This statist approach is reinforced by other facts: Many definitions of environmental security[1] were analysed in a study conducted for the US Army Environmental Policy Institute [26] and the key findings have to do, inter alia, with the identification of many environmental security "threats" over the next ten years, the importance of co-operation, and the key elements for a definition that are grouped around two basic concepts:

1.-Repairing damage to environment (a) for human life and support and (b) for the moral value of the environment itself.

2.-Preventing damage to the environment from attacks and other forms of human abuse.

But these elements provide a flavour of national security to the concept. [2]

Taking into account criticism and the limitations exposed, the link between environmental problems, that can affect all human beings and cross state borders, and human security concepts is logical. Global environmental problems apparently call into question the state as the referent subject of environmental security, its utility and thinking in national security terms.

4.The Concept of Human Security

The concept of human security has its roots in the debates on broadening the concept of security carried out before and after the Cold War. The Independent Commission on International Development (1980), the World Commission on Environment and Development (1987), the Commission on Global Governance (1995), and the United Nations Conference on the Environment and Sustainable Development (1992) were

---

[1] Few countries have an official definition of environmental security that can guide policies and actions. The international organisations where environmental issues are relevant to their goals, for instance UNEP, WHO, have not developed a working definition of environmental security.

[2] In this regard Glein and Gordon --2002—present sixteen definitions of environmental security, the following receiving the highest ratings:

1.- Environmental security is the relative public safety from environmental dangers caused by natural and human processes due to ignorance, accident, mismanagement or design and originating within or across national borders.

2.- Environmental security is the state of human-environment damaged by military actions and amelioration of resource scarcities, environmental degradation and biological threats that could lead to social disorder and conflict.

In my view the first definition with some modifications, eliminating the causes of the human processes and extending safety beyond the concept of national security can become an operational definition. The second definition is too descriptive.

influential in broadening the concept of security and the subjects of security, not only the states but also the people, the planet and the individual, the actors, the "threats", risks and security mechanisms [1,6,28,45]. The internal conflicts and the civilians that were at the epicentre of the conflicts in many parts of the world after the Cold War contributed to the reformulation of the concept of security. Many threats to human survival and well-being were internal threats. To this has to be added the awareness of global risks induced largely from the actions of people living in different states.

If Lincoln Chen coined the term Human security, Mahbub ul Haq, a respected Pakistani development economist and consultant to the UNDP, who played a key role in the construction of the Human Development Index, initially developed its content. His ideas were presented in a paper published in 1994 and entitled "New imperatives of Human security" [6] For this professor, human security has to be achieved through development, not through arms. He also established a list of threats to individual safety and well-being and a series of proposals for developing the new conception of security.

In this same year, the UN published the report on human security.

4.1. THE HUMAN DEVELOPMENT REPORT OF 1994

The United Nations Development Program introduced the concept of human security in its Human Development report [75].

According to this report, the concept of security is "people centred". The referent subject is primarily the individual or people, not the state. States are not considered irrelevant. The security of states is essential, but it is not sufficient to ensure the security of the individuals.

The concept is very comprehensive and has four essential characteristics:

1.- It is a universal concern
2.-The components are interdependent
3.- It is best ensured through early prevention rather than later intervention
4.- It is people centred.

At the global level, human security means essentially "responding to the threat of global poverty" which travels across international borders in many forms. Global threats and challenges to human security, arising more from the actions of millions of people than from the aggression of nations, include: unchecked population growth, disparities in economic opportunities, migration pressures, environmental degradation, drug trafficking, and international terrorism.

The core values to be defended are not the territory and the sovereignty of the states but the quality of life of the people or polity. The list of threats to these core values is presented under seven categories :

1.- Economic security. 2.-Food security. 3.-Health security. 4.-Environmental security. 5.-Personal security. 6.-Community security. 7.-Political security.

These seven categories imply numerous potentially interrelated and overlapping dimensions of possible threats and damage to human security, but do not provide a coherent framework for integrating them into a single concept. However this classification permits the elaboration of some measures to confront the risks and threats to human dignity, survival, well-being and freedom. The list of risks and threats is very extensive, for this very reason the establishment of priority areas is fundamental. However, as the chapter will explain, these priorities differ according to the authors.

In the United Nations Development Program several values are central: individual safety in physical terms, individual well-being and individual freedom [6]. Thus human security essentially consists of freedom from fear and freedom from want.

On environmental security, the elaboration is not sophisticated. The term "environmental threats" is used, quoting as "threats" within the states: the degradation of local ecosystems, water scarcity, pressure on land, deforestation, desertification, salinization and air pollution. At the global level, several changes that threaten global human security are quoted: air pollution, greenhouse gas emissions, loss of biological diversity and tropical deforestation, destruction of coastal marine habitats. Thus at the local level there does not exist a clear distinction between environmental security and ecological security. At the global level, the implications of the physical environmental issues for the human condition are not mentioned.

On the question of mechanisms and actors for the response, the UN report demands a wide range of actors including NGOs and co-operation among them. Vulnerability of the individual *per se* is very high and confronting and managing all possible direct or indirect threats to the individual and people in general is almost an insurmountable task. It is a question of prioritising the threats according to their importance. This explains the low priority of environmental degradation. It is not included in the core threats to human security [6].

The literature on environmental security and sustainable development put a much greater emphasis on environment than human security.

Other different elaborations of the dimensions of human security have been proposed. They only show that environmental issues are on the agenda of human security, but are not in the center of reflection.

For Jorge Nef [62] the different dimensions are centred on the idea of human dignity. He includes: environmental, personal and physical security; economic security; social security; political security; cultural security. For him, the political dimension holds the key to the safeguarding of physic-environmental, economic, social and cultural rights.

George MacLean [38] expands the laundry list to include personal security; access to the basic essentials of life, protection from crime and terrorism, pandemic diseases, political corruption, forced migration and the absence of human rights; freedom from violations based on gender; the rights of political and cultural communities; political, economic and democratic developments; preventing the misuse and overuse of natural resources; environmental sustainability; and the effort to curtail pollution.

No less broad is the recommended agenda for human security in UNESCO [78], although with little attention to environment.

Mark Halle and his collaborators [27] consider that human security and insecurity are largely a function of five security-relevant systems: economic, political, cultural, demographic and ecological. The question is how to operationalise the interactions and establish causal relationships. This has not been done yet. The establishment of links is clearly insufficient.

4.2. PLURAL DEFINITIONS AND LITTLE SUBSTANCE ON ENVIRONMENT

The different authors define the term human security including or excluding aspects that can be integrated in the definition. Depending on the school of thought, the meaning and implications of the concept are different. Thus the values and threats to these values vary.

The notion varies also according to the interests of the countries. Two leading countries, supporting research and policies of human security Japan and Canada, maintain different approaches. Canada emphasises the human security of the people and the freedom from fear resulting from violent conflicts, although it does not forget non-violent threats [13,17]. The present Japanese approach emphasises the protection of the lives, livelihoods, and dignity of individual human beings, going beyond the freedom from want that was originally emphasised [13, 82].

Regarding the definitions, nearly thirty definitions can be found. Most of them are characterisations, descriptions or enumeration of the contents, goals, objectives, relationships, linkages or outcomes. For instance: "human security is achieved when..." or "human security describes..." or "several key elements make up human security..." or "human security relates to...". Enumeration lacks explanatory power. Sometimes, instead of definitions we have tautologies or circular reasoning. Sometimes the concept is defined in a negative way [79]. Sometimes the emphasis is on threats to human beings, but labelling anything that causes a decline in human well-being as a security threat may result in the term losing any analytical usefulness [64].

Olster Hampson [28] has tried to explain the different conceptions by establishing three categories of human security:

1.-The first approach, centred on a broad definition, is a rights-based approach anchored in the rule of law and treaty based solutions to human security. In this conception, international institutions are considered central to developing new human rights norms and for bringing about a convergence of different national standards and practices

2.-The second approach is centred primarily, though not exclusively, on a humanitarian conception of human security. The "safety of peoples" is considered the paramount objective in the case of military intervention, and war is one of the principal threats to human security.

3.-The third is associated with sustainable human development and is the conception exposed in the Human Development Report of 1994.

The three conceptions differ as to which are the risks, challenges and threats, the strategies, instruments and mechanisms of response, and the institutions and governance structures to address them.

Only for the third conception, do the environmental issues have a real importance.

Edward Newman [63] has also presented four overlapping typologies of human security:

1.- Basic Human needs.

2.- Assertive/Interventionist focus.

3.- Social Welfare/Developmentalist focus.

4.-Non-traditional security and new security issues, such as AIDS, terrorism or international crime, that the processes of globalisation permit. But addressing the new security issues will imply the frequent strengthening of state capacity. The referent subject of security in this type is not only the state but the individual.

In the other three types, the referent subject is the individual or the people. Newman recognises that the types differ in focus, methodology and in the institutions, actors and policies proposed for the promotion of human security. Human security, in his opinion, is revisionist and any investigation into it will unavoidably engage in highly controversial debates regarding political and economic organisation and state sovereignty. And the issue areas that can be analysed from a human security perspective are almost endless and not uncontroversial. In this way he recognises that environmental issues can be analysed from a human security perspective. However, human security is not a theory; it is supported by theories. This complicates the analysis.

Looking at it from another angle, the differences existing among the authors can be stressed. Rasmesh Thakur [72] emphasises the quality of life of the people, society or polity as the central aspect of the concept. Astri Suhrke [71] focuses on vulnerability as the defining characteristic of the concept. Thus "the essence of human security is reduced vulnerability, and policies to this end could be aggregated into a "human security regime" designed to protect categories of extremely vulnerable persons". That is clearly insufficient.

Bajpai [6] includes the concept of protection and the concept of promotion of human development. Jennifer Leaning, M.D., S.M.H. and Sam Arie [39] maintain that human security is an underlying condition for sustainable development. This approach is not shared by Sverre Lodgaard [45] who supports a narrow or limited definition, which does not include natural disasters, humanitarian assistance, or human development. For him, human security in operational terms deals with freedom from fear of physical violence. He does not include structural or indirect violence in the concept.

In order to operationalise the concept, Gary King and Chritopher J.L. Murray [38] proposed a simple, rigorous, and measurable definition :The number of years of future life spent outside a state of generalised poverty. But this definition presupposes many things, lacks a policy orientation and is of little value in the conceptualization process.

Most recently Sabina Alkire [2] proposed the following working definition "the objective of human security is to safeguard the vital core of all human lives from critical pervasive threats, in a way that is consistent with long term fulfilment ".

Again the problem of this definition is the lack of clarity of some concepts such as the vital core of all human lives. According to the author, there is a limited core of human activities and abilities to be protected: the rights and freedoms in the vital core pertain to survival, to livelihood and to basic dignity. He adds that the task of prioritising among rights and capabilities is a question of judgement and a difficult one. At least, it can be said, environmental issues could be included as vital core values.

In a different orientation Roland Paris [65] criticises the concept of human security qualifying it as a new neologism for thinking about international security as something more than military security. He criticises the definition attempts that identify some values to be protected as more important than other players, without giving an explanation, and the need that the states and other actors have in maintaining a certain level of ambiguity in the term. For him definitional expansiveness and ambiguity are powerful attributes of human security. In this regard it can be a useful label for a broad category of security research but, as a tool for analysis, it will not be very useful.

More optimistically, Nicholas Thomas and William T. Tow [73] consider that although there exist policy limitations on the conceptual and empirical utility, human security can be considered a valid paradigm for identifying, prioritising and resolving transnational security problems, given the fact that there are too many and too diverse a range of human security issues and that it is necessary to prioritise. Their approach recommends going beyond the state in prioritising threats, "focusing on those events that transcend state borders in terms of their impact on different societies and diverse individuals", adding that "an event or crisis becomes a truly human security problem, however, when the ramifications of not overcoming it cross a state´s border and assume a truly international significance, affecting other societies and individuals". Extreme vulnerability and imminent danger are the most important considerations for establishing priorities. In this regard they establish three categories that either by themselves or in combination generate urgent and compelling humanitarian emergencies that demand attention:

1.-Victims of war and internal conflict
2.-Persons who barely subsist and are thus courting "socio-economic disaster"
3.-Victims of natural disasters.

This approach can be criticised [8] for its lack of attention to the role of states as possible agents of human insecurity, the narrow understanding of transnational security issues and the inconsistency with the normative concerns of the human security agenda. Although they maintained that freedom from fear and freedom from want are the fundamental objectives of human security, they try to define the term more narrowly, in order to increase its analytical and policy value, avoiding models "that can cope with all forms of human interaction simultaneously" [74], something impossible to achieve. Thus, they focus first on transnational trends to international norms arising from inadequacies in internal state systems that make individuals and groups within states more vulnerable; second on the effective management of such vulnerabilities; and third on the requirement of help in some form of international intervention. The environmental challenges and risks are apparently only considered in the category of natural disasters.

In conclusion, many definitions open the way to include environmental issues in the concept of human security, at least indirectly, but it is necessary to make substantial efforts to give priority in the conceptualisation of environmental issues and to operationalise the different possible interactions.

4.3. THE COMMISSION ON HUMAN SECURITY

In a recent published report, entitled "Human Security Now" [14], the Commission on Human Security tries to present a new approach to human security. Considering that security is centred on people, human security is defined as the protection of " the vital core of all human lives in ways that enhance human freedoms and human fulfilment. Human security means protecting fundamental freedoms---freedoms that are the essence of life. It means protecting people from critical(severe) and pervasive(widespread) threats and situations. It means using processes that build on people´s strength and aspirations. It means creating political, social, environmental, economic, military and cultural systems. That together give people the building blocks of survival, livelihood and dignity."

In this definition I have to stress the words " the vital core of all human lives", that is defined as a "set of elementary rights and freedoms people enjoy". The authors of the report recognise that what people consider to be vital varies according to different cultures and individuals. They explicitly renounce presenting a concrete list of the content of human security, thus avoiding the criticism that the Human Development Report of 1994 received in this regard.

For our purposes it is important to note that this report does not imply a significant departure from the UN Human Development Report of 1994. The authors consider human security as different and complementary to state security. They emphasise not only protection but the empowerment of human security, enabling people to develop their resilience to the difficult conditions they have to face. They also emphasise the interdependence of the world and that no state can assert its unrestricted national sovereignty and unilateral rights. They are concerned with violent conflict, poverty and deprivation, human development and human rights. In this regard, the report explores conflict-related aspects of human security, and poverty-related aspects of human security and proposes concrete action.

From this perspective, the interest in environmental issues affecting human security is not explored, even taking into consideration that the authors quote the Secretary General of the UN, Kofi Annan, and his explanation of building blocks of human security: Freedom from want, freedom from fear and the freedom of future generations to inherit a healthy natural environment. Only environmental issues are included as "menaces" from which people have to be protected. And in an annex box, at the end of the first chapter, hunger, water, population and environment are merely enumerated as special issues of human security. It can be said that environmental issues are not yet in the core of the thinking of the Commission on Human Security.

5. Environment and Human Security

If this panorama is not very optimistic for the inclusion of environmental security within the present core research on human security, it can be said that some clarification of the importance of environmental risks to human security has been attempted.

GECHS, in an *ad hoc* definition by Lonergan [50], has tried to include as a priority the environmental challenges to human security. Thus, human security is achieved when and where individuals and communities:

1.-Have the options necessary to end, mitigate or adapt to threats to their human, environmental and social rights.

2.-Have the capacity and freedom to exercise these options

3.-Actively participate in attaining these options.

According to this definition, human security deals with a secure livelihood and the absence of (armed) conflict within and between societies.

In another definition proposed by GECHS, human security is having the capacity to overcome vulnerability and to respond positively to environmental change. This is an almost eschatological definition.

John Barnett [7] for his part, links environmental security and human security, defining environmental security as "the process of peacefully reducing human vulnerability to human-induced environmental degradation by addressing the root causes of environmental degradation and human insecurity". This definition assimilates environmental degradation with human induced degradation, considering that human beings are not opposed to nature, but are constitutive of nature. Human beings are the principal referent of security, not the biosphere. The problem with this definition is that it puts aside environmental issues not caused by human activities which, like natural disasters, have a tremendous impact on human security. An environmental issue that lies outside human behaviour may become a security issue when conservation, mitigation, adaptation, and risk reduction measures can be applied. If these kinds of measures can not be applied to environmental phenomena, because they are impossible or irrelevant, then they can not be integrated in the category of security issues.

What is under discussion are not these kinds of problems that lie outside human behaviour but the problems of development and growth that undermine long-term sustainability. Even in this field there are significant problems to be clarified. Steve Lonergan [49,50,51] has presented the main components of human security related to sustainable development: access to economic resources (income), access to natural resources (energy, food and water), access to social resources (education, healthcare), and the availability of adequate institutional arrangements (self-determination, democracy). He recognises that sustainable development and human security are closely interlinked and have a variety of indicators in common, but in the social and institutional domains little progress has been made through modelling exercises.

I can not develop in this chapter the research done on sustainable development and human security. What I stress here is the process of securitization. The specialists working at GECHS pay special attention to the environmental component of human security placing it in the context of social processes, poverty alleviation, institutions, governance and regional conflict resolution [58]. By doing this, environmental security is considered as a process and includes monitoring environmental changes, and the adoption of conservation, mitigation and adaptation programmes and policies in order to address environmental degradation and scarcities and people´s vulnerabilities. These programs and policies will try to bring human behaviour into line with physical phenomena and processes of environmental change. Once established the risks that have to be confronted, it is necessary to focus on vulnerabilities and resilience of the referent subjects of environmental security. That is normal in traditional security studies, where states devise different possible responses to threats and risks, mechanisms of adaptation and adjustment and threat and risk reduction. Apparently this was not sufficiently taken into consideration when dealing with environmental change [58].

In fact, the adaptive mechanism of human beings throughout history has been remarkable although it is difficult to assess the magnitude of human-generated environmental change and its implications for the human species [58].

Types of social responses to environmental change going from non-adaptation to adaptive measures and policies that include migration, innovation and evolution have also been presented [57]. But migration is declining as a viable strategy and evolution depends on the rate of evolutionary change. It can be too slow to accommodate the rate of environmental change.

In this regard, several response strategies to reduce human vulnerability to environmental change can be mentioned:
-conservation,
-adaptation,
-preventing or limiting environmental changes (mitigation),
-avoidance of potential impacts,
-defence against adverse impacts [70].

These response strategies, that are co-operative strategies, imply several preconditions. The costs of the different alternatives have to be known, the priorities have to be established, and the implications of the possible responses have to be assessed. All this is central in the state political economy and political security.

Regarding risk reduction strategies and policies, it is abundantly clear that national strategies and policies are insufficient to deal with global environmental changes. International efforts are necessary in order to limit, reduce or even eliminate environmental risks, overcoming narrow national interests or free riding. Co-operation among states is essential.

6. Adaptation

Focusing on adaptation as a particular research field, Richard Matthew [58] has presented two new hypotheses about environmental degradation and security:
1.- The environmental stress only becomes an important explanatory variable when its level exceeds the adaptive capacity of a state
2.-This rarely happens because societies almost always find ways to cope.

He proposes broadening the theoretical and temporal scope of environmental and security research. His point is that "adaptation is, by definition, an ongoing and dynamic process, and that only a research project that focuses on a broad time scale can capture the action-response- reaction nature of this phenomenon. By contrast, a research project that focuses on a short time frame will only get a snapshot of these processes--a snapshot that is likely to yield little explanatory and predictive power". "Broadening the temporal scope of research will highlight the dynamic interaction between social and ecological systems and reveal that the human adaptive capacity is more resilient than has been alleged in most studies of environmental stress and conflict".

But the concept of adaptation has become a mode. For instance, regarding climate change, in the last few years the number and the increasing sophistication of climate models and impact assessments have provided the basis for more concrete statements on vulnerabilities and adaptation measures. The Marrakech agreements to the United Nations Convention on Climate Change of 2001 implied the provision of new funding opportunities for the promotion and support of adaptation policies and programmes. However, the concept of adaptation differs depending on the subject of study, on how the concept is understood and which methodologies are used. The types also differ. Most of the recent research has been done focusing on climate change. Several important efforts can be quoted:

- The United Nations Framework Convention on Climate Change (UNFCC) signed in Rio de Janeiro in 1992 used the term but the interest in adaptation was not very high. It was considered a long-term option and its role in reducing vulnerability to climate change small. The Conference of the parties to the framework convention (COP-1) held in Berlin in 1995 changed this perception and a decision was taken to approach adaptation in three stages: planning, measures which may be taken to prepare for adaptation, and measures to facilitate adequate adaptation.. The IPCC third assessment report pays particular attention to impacts, adaptation and mitigation. [34,35].
- UNDP/GEF has developed an Adaptation Policy Framework that is now (2003) in the final stages of elaboration [76]. Its objective is to assist in the process of incorporating adaptation concerns into national, sector-specific, and local development planning processes.
- The World Bank [83] has also developed a methodology for rapid assessments and has launched its own National Adaptation Strategy Studies

Many other initiatives from Institutes and Universities and NGOs have also been developed in the last few years.

Broadening the field, not focusing only on climate change, several of the general strategies of adaptation proposed can be mentioned:

1.-Coping strategies: information collection, shielding, coalition, specialisation, diversification, rationalisation.

2.-Transforming strategies: bargaining, institutionalisation, coercion, inducements, education, systemic change [40].

It is recognised that the selection of the adequate strategy depends on numerous factors. In this regard several approaches were proposed taking into consideration three different perspectives:1.-preconditions that may shape the universe of subsequent options, for instance the type of uncertainty that has to be faced, the range of policy instruments available to the actor, the nature of the environmental problem;2.-the actors preferences and calculations;3.-the assessment of the danger and selection of strategies where psychological variables, cultural and historical variables, domestic variables, and systemic variables are to be taken into consideration. The selection of a particular strategy needs extensive case studies given the types of uncertainty, the specific environmental problems, and the characteristics of the different states [40].

Given that adaptation is a multifaceted term, other authors prefer to focus on the effects of adaptation, distinguishing macro level effects that include technological innovation, environmental regulation, building state capacity through regional regimes, greater political openness, economic exchange, and value change fostered by activities of international organisations and other non-state actors. And micro effects such as individual searching for new livelihood, farmers experimenting with different crops and people relocating [59].

In general, recognition is given to the importance of closer collaboration among different disciplines, such as disaster risk management, resource management and vulnerability reduction, conceptual frameworks such as natural hazards approach, vulnerability-based approach, policy analysis approach, adaptive capacity approach and, among agencies, sectors and policies. According to the World Bank [83] many adaptation mechanisms will be strengthened by making progress in areas such as good governance, human resources, institutional structures, public finance, and natural resource management. It supports the integration of adaptation in developing strategies. In this regard other initiatives on capacity building follow this direction [76].

In conclusion it can be said that adaptation strategies, policies and measures try to reduce vulnerability at the local, national, regional and global level, enhancing the capability to change and accommodate. It is a significant aspect for achieving a secure livelihood. The strategy and the process are equally important.

7. Actors and Mechanisms of Response

Another problem to be resolved for clarifying the inclusion of environmental security in the context of human security, as proposed by GECHS, has to do with the actors to be mobilised in response to the environmental risks and hazards. It is clear that the actors to be mobilised for the mitigation, conservation and adaptive responses are not only the referent subjects of human security: the individuals and the people. A good number of the present environmental problems have given place to conventions and agreements among states and the creation of international organisations or agencies to manage these kinds of problems. Many environmental problems cross state borders and no single state can manage problems such as pollution or the effects of climate change.

The principal players in this endeavour are the states (national governments and ministries, local authorities, planning bodies) Other actors such as the transnational enterprises can decide to operate with more environmental restrictions, can provide some ingenuity[3], and can agree to environmental liabilities, but there is a long way to go before they can be accepted as genuine actors in providing responses, and can avoid being perceived as the principal suppliers of environmental problems. NGOs, epistemic communities, and people affected have a supporting role in developing awareness of the problems, technical and financial capacities—in some cases substantial--, mobilisation, and lobbying.

International networks of NGOs and non-state actors are well suited for different initiatives, in particular transnational risks. In fact they have tried to influence actions and beliefs of governments, shaping and influencing policies and international law [53].

For this reason NGO networks can play an invaluable role, make a long term contribution, apply bottom-up approaches and supply ingenuity. Their activities, however, are often dispersed, lacking consensus on the lines to be taken [55]. NGOs have also a limited action capability in numerous developing countries where democracy is lacking and freedom of association is strictly regulated. The idea of an increasing civil society in competition with states is still an illusion. Global citizenship is still a debatable issue with few definite answers. The existing institutions in the world do not reflect this global citizenship. Cosmopolitan democracy and global governance that fall short of world government have a long way to go to become a reality [21]. It is

---

[3]Homer -Dixon [33] considers that ingenuity "consists not only of ideas for new technologies like computers or drought-resistant crops but, more fundamentally, of ideas for better institutions and social arrangements, like efficient markets and competent governments". I use this term with this meaning.

illustrative to state the increasing barriers to migration, the distinction between nationals and foreigners or to study the financial sources of numerous NGOs in developed and developing states.

It is also important to stress the difference between both types of states. In the developed states the idea of sovereignty is losing strength because of the importance of other non-state actors and the globalisation process. In the developing states the first priority is the achievement of an effective state in order to attain a real security. The contraposition is clear [3].

From this perspective, the idea of constructing environmental security within the framework of human security, clearly differentiated from the security—and policies-- of states, which are considered less effective agents of security because of globalisation, is still a pious desire. Environmental and human security can not yet be considered as opposed to national security.

8. Conclusion

Environmental problems are diverse in nature and in the consequences they induce. They can affect different security subjects: individuals, local communities, states, regional systems, and the whole planet and people. In some cases the survival of entire communities and peoples may be in question if environmental degradation, scarcity and environmental stress is not contained or physical disasters cannot be managed.

The process of environmental degradation, scarcity and environmental stress has a temporal dimension affecting not only the lives of the present generation but the lives and the well-being of future generations. Thus the process of conservation, adaptation, mitigation and risk reduction and the policies and measures to be adopted have to take into consideration the different timing existing between the political and social interests and process—short term many times- and the environmental process—medium long term consequences--. This is probably an insurmountable task for conflict prevention if a profound change in the political culture and approaches of many governments and political parties does not occur.

Another significant difficulty is the analysis and evaluations of the tendencies and causes of environmental degradation, scarcity and environmental stress. Presenting only possible tendencies and links is very little and probably an irrelevant exercise if the root causes of environmental degradation, scarcity and environmental stress cannot be clearly determined and measured. The effects of human activities on environment are not linear and simple. They are combined in multi-scale and multi-subject complex processes difficult to assess. What is abundantly clear is that the possible consequences of the environment degradation, scarcity and stress cannot be studied as consequences of a more or less mechanical process. It is difficult to assess the magnitude of human generated environmental change.

On the other hand, environmental security is a normative concept and can be integrated in the broader concept of human security, however, the different approaches, conceptualisations and definitions already made pay, in general, little attention to the environmental component of human security. GECHS has made a significant effort in this regard, but more research and studies are needed to operationalise the conceptualisation. Human security focuses on people not on states. The states have the fundamental responsibility to provide security to the people. Though they often fail to fulfil their obligations, the role of states as security providers is not diminished, even taking into account the involvement of more stakeholders in the processes of conservation, adaptation, mitigation and risk reduction.

References

1. Acharya, A. (2002) Human security: What kind for Asia-Pacific? What options? in Hassan, M.J., Leong, S., Lim, V. (eds.) *Asia-Pacific Security. Challenges and Opportunities in the 21ˢᵗ Century*, ISIS Malaysia, Kuala Lumpur, pp. 23-51.
2. Alkire, S. (2002) *Conceptual framework for Human Security.* http://www. humansecurity-chs.org/doc/ frame.pdf.
3. Ayoob, M. (1995) *The Third World Security Predicament. State Making,, Regional Conflict, and the International System*, Lynne Rienner, Boulder.
4. Bächler, G., Spillman, K. R.. (1996) *Environmental Degradation as a Cause of War,* 3 V., Rüegger Verlag, Zurich-Chur.
5. Bächler, G. (1999) *Violence Through Environmental Discrimination.* Kluwer Academic Publishers, Dordrecht.
6. Bajpai, K. (2000) *Human Security: Concept and Measurement*, Kroc Institute Occasional Paper n.19, Notre Dame.
7. Barnett, J. (2001) *The Meaning of Environmental Security*, Zed Books, London .
8. Bellamy, A. J. (2002) "The Utility of Human Security": Which Humans? What Security? A Reply to Thomas & Tow, *Security Dialogue*, 33,3, 373-377.
9. Brock, L. (1991) Peace through parks: the environment on the peace research agenda, *Journal of Peace Research*, 28,4, 407-423.
10. Brown, L. (1977) Redefining National Security, Paper n.14, World Watch Institute, Washington D.C.
11. Buzan, B. (1991) *People, States and Fear. An Agenda for International Security Studies in the Post-Cold War Era*, London, Harvester Wheatsheaf.
12. Buzan, B., Waever, O., de Wilde, J. (1998) *Security. A New Framework for Analysis*, Lynne Rienner, Boulder.
13. Caballero-Anthony, M. (2002) Human Security in the ASIA-Pacific: Current Trends and Prospects in Hassan, M.J.,Leong, S.,Lim,V.(eds.) *Asia-Pacific Security. Challenges and Opportunities in the 21ˢᵗ Century*, ISIS Malaysia, Kuala Lumpur, pp. 53-73
14. Commission on Human Security (2003) *Human Security Now*, Commission on Human Security, New York.
15. Dalby, S. (1999) Threats from the South? Geopolitics, Equity and Environmental Security, in Deudney, D.H. and Matthew, R. A., *Contested Grounds. Security and Conflict in the new Environmental Politics*, Albany, State University of New York, pp. 155-185.
16. Dalby, S. (2002) *Environmental Security*, University of Minnesota Press, Minneapolis.
17. Department of Foreign Affairs and International Trade (2003) *Human Security Program* http://www.humansecurity.gc.ca/freedom_from_fear-en.asp.
18. Deudney D. (1991) Environmental Security: Muddled Thinking, *Bulletin of Atomic Scientists*, 47,3, 22-28.
19. Deudney, D.H., and Matthew, R. A. (1999) *Contested Grounds. Security and Conflict in the New Environmental Politics*, State University of New York Press, Albany.
20. Diehl, P.F., and Gleditsch, N.P. (eds) (2001) *Environmental Conflict*, Westview Press, Boulder.
21. Dower, N., Williams, J. (2002) *Global Citizenship. A Critical Reader*, Edinburgh University Press, Edinburgh.
22. Dubois-Maury, J. (2001) *Les risques naturels: quelles réponses?*, La Documentation Française, Paris.
23. GEF (2001) *Proposed Elements for Strategic Collaboration and a Framework for GEF Action on Capacity Building for the Global Environment*, a briefing document, htpp://www.gefweb.org/Outreach/ outreach-Publications/outreach-publications.html.
24. Gleditsch, N. P. (1997) *Conflict and Environment*, Kluwer Academic Publishers. Dordrecht.

25. Gleick, P. H. (1989) The Implications of Global Climate Changes for International Security, *Climate Change*, 15, 303-325.
26. Glenn, J.C., Gordon.T.J. (2002) *2002 State of the Future*. American Council for the United Nations University. Washington D.C.
27. Halle, M., Dabelko, G., Lonergan, S., Matthew, R. (2000) *State-of-the-Art. Review on Environment, Security and Development Co-operation*, http://www.1.oecd. org/dac/pdf/ envsec.pdf.
28. Hampson, F.O. with Daudelin, J., Hay, J.B., Martin, T., Reid, H. (2002) *Madness in the multitude. Human Security and World Disorder*, Oxford U.P., Toronto.
29. Harvard University, *Concepts of Human Security*, http://www.hsph.harvard.edu/hpcr/events/ hsworkshop/list_definitions.pdf
30. Harvard Kennedy School (2001) *Summary of Deliberations*, Workshop: Measurement of Human Security. November.
31. Homer-Dixon, T. F., and Blitt, J. (1998) *Ecoviolence. Links among environment, Population, and Security*, Rowman and Littlefield Publishers, Lanham.
32. Homer-Dixon, T. F. (1999) *Environment, Scarcity and Violence*, Princeton University Press, Princeton.
33. Homer-Dixon, T. (2001) *The Ingenuity Gap*, Vintage, London.
34. IPCC (2001) *Climate Change 2001. Impacts, Adaptation and Vulnerability Mitigation*, Cambridge University Press, Cambridge.
35. IPCC (2001) Climate Change 2001. Mitigation, Cambridge University Press, Cambridge.
36. IUCN, Worldwatch Institute, ISSD, SEI-B (2002*) Adapting to Climate Change: Natural Resource Management and Vulnerability Reduction Background Paper to the Task Force on Climate Change, Adaptation and Vulnerable Communities*, htpp://www. issd.org/pdf/2002/envsec_cc_bkgd_paper.pdf.
37. Kasperson J.X., and Kasperson R.E.. (2001) *Global Environmental Risk*, The United Nations University Press. Tokyo.
38. King, G., Murray, Ch. (2001-2002) Rethinking Human Security, *Political Science Quarterly,* 116,4, 585-610.
39. Leaning, J., M.D., S.M.H., and Sam, A.. (2000) *Human Security. A Framework for Assessment in Conflict and Transition. CERTI Crisis and Transition Tool Kit*, http://www.certi.org/publications/policy/ human%20security-4.htm.
40. Le Pestre, Ph. (1999) Adapting to Environmental Insecurities: A Conceptual Model, in Lonergan, S.C. (ed.) *Environmental Change, Adaptation, and Security*, Kluwer Academic Publishers, Dordrecht.
41. Lee Peluso, N., and Watts, M.. (2001) *Violent Environments*, Cornell University Press, Ithaca.
42. Levy, M. (1995) Is the Environment a National Security Issue? *International Security*, 20,2, 35-62.
43. Lietzmann, K. M., Vest, G.D. (1999) *Environment and Security in an International Context*, NATO, Brussels.
44. *Livelihoods and Climate change. Combining disaster, risk reduction, natural resource management and climate change adaptation in a new approach to the reduction of vulnerability and poverty*, International Institute for Sustainable Development, International Union for Conservation of Nature and Natural Resources and Stockholm Environmental Institute, Winnipeg. http://www.iisd.org.
45. Lodgaard, S. (2000) *Human security: Concept and Operationalization*, http://www.hsph.harvard.edu/ hpcr/events/hsworkshop/lodgaard.pdf.
46. Lomborg, B. (2001) *The Skeptical Environmentalist. Measuring the Real State of the World*, Cambridge University Press, New York.
47. Lonergan, S.C. (ed.) (1999) *Environmental Change, Adaptation, and Security* , Kluwer Academic Publishers, Dordrecht.
48. Lonergan, S.C., Gustavson, K., and Harrover, M. (1999) Mapping Human Insecurity, in Lonergan, S.C. (ed.) *Environmental Change, Adaptation, and Security*, Kluwer Academic Publishers, Dordrecht, pp 397-413.
49. Lonergan, S., Gustavson, K., Carter, B. (2000) The Index of Human Insecurity, *Aviso* 6, 1-9.
50. Lonergan, S., Langeweg, F., Hilderingk, H. (2002) Global environmental change and human security: what do indicators indicate? In Page, E.A., Redclift, M., *Human Security and the Environment*, Edward Elgar, Cheltenham.
51. Lonergan, S. Langeweg, F., Hilderink, H. (2002) Environmental Change and Human Security: Indicators and Trends in Matthew, R. A., Fraser, L. *Global Environmental Change and Human Security: Conceptual and Theoretical Issues*. GECHS   Professional Report, http://www.gechs.uci.edu/ publications.htm.
52. Lowi, M.R., and Shaw, B.R. (2002) *Environment and Security. Discourses and Practises*, MacMillan Press Ltd, Basingstoke.

53. Luterbacher, U. (2000) *International Relations and Global Climate Change*, MIT, Cambridge.
54. Mack, A. (2002) *Report on the feasibility of creating an Annual Human Security Report*. Program on Humanitarian Policy and Conflict Research, Harvard University, http://www.hsph.harvard.edu/hpcr/ Feasibility Report.pdf.
55. Martínez-Alier,J.(2002) *The Environmentalism of the Poor. A Study of Ecological Conflicts and Valuation*, Edward Elgar, Cheltenham.
56. Matthew, R. A. (1997) Rethinking Environmental Security, in Gleditsch, N.P., *Conflict and Environment*, Dordrecht, Kluwer Academic Publishers, pp. 71-90.
57. Matthew, R. A. (1999) Social Responses to Environmental Change, in Lonergan, S.C. (ed.) *Environmental Change, Adaptation, and Security*, Kluwer Academic Publishers, Dordrecht, pp 17-39.
58. Matthew, R. A., Fraser, L. (2002) *Global Environmental Change and Human Security: Conceptual and Theoretical Issues*. GHECS Professional Report. http/www.gechs.uci.edu/publications.htm.
59. Matthew, R. A., Gaulin, T., McDonald, B. (2002) The Elusive Quest: Linking Environmental Change and Conflict, in Matthew, R. A., Fraser, L. *Global Environmental Change and Human Security: Conceptual and Theoretical Issues*. GHECS Professional Report, http/www.gechs.uci.edu/ publications. htm.
60. Mathews, J.T. (1989) Redefining Security, *Foreign Affairs*, 68,2, 162-177.
61. Myers, N. (1993) *Ultimate Security. The Environmental Basis of Political Stability*, Norton, New York.
62. Nef, J. (1999) *Human Security and Human Vulnerability. The Global Political Economy of Development and Underdevelopment*, International Development Research Centre, Ottawa.
63. Newman, E. (2001) Human Security and Constructivism, *International Studies Perspectives*, 2,3, 239-251.
64. Page, E.A., and Redclift, M. (2002) *Human Security and the Environment. International Comparations*, Edward Elgar, Cheltenham.
65. Paris R. (2001) Human Security. Paradigm Shift or Hot Air?, *International Security*, 26,2, 87-102.
66. Redclift, M. (1999) Environmental Security and Competition for the Environment, in Lonergan, S.C. (ed) *Environmental Change, Adaptation, and Security*, Kluwer Academic Publishers, Dordrecht, pp. 3-16.
67. Rotchild, E. (1995) What is Security?, *Daedalus*, 124, 3, 53-98.
68. Soroos, M. (1992) Why I am insecure about environmental security, Geopolitical Perspectives on Environmental Security, Cahier 92-05.
69. Soroos, M. S. (1999) Strategies for enhancing Human Security in the face of Global Change, in Lonergan S.C. (ed.) *Environmental Change, Adaptation, and Security*, Kluwer Academic Publishers, Dordrecht, pp. 41-55.
70. Soroos, M. (2002) Approaches to Enhancing Human Security in Matthew, R. A., Fraser, L. *Global Environmental Change and Human Security*. GHECS Professional Report, http://www.gechs.uci.edu/ publications.htm.
71. Suhrke, A. (1999) Human Security and the Interests of the States, *Security Dialogue*, 30,3, 265-276.
72. Thakur, R. (1999) The UN and Human Security, *Canadian Foreign Policy*, 7,1, 51-59.
73. Thomas, N., Tow, W. T. (2002) The Utility of Human Security: Sovereignty and Humanitarian Intervention, *Security Dialogue*, 33,2, 177-192.
74. Thomas, N., Tow, W. T. (2002) Gaining Security by Trashing the State? A Reply to Bellamy & McDonald, *Security Dialogue*, 33,3, 379-382.
75. UNDP (1994) Human Development Report, Oxford University Press, Oxford.
76. UNFCCC Workshop on Methodologies on Climate Change Impact and Adaptation. Hotel Mont Gabriel, Ste Adele, Nr. Montreal 11-14 June 2001, http://unfccc.int/ int/sessions/workshop.
77. Ullman, R. (1983) Redefining Security, *International Security*, 8,1, 129-153.
78. UNESCO (2001) *What Agenda for Human Security in the Twenty-first Century?* UNESCO, Paris.
79. Vigilante, A.(1996) El Concepto Humano de Seguridad, *Claves*, 6,96.(Annex: The Human Concept of Security in Bulgaria National Human Development Report 1998), http://ftp.online.bg/pub/pdf/undp/ publications/security/security-eng/annex. pdf
80. Walt, S.M. (1991) The Renaissance of Security Studies, *International Studies Quarterly*, 35, 211-239
81. Wolfers, A. (1962) *Discord and Collaboration*, Johns Hopkins University Press, Baltimore.
82. The Ministry of Foreign Affairs of Japan (2002) *Diplomatic Bluebook*, Ministry of Foreign Affairs, Tokyo.
83. The World Bank (2003) *Poverty and Climate Change. Reducing the Vulnerability of the Poor through Adaptation*, The World Bank, Washington DC.

Section II

Climate Change, Scenarios and Modelling

Chapter 2

Solar Variability and Climate Change

M. Vázquez[1]
Instituto de Astrofísica de Canarias, E-38205 La Laguna, Tenerife, Spain

The variability of solar radiation is a key factor influencing the terrestrial climate. In the last few decades space-borne measurements of solar irradiance have revealed evidence of a correlation with the 11 yr solar activity cycle, although the amplitude of these changes is too small to produce significant climate variation through direct heating. However, solar activity also shows variations on larger scales that are clearly related to several climate indicators. Various feedback mechanisms are currently being explored that could amplify different internal patterns of the climate system. Finally, the influence of solar variability on the terrestrial climate is compared to that originating from other natural and anthropogenic sources, and predictions for the future are outlined.

1.Introduction

Throughout the Earth's history, the climate has suffered dramatic changes [60,73] characterized by different time scales corresponding to the dissipation rates of the energies involved (Table 1). Although undergoing important crises, climatic conditions have always been compatible with the existence of life on our planet.

Table 1. Temporal Variation of Energy sources producing Changes in Terrestrial Climate.

PLANETS	Time Scales (years)	Effects
Gravitational	$10^4 - 10^5$	Earth's Orbit (ice – ages)
	10^8	Impacts > 1 km
SOLAR		
Gravitational	10^7	Solar Luminosity
Nuclear	10^9	Solar Luminosity
Rotation	10^9	Magnetic Energy
Magnetic	10^3	?
TERRESTRIAL		
Tectonic activity	$10^8 - 10^9$	Atmospheric composition
Volcanic eruptions	10^1	Aerosol emission
Life		Atmospheric composition

[1] I would like to thank the organizers for their kind invitation to participate in the NATO Workshop. We thank M. Schüssler and S. Sofia for a critical reading of the paper and D.T. Shindell for supplying some figures for the talk. This work was partly financed by the Spanish PNAYA project 2001-1649. The paper has been revised for English and style by the Scientific Editorial Service of the IAC

A. Marquina (ed.), Environmental Challenges in the Mediterranean 2000-2050, 29–46.
© 2004 *Kluwer Academic Publishers. Printed in the Netherlands.*

Nowadays, we are confronted with an increase in global temperatures (Figure 1) that characterizes the present climate change. We can see that the warming shows different phases. Starting with some fluctuations during the XIX century, we have a growth at the beginning of XX century, followed by a period of stabilization, including a slight cooling in the northern hemisphere (NH). Finally, a steep increase is noticeable during the last 30 years of the last century. Several monographs give the necessary background of this process [38,39,67].

Figure 1.Hemispheric and global average temperatures of the Earth for the last 150 years expressed as anomalies from the mean of the period 1961--1990

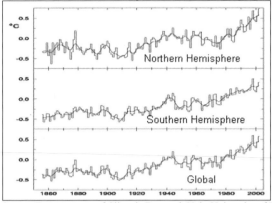

Source: Data courtesy of Climatic Research Unit, University of East Anglia, UK.

On this time scale, hundreds of years, a change in the terrestrial climate can be produced by variations in natural and anthropogenic factors (see scheme of Figure 2), namely: a) incoming solar radiation, and b) the amount of reflected solar radiation, also called albedo. This include natural changes (clouds, volcanic aerosols) and anthropogenic ones, derived from the emission of industrial aerosols and from changes in land use and deforestization, and c) the emission of greenhouse gases (CO_2, CH_4, N_2O), mainly of anthropogenic origin. Finally, we must also consider the internal variability of the climate that also occurs in the absence of external influences. Its quantitative determination is a crucial task for the future.

Figure 2. Diagram showing the different factors related with the change in the Earth's surface temperatures. Dashed lines separate the different environments involved in the process: Sun, Earth's atmosphere and Earth's surface.

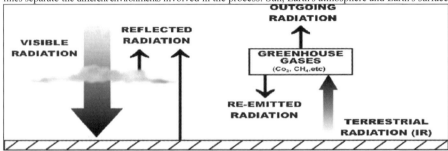

In this paper we are mainly interested in the solar contribution to global warming. Of all the forms of solar energy [see 24 for a summary] magnetic energy is the most suitable for this time scale of hundreds of years. Various review papers [10,21,29,50,66,94] reports and proposals [45,61,62], monographs [11,16,37,40,64,94] and conference proceedings [25,63,65,82,96,98] provide an evolutionary view of our knowledge of this topic.

2.Channels of Sun--Earth Interaction

The solar magnetic field emerges at the solar surface, forming, depending on its size, dark sunspots and bright faculae (see Figure 3). These constitute the active regions, characterized by the closed topology of their magnetic field lines.

Figure 3: Image of the solar surface, close to the limb, obtained at the Swedish Vacuum Solar Tower at the Roque de los Muchachos Observatory (La Palma). It shows large dark sunspots and small bright faculae.

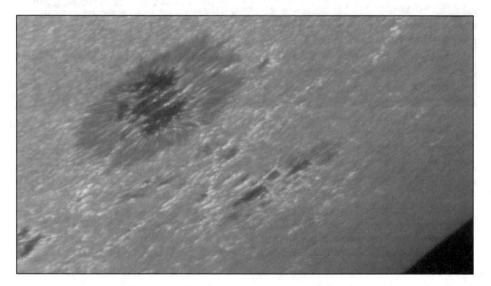

Large-scale magnetic regions have field lines open toward the interplanetary medium. They are visible in X-ray images as coronal holes (Fig. 4) and are the main source of a continuous outward flow of charged particles (protons, electrons and He nuclei) known as the solar wind.

The solar magnetic field of the open regions (OMF) is frozen in this wind, configuring the interplanetary magnetic field (IMF), which produces a huge magnetic region, the heliosphere, that fills practically the whole Solar System. Galactic cosmic rays (hereafter GCRs) are high-energy particles (mainly protons, with energies in the range 1---20 Gev), originating outside our planetary system and striking the Earth from all directions. Both the flux and energy spectrum of GCRs are modulated by the strength of the heliosphere, being stronger when the IMF is weaker.

Figure 4: X-ray image of the Sun obtained by the Yohkoh satellite. At one of the poles we see a coronal hole: a region of reduced X-ray emission.

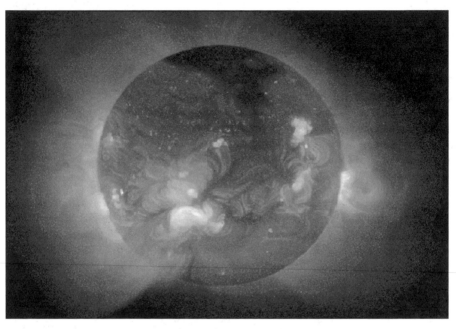

Source: T. Shimizu, The Institute of Space and Astronautical Science.

Table 2 shows the energy transmitted in the different channels. Although the flux is much less than in the visible, the high-energy radiation and the particle flux are interesting for their capacity to ionize matter.

Table 2. Energy densities in the three main channels for the Sun--Earth connection

Visible radiation	1366 Watt / m^2
UV and X – ray Radiation	15.5 Watt/ m^2
Solar Wind	0.0003 Watt/m^2

Fortunately, we are protected on the terrestrial surface from their actions by the absorbing properties of atmospheric gases and by the magnetosphere, respectively. Therefore, let us start by considering first the influence of the visible radiation.

3.The measurement of solar irradiance

In principle, we can expect that when a large group of sunspots crosses the solar disc, carried by the solar rotation, we will get a decrease in the amount of flux arriving from the Sun, the solar irradiance S_S. Moreover, the sunspot number changes with a 11 yr cycle (Figure 5) and therefore we can also expect to measure a variation in solar flux correlated with the level of solar activity.

Figure 5: Temporal variation of the number of sunspots since 1700. A predominant 11 yr cycle is clearly visible, although an 80--90 yr period is also detected.

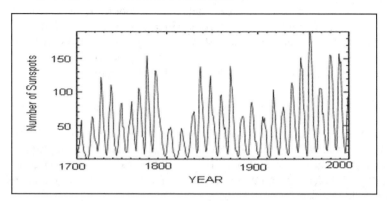

Fluctuations in atmospheric transparency hamper ground-based observations of solar irradiance [1,40]. Only by space observations was it possible to reach the necessary accuracy to detect temporal changes in SS. Over the past two decades, satellite observations have revealed that the total solar irradiance, S_S, changes both on short time scales [99,101] and on the time scale of the solar cycle [102]. Classically, this has been interpreted as being due to the above-mentioned structures (sunspots and faculae) visible at the surface. However, we should also consider a contribution to the observed ΔS_S fluctuations from a mechanism operating deeper inside the Sun, owing to the larger energetic content (e.g 78]

Figure 6. Variation in solar irradiance during the last two solar cycles.

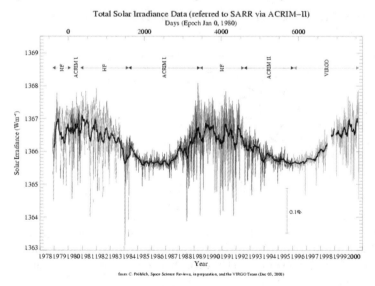

Source: Fröhlich, C., Lean, J. (1998) The Sun´s irradiance: Cycles, trends, and related climate change uncertainties since 1976, *Geophysical Research Letters 25*, 4377—4380 [32].

Fröhlich & Lean [32] have combined the data sets from different experiments into a common temporal series (Figure 6). These data are periodically updated at *http://www.pmodwrc.ch/pmod.php?topic=tsi/composite/SolarConstant.html* and do not show any noteworthy difference between the last two activity cycles. However, we should note that the procedure is not trivial, involving detrending methods, which could eliminate slight long-term variations [see, e.g.100 for a different analysis]

3.1. DIRECT INFLUENCE OF SOLAR HEATING

We establish a simple energy balance between the received energy per unit time and area, $E_{in} = S_S(1 - a)$, and that emitted by the Earth, E_{out}, under the simplifying assumption that there is no greenhouse effect. The terrestrial cross-section exposed to solar energy is πR_E^2, whereas the total surface of the Earth is $4 \pi R_E^2$ (Figure 7). Therefore, balancing the flux densities,

$$\pi R_E^2 E_{in} = 4 \pi R_E^2 E_{out} = F_E$$

Assuming now that the Earth radiates as a black body ($F_E = \sigma T^4$), then

$$T = [S_S(1 - a)/4 \sigma]^{1/4}$$

where a is the terrestrial albedo, R_E the Earth's radius, σ the Stefan--Boltzmann constant and T the mean equilibrium temperature of our planet.

Figure 7. The cross-section exposed to the solar energy corresponds to an area of πR_E^2, whereas the total surface area of the Earth is $4 \pi R_E^2$. }

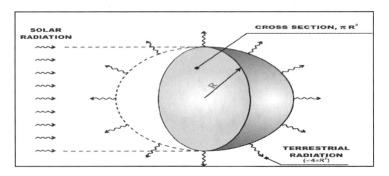

We see that only a fraction, $Q = S_S(1 - a)/4$, of the energy received is effective for climate changes. Taking $S_S = 1368$ W/m^2 we get $Q = 239.4$ W/m^2. Climate models indicate the existence of a linear relationship between the radiative forcing[2], ΔF, and the

---

[2] Climate forcing is a perturbation of the Earth's radioactive energy balance, with the convention that a positive forcing leads to warming, and a negative forcing to cooling.

resultant temperature change, ΔT, expressed by $\Delta T = \lambda \Delta F$, where λ is the climate sensitivity. For example, a doubling of atmospheric CO_2 from pre-industrial levels will produce $\Delta F = 4$ W / m^2 and a ΔT between 1.5 and 4.5 °C (Houghton et al., 2001). Therefore, we will have $\lambda = (3 \pm 1.5)/4 = (0.75 \pm 0.4)$ °C/W m^{-2}. This value can be compared with the response of the Earth if it were to act as a black body ($\lambda = 0.3$ °C/W m^{-2}).

Simple energy-balance models [17,74] or more sophisticated global circulation models [18,71,72] have been used to estimate the terrestrial response to variable solar forcing and give values of λ in the range 0.5--1.0. A simple calculation indicates that the influence of solar radiation changes during the solar cycle (0.1% of 239.4 (0.24 W / m^2) on terrestrial climate should be negligible ($\Delta T = 0.07$ °C for $\lambda = 0.3$).

However, there are several facts indicating that we have an influence well above this simple estimate. Friis-Christensen & Lassen [31] show an apparently clear correlation between the length of solar cycles and terrestrial temperatures, with shorter (longer) cycles related to episodes of warming (cooling).

Lassen & Friss-Christensen [49] extended this link to the past five centuries. Other correlations of the solar cycle length with different climate parameters have also been found [15,36].

Moreover, the idea of static equilibrium and instantaneous climatic response represents an unrealistic situation. The response of the system depends also on its heat capacity, C. This will produce phase lags between cause and effect. White et al. [97] determined that changes in the global temperature of upper oceanic layers fluctuate in phase with solar irradiance on decadal, interdecadal and centennial time scales.

The next advance was achieved by White, Lean & Cayan [97], who showed how the fluctuations, S, on interdecadal scales are capable of exciting an interdecadal mode of variability in the terrestrial climate. This idea was quantified by White et al. [97], who estimate that a factor of 2 or 3 in the amplification of the solar forcing is caused by this effect.

4.Long-term variability of solar activity

4.1.THE MAUNDER MINIMUM

Although Spörer [83] and Maunder [58] had already noticed the existence of a period of reduced activity during the XVII century, it was Eddy [27] who brought the attention of the scientific community to the lack of sunspots during the second half of the XVII century. In his seminal paper Eddy also showed the coincidence of this event, hereafter called MM, with a period of important cooling in London and Paris.

The objective verification of the existence of this event comes from the above-mentioned relation between GCRs and solar activity through a process depicted in Figure 8. Cosmic rays react with the main constituents of the terrestrial atmosphere (N,O) producing such isotopes as ^{14}C and ^{10}Be. After a certain residence time in the atmosphere, these elements are incorporated into trees and ice-cores, respectively. There, we can measure their concentration in age-calibrated layers, which are a good indicator of the time variability of solar activity.

Figure 8. The steps between the production of cosmogenic isotopes by the action of cosmic rays and their deposition in terrestrial archives.

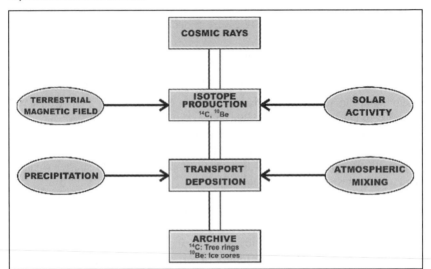

Periodicities of 2300, 212 and 87 years have been found in the ^{14}C records, showing that the MM is not a unique event, although other factors such as changes in the geomagnetic magnetic field and in the carbon cycle could also play a role [84,86]

On longer time scales measurements of isotopic ratios, such as ^{18}O/^{16}O, are sensitive indicators of ocean temperatures. Measurements in the North Atlantic show periods of abrupt climate change. Examples of abrupt cooling, during the Holocene, are a short event at 10,300 yr BP, which lasted less than 200 yr [13] and many others with a mean frequency of 1500 years [12]. These are mainly explained by the invasion of fresh water, which alters circulation in the Atlantic Ocean. However, modelling and correlations with ^{14}C and ^{10}Be concentrations show how the solar influence could trigger and sustain such periods.

Solar-like stars provide us with information about variations in magnetic activity on even longer time scales. The archive of Mount Wilson [6] constitutes a stellar analogue of the sunspot number records. Rotation periods, P_{rot}, and cycle lengths, (P_{cyc}), are the most relevant outputs. Baliunas & Soon [8] verified the relation between brightness variations and the length of the activity cycle, using stellar observations.

This long-term monitoring has permitted the verification that approximately one third of the sampled stars are in a situation analogous to the Maunder Minimum [6]. Stars of a similar age to the Sun exhibit the same behavioural pattern, with maximum brightness at the cycle maximum. However, for young stars the situation is reversed [70]. This can be explained in terms of a greater contribution coming from sunspots instead of the dominant long-term contribution from plages in Sun-like stars. The level of irradiance variations in the Sun over the solar cycle seems to be low compared to the average activity level [53].

4.2.SIMULATION OF LONG-TERM TRENDS IN SOLAR IRRADIANCE

Simulating modern irradiance observations by means of an adequate combination of different kinds of proxies, it could be possible to extrapolate the irradiance changes back in time to periods for which no satellite measurements are available. In other words, *the present could be the key to the past*. What we learn from the past could be applied to the forecasting of future behaviour: *the past could be the key to the future*. However, the validity of these principles for the solar activity remains to be demonstrated.

Figure 9 shows one of these simulations [52]. As expected, we have reduced fluxes of solar radiation at the Maunder and Dalton minima and an increase since the middle of XIX century.

Figure 9. Simulation of solar irradiance variations since 1600.

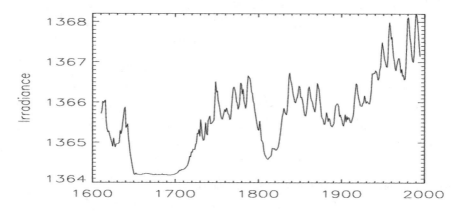

Source: Lean, J., Beer, J., Bradley, R. (1995) Reconstruction of solar irradiance since 1610: Implications for climate change, *Geophysical Research Letters 22*, 3195 – 3198 [52].

Rind et al. [72] have used these ΔS_S reconstructions to model the climate response of the oceans. They found an increase in temperatures $\Delta T = 0.5$ °C since 1600, and 0.2 over the last century. Ocean temperatures lagged solar fluctuations, ΔS_S, by up to 10 years.

4.3.EUROPEAN CLIMATE DURING PERIODS OF REDUCED SOLAR ACTIVITY

Figure 10 shows estimates of the temperature variation over the last millenium. The Maunder and Dalton minima are clearly identified as the coldest part of the so-called "Little Ice-Age" [28]. We should note that the last event coincides with the eruption of the Tambora volcano, in 1816, producing `the year without a summer' [22]. Especially impressive in this plot is the steep increase of temperatures in the XX century, the warmest epoque of this millenium.

Figure 10: Global temperature estimate (°C anomalies from the observed 1961--1990 mean) in the last millenium. (solid) Jones et al. 2001; (dotted) Mann et al., 1998; (dashed) instrumental records.

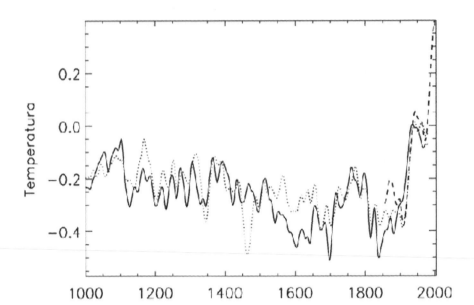

Source: Data courtesy of World Data Center for Paleoclimatology, Boulder.

The European climate is clearly dominated by fluctuations in the North Atlantic Oscillation (NAO), especially during the winter [42,92,103]. The NAO index is defined as the difference in sea-level pressure between two stations situated close to the ``centres of action'' over Iceland and the Azores. Strong positive phases of the NAO tend to be associated with above-normal temperatures in northern Europe and below-normal temperatures across southern Europe and the Middle East. They are also associated with above-normal precipitation over northern Europe and Scandinavia and below-normal precipitation over southern and central Europe. Opposite patterns of temperature and precipitation anomalies are typically observed during strong negative phases of the NAO. The NAO is clearly associated with the Arctic Oscillation (AO), defined by its effects on the stratosphere [90]. During its positive phase, there is a strengthening of the polar vortex and a subsequent cooling of the stratosphere.

During the MM important changes occur in the pressure distribution in the North Atlantic, triggering cold episodes in central and southern Europe [56,77]. It seems that the reduced solar forcing amplifies one of the natural patterns of variability of the climate, the negative phase of the NAO, producing cold and wet weather in central and southern Europe. In modern times, evidence for solar influence on sources of internal variability of the terrestrial climate system (Indian Monsoons, El Niño/La Niña, the North Atlantic Oscillation) has also been presented [2, 59]

Drijfhout et al. [26] Tobias & Weiss [91] and Ganopolski & Ramhstof [33] have studied this type of climate response produced by the interaction between an external forcing and a stochastic process representing the internal variability.

5.Feedback mechanisms

The upper layers of the Earth's atmosphere suffer directly the action of the high-energy radiation and flux of GCRs. Physical parameters, such as the temperature, are clearly correlated with the activity level, producing, for example, the change of altitude in satellites placed in orbits below 300 km [9].However, to transfer this influence to the troposphere, and so to the climate, we need to discover a suitable mechanism.

5.1.THE UV--OZONE LINK

The spectral radiation longward of 300 nm accounts for 99 % of the total solar irradiance. However, the amplitude of the variability from maximum to minimum of the solar cycle increases with decreasing wavelength. As a result, solar UV radiation below 300 nm accounts for up to 14% of the total variation of the irradiance [30].

The Berlin data set supplies continuous monitoring of geopotential heights[3] at different pressures in the lower stratosphere since 1957 [46]. Measurements show a clear correlation with solar activity over a large vertical and latitudinal range [46,47].

The interaction of UV radiation with tropospheric ozone via the process of formation and decay (Figure 11) seems to be the most promising mechanism for explaining this correlation. The net effect over the solar cycle is an increase of a few per cent in ozone [20,21]. The result is stratospheric heating in sunlit areas, which is balanced by subsequent infrared cooling, producing almost negligible radiative forcing [34,35,48] However, it represents the dynamical channel to transfer the perturbation to lower layers and so to the terrestrial climate.The agent could be planetary waves, which are formed in the troposphere by orographic obstacles (the Tibet plateau, the Rocky mountains, etc.) and land--sea contrasts, and are clearly stronger in the northern hemisphere. Planetary waves propagate upwards into the stratosphere and provide a way to connect stratospheric influences to the lower layers [4,5].

Figure 11. Role of solar ultraviolet radiation in controlling the balance of the ozone in the stratosphere.

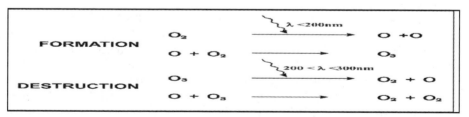

[3] Geopotential is a measure of the energy required to lift a mass of air from the surface to a given altitude, working against gravity. Geopotential height (GH) is the geopotential divided by g0, the globally averaged value of terrestrial gravity. Usually, it is considered as the GH at which atmospheric pressure has a particular value, for example 70mb at 18.5 k m. If the temperature in the column below increases, the level of 70 mb will be found to be higher.

In modelling this interaction, special attention has been given to behaviour in the winter northern hemisphere. The corresponding polar region is in total darkness, so temperatures will not change, inducing a latitudinal temperature gradient and a zonal wind and affecting the propagation of planetary waves [44]. Shindell et al. [75] have developed a code to model climate response to these UV variations.

In MM conditions [77] they found that the global ΔT produced is small (0.3 - 0.4 °C). However, a shift is produced towards the low index state of the NAO/AO, leading in winter to strong coolings (1--2 °C) in central Europe, in clear agreement with the historical data and proxy data for surface temperatures. In contrast, the effect of volcanoes leads only to global cooling.

Shindell et al. [77] have used the same climate code to study the impacts of increasing greenhouse gases, polar ozone depletion, volcanic eruptions and solar-cycle variability on climate. They show that, while ozone depletion has a significant effect, greenhouse gas forcing is the only factor capable of causing the large, sustained increase in the AO over recent decades.

Other models [34,35] have tried to reproduce the behaviour in the summer hemisphere, finding a broadening of the Hadley cells and a poleward movement of subtropical jets and mid-latitude storm-tracks.

5.2. THE COSMIC RAY--CLOUD CONNECTION

We can now return to GCRs, but as direct agents of climate change. Svensmark & Friis-Christensen [87] and Svensmark [88] proposed the existence of a correlation between such a GCR flux and the level of terrestrial cloudiness.

When more data were accumulated, it was found that the correlation was only valid for low-lying clouds[4], producing a net effect of cooling on the climate [68,88]. Surprisingly, these last authors found that the tropics are the regions where the correlation between GCR flux and cloud-top temperatures is higher.

Lockwood et al. [54] derived the intensity of the interplanetary magnetic field (IMF) from the *aa* geomagnetic index[5], a record that extends back to 1868 (Figure 12). They found that the average strength of the solar magnetic field has doubled over the last 100 years, implying a decrease in the GCR flux, which has in fact been observed [85]. Lockwood & Stamper [55] deduced that this increase corresponds to a rise in the average total solar irradiance of 1.65 ± 0.23 W/m^2 (0.29 ± 0.04 of the effective irradiance, Q).

This last value can be compared with the thermal effect (0.8 W/m^2) of decreasing the expected amount of low cloud [45]. However, this should imply a corresponding decrease in the fraction of low cloud over the same period, a hypothesis that is difficult to verify. In fact, an increase in total cloud has been detected, although using data limited to some locations and using such proxies as sunshine [68].

Solanki & Schüssler [79] have simulated the variations in the OMF (the dominant factor of the IMF) with a model based on the emergence and decay rates of active regions that fits the ^{10}Be records reasonably well.

---

[4] These are below an altitude of 3 km and include stratocumulus and stratus cloud types.
[5] The *aa* index is a measurement of the geomagnetic field at the Earth's surface by two antipodal stations.

Figure 12: Variation in the interplanetary magnetic field, expressed by the geomagnetic index aa, showing a long-term increase during the XX century.

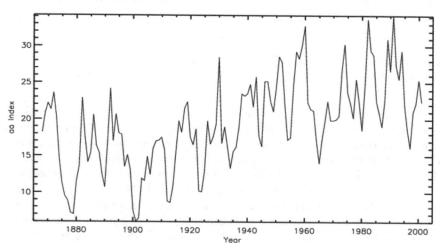

The model has been improved by Solanki et al. [80] by additionally taking into account the contribution of the ephemeral magnetic regions[6], whose overlapping between consecutive activity cycles together with a long lifetime of large-scale magnetic patterns explains the secular variation in open magnetic flux. On the basis of this model, Usoskin et al. [93] have reconstructed the cosmic ray intensity since 1610.

6.The role of solar changes in global warming

The separation of the solar and anthropogenic components in the temperature record is not a trivial task. In principle, greenhouse gases will produce cooling in the stratosphere, and solar radiation will cause heating, but the limited length of the records and the influence of other events (volcanic eruptions) prevents clear conclusions from being established. Generally, the discrimination between natural and man-made sources of global warming has been made by fitting the response of the climate to different forcings to the measured ΔT [3,89].

The IPCC has recently published its last report [39] and in Table 3 are summarized the different contributions to the global radiative forcing of the climate from pre-industrial times to the present. The net effect of the anthropogenic effects (2.9 - 0.2 - 1.4 = 1.3 W/m^2) with respect to the solar (0.4) is significant, although our knowledge of aerosols is minimal. However, if we include the contribution from the GCR--cloud interaction, estimated at 0.8 W/m^2, the difference (1.3 vs. 1.2) should be really marginal.

---

[6] Small-scale magnetic regions located outside the active regions.

Table 3. Radiative forcing from pre-industrial times to the present.

Forcing	W / m ^{2}
Greenhouse Gases	2.9
Land Use	- 0.2
Aerosols	- 1.4
Solar Irradiance	0.4

Source: Houghton, J.T., Ding, Y., Griggs, D.J., Noguer, M., Van der Linden, P.J., Dai, X., Maskell, K., Johnson, C.A. (2001) *Climate Change 2001: The Scientific Basis*, Cambridge University Press, Cambridge [39].

The long-term prediction of solar activity is a very difficult task, because the underlying physical mechanism has a strong chaotic component. A combination of the periodicities observed in the ^{14}C record leads us to expect a slight increase until 2050, followed by a reduction, which could perhaps lead to another event similar to the Maunder or Dalton ones, at the end of this century. Lean [49] estimates the solar forcing between the cycle minimum in 1996 and 2016 to be ~ 0.1 W/m^2. For comparison, the expected anthropogenic forcing in the same period is in the range 0.5--0.9 W/m^2.[For a dissenting view on this conclusion, see 81]. In any case, the action of solar *oscillatory* forcing on climate will be obscured by the *growing* influence of the anthropogenic emission of greenhouse gases.

7.Conclusions

• Statistical studies and models indicate that much of the variability in the climate over the last millenium can be accounted for by variations in solar output, volcanic eruptions and the internal variability of the climate system.

• XX century warming cannot be explained by these mechanisms. Only increased levels of greenhouse gases---of anthropogenic origin---are able to explain the unusual warming.

• The climate in the XXI century will be dominated by the anthropogenic emission of greenhouse gases.

References

1. Abbot, C.G. (1935) Solar Radiation and weather studies, *Smithsonian Miscellaneous Collections*, *94,10*, 1-89,

2. Agnihotri, R., Dutta, K., Bhushan, R., Somayajulu, B.L.K. (2002) Evidence for solar forcing on the Indian monsoon during the last Millenium, *Earth and Planetary Science Letters 198*, 521 - 527,.

3. Andronova, N.G., Schlesinger, M.E. (2000) Causes of global temperature changes during the 19th and 20th centuries, *Geophysical Research Letters 27*, 2137 - 2140.

4. Arnold,N.F., Robinson, T.R. (1998) Solar cycle changes to planetary wave propagation and their influences on the middle atmosphere circulation, *Ann. Geophysicae 16*, 69 - 76.

5. Arnold, N.F., Robinson, T.R. (2000) Amplification of the influence of solar flux variations on the middle atmosphere by planetary waves, *Space and Science Reviews 94*, 279 - 286.

6. Baliunas, S.L., Jastrow, R. (1990) Evidence for long-term brightness changes of solar-type stars, *Nature 348*, 520—523.

7. Baliunas, S.L. et al. (1995) Chromospheric variations in main-sequence stars, *Astrophysical Journal 438*, 269 – 287.

8. Baliunas, S.L., Soon, W. (1995) Are variations in the length of the activity cycle related to changes in the brightness in solar-type stars?, *Astrophysical Journal 450*, 896 -901.

9. Banks, P.M., Kockarts, G. (1973) *Aeronomy*, Academic Press, New York.

10. Beer, J., Mende, W., Stellmacher, R. (2000) The role of the Sun in climate forcing, *Quaternary Science Reviews 19*, 403-415.

11. Benestad, R.E. (2002) *Solar Activity and Earth's Climate*, Springer, Heidelberg.

12. Bond, G.C. (2001) Persistent solar influence on North Atlantic climate during the Holocene, *Science 294*, 2130-2136.

13. Björck, S., et al. (1990) High-resolution analyses of an early Holocene climate event may imply decreases solar forcing as an important climate trigger, *Geology 29*, 1107-1110.

14. Bradley R..S. and Jones P.D. (1993) Little Ice Age summer temperature variations: their nature and relevance to recent global warming trends, *The Holocene 3/4*, 367-376.

15. Butler, C.J. (1994) Maximum and Minimum temperatures at Armagh Observatory, 1884 - 1922, and the length of the Sunspot cycle, *Solar Physics 152*, 35-42.

16. Calder, N. (1997) *The Manic Sun*, Pilkington Press, London.

17. Crowley, T.J., Kim, K.Y. (1996), Comparison of proxy records of climate change and solar forcing, *Geophysical Research Letters 26*, 1901—1904.

18. Cubasch, U., Voss, R., Hegerl, G.C., Waszkewitz, J., Crowley, T. (1997) Simulation of the influence of solar radiation variations on the global climate with an ocean - atmosphere general circulation model, *Climate Dynamics 13*, 757 – 767.

19. Chandra ,S. (1991) The solar UV related changes in total ozone from a solar rotation to a solar cycle, *Geophysical Research Letters 18*, 837 – 840.

20. Chandra, S., Ziemke, J.R., Stewart, R.W. (1999) An 11 - year solar cycle in tropospheric ozone from TOMS measurements, *Geophysical Research Letters 26*, 185 - 188,.

21. Chapman, G.A. (1987) Variations of solar irradiance due to magnetic activity, *Annual Review of Astronomy and Astrophysics 25*, 633 - 667.

22. Chenoweth, M. (2001) Two major volcanic cooling episodes derived from global marine air temperature, AD 1807 - 1827, *Geophysical Research Letters 28*, 2963 -2966.

23. Damon, P.E., Jirikowic, J.L. (1992) *Solar forcing on global climate change*, in R.E. Taylor, A. Long, R.S. Kra (eds.) *Radiocarbon after four decades*, Springer, Heidelberg.

24. De Jager C. (1972) Solar energy sources, in Dyer E.R. (ed) *Solar Terrestrial Physics:* Part I, Reidel, Dordrecht, 1-8.

25. Donnelly, R.F.(ed) (1992) *Workshop on the Solar Electromagnetic Radiation Study for Solar cycle 22*, US Department of Commerce, Washington DC.

26. Drijfhout, S..S., Haarsma, R..J., Opsteegh, J..D., Selten, F..M. (1999) Solar-induced versus internal variability in a coupled climate model, *Geophysical Research Letters 26*, 205-208.

27. Eddy, J.A. (1976) The Maunder Minimum, *Science 192*, 1189-1202.

28. Fagan, B. (2000) *The Little Ice Age*, Basic Books.

29. Fligge, M., Solanki, S..K., Pap, J..M., Fröhlich, C., Wehrli, C. (2001) Variations of solar spectral irradiance from near UV to the infrared - measurements and results, *Journal of Atmospheric and Solar-Terrestrial Physics 63*, 1479-1487.

30. Floyd, L., Tobiska, W..K., Cebula, R.P. (2002) Solar UV Irradiance, its variation, and its relevance to the Earth, *Advances in Space Research. 29, 10,* 1427-1440.
31. Friis-Christensen, E., Lassen, K. (1991) Length of the solar cycle: an indicator of solar activity closely associated with climate, *Science 254*, 698.
32. Fröhlich, C., Lean, J. (1998) The Sun´s irradiance: Cycles, trends, and related climate change uncertainties since 1976, *Geophysical Research Letters 25*, 4377-4380.
33. Ganopolski,A..Rahmstof, S. (2002) Abrupt glacial climate changes due to stochastic resonance, *Physical Review Letters 88*, 038501.
34. Haigh, J. (1996) The impact of solar variability on climate, *Science 272*, 981-984.
35. Haigh, J.,D. (1999) Modelling the impact of solar variability on climate, *Journal of Atmospheric and Solar-Terrestrial Physics 61*, 63 –72.
36. Hameed, S., Gong, G. (1994) Variation of spring climate in lower-middle Yangtse river valley and its relation with the solar cycle, *Geophysical Research Letters 21*, 2693-2696
37. Herman, J.R., Goldberg, R.A. (1978) Sun, weather, and climate, *NASA SP* ,426.
38. Houghton, J. (1997) *Global warming: The complete briefing*, Cambridge University Press, Cambridge.
39. Houghton, J.T., Ding, Y., Griggs, D.J., Noguer, M., Van der Linden, P.J., Dai, X., Maskell, K., Johnson, C.A. (2001) *Climate Change 2001: The Scientific Basis*, Cambridge University Press, Cambridge.
40. Hoyt, D.V. (1979) The Smithsonian Astrophysical Observatory solar constant program, *Review of Geophysics and Space Physics 17,* 427-458.
41. Hoyt, D.V., Schatten, K. (1998) *The role of the Sun in Climate Change*, Oxford University Press, Oxford.
42. Hurrell, J.W. (1995) Decadal trends in the North Atlantic Oscillation and relationships to regional temperature and precipitation, *Science 269*, 676-679.
43. Jones, P.D., Osborn, T.J., Briffa, K. (2001) The evolution of the climate over the last millenium, *Science 292*, 292.
44. Kodera, K. (1995) On the origin and nature of the interannual variability of the winter stratospheric circulation in the northern hemisphere, *Journal of Geophysical Research 100*, 14077-14087.
45. Kirkby, J., Cloud (2001) A particle beam facility to investigate the influence of cosmic rays on clouds, *Proc. Workshop on Ion - Aerosol-Cloud Interactions*, ed. J. Kirkby, CERN, Geneva, available at http://cloud.web.cern.ch/cloud/index.html.
46. Labitzke, K., van Loon, H. (2000) *The Stratosphere*, Springer, Heidelberg.
47. Labitzke, K. (2001) The global signal of the 11 - year sunspot cycle in the stratosphere: Differences between solar maxima and minima, *Meteorologische Zeitschrift 10, 2*, 83-90.
48. Larkin, A., Haigh, J.D., Djavidnia, S. (2001) The Effect of Solar UV Irradiance Variations on the Earth's Atmosphere, *Space Science Reviews 94*, 199-214.
49. Lassen, K., Friis - Christensen, E. (1995) Variability of the solar cycle during the last five centuries and the apparent association with terrestrial climate, *Journal of Atmospheric and Terrestrial Physics 57*, 835—845.
50. Lean, J. (1997) The Sun variable radiation and its relevance for Earth, *Annual Review of Astronomy and Astrophysics 35*, 33 - 67.
51. Lean, J. (2001) Solar irradiance and climate forcing in the near future, *Geophysical Research Letters 28*, 4119-4122.
52. Lean, J., Beer, J., Bradley, R. (1995) Reconstruction of solar irradiance since 1610: Implications for climate change, *Geophysical Research Letters 22*, 3195-3198.
53. Lockwood,G.W., Skiff, B.A.,Baliunas, S..L., Radick, R..R. (1992) Long-term brightness changes estimated from a survey of Sun-like stars, *Nature 360*, 653-655.
54. Lockwood, M.., Stamper, R. (1999) A doubling of the Sun´s coronal magnetic field during the past 100 years, *Nature 399*, 437-439.
55. Lockwood, M., Stamper, R. (1999) Long-term drift of the coronal source magnetic flux and the total solar irradiance, *Geophysical Research Letters 26*, 2461-2464.
56. Luterbacher, J., Rickli, R., Xoplaki, E., Tinguely, C., Beck, C., Pfister, C., Wanner, H. (2001) The late Maunder Minimum (1675-1715) - a key period for studying decadal scale climatic change in Europe, *Climatic Change 49*, 441-462.
57. Mann, M..E., Bradley, R..S., Hughes, M..K. (1998) Global-scale temperature patterns and climate forcing over the past six centuries, *Nature 394*, 779-787.
58. Maunder R.W. (1894) A prolonged sunspot minimum, *Knowledge 17*, 173-176.
59. Mehta, V., Lau, K.M. (1997 Influence of solar irradiance on the Indian monsoon-ENSO: Relationship at decadal-multidecadal time scales, *Geophysical Research Letters 24*, 159-162.

60. Mitchel, J.M. (1982) An overview of climatic variability and its causal mechanisms, *Quaternary Research* 6, 481-493, 1973
61. National Research Council (1982) *Solar variability, weather and climate*, National Research Council, Washington.
62. National Research Council (1994) *Solar influences on global change*, National Research Council, Washington.
63. Nesme-Ribes, E. (ed.) (1994) *The solar engine and its influence on Terrestrial Atmosphere and Climate*, Springer , Heidelberg.
64. Nesme - Ribes, E., Thuillier, G. (2001) *Histoire solaire et climatique*, Éditions Belin, Paris.
65. Pap, J., Fröhlich, C., Hudson, H..S., Solanki, S.K. (eds.) (1994) *The Sun as a Variable Star: Solar and Stellar Irradiance Variations*, Cambridge University Press, Cambridge.
66. Pap, J., Fröhlich, C. (1999) Total solar irradiance variations, *Journal of Atmospheric and Solar Terrestrial Physics 61*, 15.
67. Philander, S.G. (1999) *Is the temperature rising?*, Princeton University Press. Princeton.
68. Palle, E., Butler, C.J (2002) The proposed connection between clouds and cosmic rays: Cloud behaviour during the past 50-120 years, *Journal of Atmospheric and Solar-Terrestrial Physics 64*, 327--337.
69. Pecker, J.C., Runcorn, S.K. (eds.) (1990) *The Earth's climate and variability of the Sun over recent Millenia*, Cambridge University Press, Cambridge.
70. Radick, R.R., Lockwood, G.W., Skiff, B.A., Baliunas, S.L. (1998) Patterns of variation among sun-like stars, *Astrophysical Journal Supplement Series 118*, 239-258.
71. Rind, D., Overpeck, J. (1993) Hypothesized causes of decadal - to century climate variability: climate model results, *Quaternary Science Reviews 12*, 357-374.
72. Rind, D., Lean, J., Healy, R. (1999) Simulated time-dependent climate response to solar radiative forcing since 1600, *Journal of Geophysical Research 104*, D2, 1973 - 1990.
73. Rudiman, W.F. (2001) *Earth's climate: Past and Future*, W.H. Freeman and Company, New York.
74. Schlesinger, M.E., Ramankutty, N. (1992) Implications for global warming of intercycle solar irradiance, *Nature 360*, 330 - 333.
75. Shindell, D., Rind, D., Balachandran, N., Lean, J., Lonergan, P. (1999) Solar cycle variability, ozone and climate, *Science 284*, 385 -- .
76. Shindell, D.T., Schmidt, G.A., Mann, M.E., Rind, Waple, A. (2001) Solar forcing of regional climate change during the Maunder Minimum, *Science 294*, 2149-- 2152.
77. Shindell, D.T., Schmidt, G.A., Miller, R.L., Rind, D. (2001) Northern Hemisphere winter climate response to greenhouse gas, ozone, solar and volcanic forcing, *Journal of Geophysical Research 106*, D7, 7193 --, 7210.
78. Sofia, S., Li, L.H. (2001) Solar variability and climate, *Journal of Geophysical Research* 106, A7, 12969-12974.
79. Solanki, S.K., Schüssler, M. (2000) Secular variation of the Sun's large-scale magnetic field, *Nature 408*, 405-407.
80. Solanki, S.K., Schüssler, M., Fligge, M. (2002) Secular variation of the Sun´s magnetic flux, *Astronomy and Astrophysics 383*, 706- 712.
81. Soon, W., Posmentier, E., Baliunas, S. (2000) Climate hypersensitivity to solar forcing, *Ann. Geophysicae 18*, 583-588.
82. Sonett, C.P., Giampapa, M.S., Mattews, M..S. (eds.) (1991) *The Sun in Time*, University of Arizona Press, Tucson.
83. Spörer, G., Über, (1889) Die periodizität der Sonnenflecken seit dem jahre 1618, *Nova Acta der Kaiserliche Leopold- Caroline Deutschen Akademie der Naturfoscher 53*, 283-324.
84. Stephenson, F.R., Wolphendale, A.W. (eds) (1998) *Secular Solar and Geomagnetic Variations in the last 10.000 years*, Kluwer Academic Publishers, Dordrecht.
85. Stozhkov, Y.I., et al. (2000) Long - term negative trend in cosmic ray flux, *Journal of Geophysical Research 105*, 9.
86. Stuiver, M., Braziunas, T.F., Becker, B., Kromer, B. (1991) Climatic ,solar, oceanic and geometric influences on late-Glacial and Holocene atmospheric 14C/12C change, *Quarternary Research 35*, 1-24.
87. Svensmark, H., Friis - Christensen, E. (1997) Variation in cosmic ray flux and global cloud coverage - a missing link in solar - climate relationships, *Journal of Atmospheric and Solar - Terrestrial Physics 59*, 1225.
88. Svensmark, H. (1998) Influence of Cosmic Rays on Earth´s Climate, *Physical Review Letters 81*, 5027-5030.

89. Tett, S.F.B. (1999) Causes of Earth's near-surface temperature change in the twentieth century, *Nature 399*, 569-572.

90. Thompson, W.J., Wallace, J.W. (1998) The Artic Oscillation signature in the wintertime geopotential height and temperature fields, *Geophysical Research Letters 25*, 1297-1300.

91. Tobias, S.M., Weiss, N.O. (2000) Resonance in a coupled solar - climate model, *Space Science Reviews 94*, 153-160.

92. Trigo, R.M., Osborn, T.J., Corte-Real, J. M. (2002) The North Atlantic Oscillation influence on Europe: climate impacts and associated physical mechanisms, *Climate Research 20*, 9-17.

93. Usoskin, I.G., Mursula, K., Solanki, S.K., Schüssler, M., Kovaltsov, G.A. (2003) A physical reconstruction of cosmic ray intensity since 1610, *Journal of Geophysical Research.*.

94. Vázquez, M. (1998) Long-term variation in the total solar irradiance, in A. Hanslmeier & M. Messerotti (eds.), *Motions in the Solar Atmosphere*, Kluwer Academic Publishers, Dordrecht.

95. Vázquez, M. (1998) *La Historia del Sol y el Cambio Climático*, Mc Graw-Hill Interamericana, Madrid.

96. White, O.R. (ed.) (1977) *The solar output and Its Variation*, Colorado Associated University Press.

97. White, W.B., Lean, J., Cayan, D.R., Dettinger, M.D. (1997) A response of global upper ocean temperature to changing solar irradiance, *Journal of Geophysical Research 102*, 3255-3266.

98. Wilson, A. (ed.) (2000) *The Solar Cycle and Terrestrial Climate*, ESA SP-463.

99. Willson, R.C. (1982) Solar Irradiance variations and Solar Activity, *Journal of Geophysical Research 87*, 4319-4326.

100. Willson, R.C. (1997) Total solar irradiance trend during solar cycles 21 and 22, *Science 277*, 1963-1965.

101. Willson, R.C., Gulkis, S., Janssen, M., Hudson, H..S., Chapman, G.A. (1981) Observations of solar Irradiance Variability, *Science 211*, 700-702.

102. Willson, R.C., Hudson, H.S., Fröhlich, C., Brusa, R.W. (1986) Long-term downward trend in total solar irradiance, *Science 234*, 1114-1117.

103. Zorita, E., Kharin, V., Von Storch, H. (1992) The atmospheric circulation and Sea Surface Temperature in the North Atlantic Area in Winter: Their Interaction and Relevance for Iberian Precipitation, *Journal of Climate 5*, 1097-1108.

Chapter 3

Qualitative-quantitative scenarios as a means to support sustainability-oriented regional planning

P. Döll

Center for Environmental Systems Research, University of Kassel
Kurt-Wolters-Str. 3, 34109 Kassel, Germany

Scenario analysis facilitates a better understanding of future environmental change and is a useful means to support a regional planning that is geared towards sustainable development. In particular if developed by close cooperation of decision-makers and scientists, scenarios help to understand the consequences of today's decisions in a quite distant and uncertain future. Scenarios describe a range of plausible futures in an integrated manner, considering the most important driving forces of the socio-environmental system of interest. Ideally, they combine qualitative and quantitative elements, i.e. narratives (storylines) with mathematical modeling. In this paper, a methodology to develop environmental scenarios is described, considering, in particular, the issues of participation and scale. Examples for scenarios which explore the impact of climate change on water resources and irrigation requirements in the whole Mediterranean region are provided. Due to the high uncertainty of the precipitation changes computed by global climate models, it is not possible to determine, on the river basin scale, how different greenhouse gas emission scenarios will translate to changes in water resources and irrigation requirements.

1. Scenario analysis

How can people, institutions and businesses plan for the future when they do not know what tomorrow will bring? Most of us prefer to only think of what we believe is the likely future, instead of really considering that the future can look very different. However, "to operate in an uncertain world, people need to be able to question their assumptions about the way the world works" [18]. Scenario analysis is a methodological approach to deal with the uncertain future, and it helps to arrive at better decisions. An example from the past is the "oil crisis" of the 1970s when the oil price increased drastically, after being more or less constant since World War II. Of the big oil companies, only Shell was prepared as they had worked before with scenarios. Some of these assumed a constant oil price, while others considered the impacts of an oil price hike by OPEC. As a consequence, Shell's position among the big oil companies improved strongly in the following years [18]. Nowadays, scenarios are not only well used in the business world but are widely applied to assess environmental problems, in particular those related to global change.

A. Marquina (ed.), Environmental Challenges in the Mediterranean 2000-2050, 47–60.
© 2004 *Kluwer Academic Publishers. Printed in the Netherlands.*

Scenarios are consistent and plausible images of alternative futures, i.e. they show various possibilities of how the future might look and how it will be reached. They rely on a rather high degree of systems (or process) knowledge, i.e. knowledge about the interdependencies and feedbacks between various system components. Scenarios focus on laying out complex causal relations and try to answer "what if" questions. They are not predictions of the future, and should not be qualified by a probability. Predictions are only possible if system complexity is rather low and process knowledge and data availability is high. Prediction of engineered systems is often possible, while prediction of natural or human systems mostly is not, unless the time horizon is short.

Scenarios can be qualitative, quantitative or qualitative-quantitative. Purely qualitative scenarios are, in most cases, narratives of the futures. These describe, in the form of storylines, how relevant aspects of the system under consideration will develop in the future. Purely quantitative scenarios are generated, for example, by running mathematical models of the system of interest such that future states of the system are computed. State-of-the-art environmental scenarios combine qualitative with quantitative elements, i.e. storylines with model calculations. The storylines are the basis for quantifying the driving forces and thus the input for the models that are then applied to compute numerical estimates of environmental indicators. The quantitative modeling part may also help to improve the consistency of the storylines, e.g. if the computed impact of the change of a certain system component on another one proves to be quite different from what was thought before the quantitative modeling. Examples for global-scale qualitative-quantitative environmental scenarios are:

- the greenhouse gas (plus sulphate) emission scenarios developed for the Intergovernmental Panel on Climate Change IPCC [15], which serve as input to climate models but also include scenarios of population and Gross Domestic Product until 2100,
- the scenarios of the water situation in 2025 developed in the framework of the World Water Vision [3, 5] and
- the scenarios of the Global Environmental Outlook 3 of the United Nations Environment Program [19].

Qualitative-quantitative scenarios can be a good tool for supporting regional planning, as they show the consequences of today's decisions in the distant future. The robustness of a certain policy can be tested by assessing its impact in alternative futures. A robust policy measure is one that performs reasonably well in all plausible futures. In the context of regional planning, scenario development should be done by scientists and stakeholders (policy makers, representatives of the civil society, etc.) together. A close cooperation, or participation, leads to the "ownership" of the scenarios by the stakeholders that is necessary for the scenario process to become relevant for actual decision making. Scenario development can be an ideal means for establishing the often desired communication between stakeholders and scientists, and among scientists of different disciplinary backgrounds; it requires, and thereby encourages, interdisciplinary communication and cooperation. Environmental scenarios are always interdisciplinary, as it is not possible to derive images or stories of the future state of the environment that do not include demographic, economic and technological aspects. They are integrated in the sense that disciplinary knowledge must be made consistent.

In Section 2, the most important steps involved in the development of qualitative-quantitative scenarios are described, and the questions of participation and scale issues in scenario development are discussed. In Section 3, the implications of global-scale qualitative-quantitative scenarios for water resources in the Mediterranean region are shown. Finally, some conclusions are drawn.

2. Methodology

For developing qualitative-quantitative scenarios which are to support sustainability-oriented regional planning, we propose the following steps:

1. Identification of the problem field and of the participants of the scenario process (stakeholders and scientists)
2. System definition including driving forces as well as temporal and spatial resolution and extent (base year, time horizon and time step, scenario regions)
3. Definition of indicators of the system state (related to the mathematical models that are available to compute indicators)
4. Development of qualitative reference scenarios in the form of storylines (narrative descriptions of alternative futures)
5. Development of quantitative reference scenarios
   a) Quantification of the driving forces
   b) Computation of indicators using mathematical models
6. Development of intervention scenarios
   a) Identification of interventions of interest
   b) Modification of selected driving forces or parameters of reference scenarios
   c) Computation of indicators using mathematical models
7. Evaluation of the scenarios (based on defined targets, e.g. by multi-criteria analysis, cost-benefit analysis, equity analysis)

Generally, intervention scenarios serve to test the robustness of an intervention (or policy measure) on the background of all reference scenarios, i.e. on the background of alternative plausible futures. They are defined by modifying one or more driving forces or parameters of the reference scenarios.

The steps listed above reflect those that were taken during scenario development for two federal states of Northeastern Brazil, in the framework of the German-Brazilian research cooperation program WAVES [6]. The WAVES scenarios cover the problem fields of water scarcity, agriculture and migration up to the year 2025, taking into account climate change as well as demographic and socioeconomic changes (step 1). Participants in the scenario process were a group of about eight scientists and a larger group of policy makers and technicians of Brazilian authorities dealing with regional planning. The involvement of the policy makers and technicians occurred during three workshops in the final year of the program. The indicators were calculated using an integrated model which combines the following modules: water use, hydrology, agricultural productivity, agricultural income and migration. They were computed for each of the 332 municipalities of the two federal states, while for the storylines and the quantification of the driving forces, eight scenario regions were distinguished. These eight scenario regions were defined based on similar (agro)economic and natural (sedimentary vs. crystalline subsurface, location with river basin, precipitation) conditions (steps 2 and 3).

Two different reference scenarios were developed which served as the background to test the impact of various policy interventions. Alcamo [2] recommended producing two or four reference scenarios with quite distinct storylines, in order to avoid that one scenario is considered as the "most probable", while the others are not given full consideration. The two storylines were formulated for the whole of the study region, distinguishing the events and developments occurring in each of the eight scenario regions. Each of the reference scenarios continues certain existing trends. The "Coastal boom and cash crops" scenario carries on the current trend of increased cash crop production for Southern Brazilian and international markets, the efforts to promote tourism mainly along the coast and the fast economic development of the capital of one of the two states. The "Decentralization – integrated rural development" scenario takes up the strengthening of regional centers e.g. by the establishment of universities, and thus increasing regional demand for agricultural products (step 4).

Based on the storylines, driving forces and indicators the reference scenarios are quantified (step 5). Here, those variables are defined as driving forces that cannot be computed by the applied models but are needed as model input. Both the quantified driving forces and the computed model output (indicators) are part of the quantitative scenario. In order to make assumptions about the future development of certain driving forces, their historical development is first analyzed. Then, numerical values of the driving forces that reflect the respective qualitative scenario are defined for future time periods. Care must be taken to guard consistency in quantifying driving forces that are known to be correlated. When all input necessary for the various models is quantified, the model(s) can be used to compute the system indicators of the reference scenarios. Examples of interventions considered in WAVES were changed water prices or reduced investments in reservoir construction (step 6). To test the robustness of an intervention, its impact was generally assessed on the background of both reference scenarios.

A formal evaluation (not done in WAVES) helps to analyze the results of a scenario analysis (step 7). Depending on the system of interest and the applied models, each scenario will include a large number of indicators, and it might be difficult to recognize which intervention performs best in alternative futures. Any "best" performance depends, of course, on the ideas, goals and visions we have. Different stakeholders (and scientists) are likely to have different ideas, and it is therefore recommended to discuss these openly. With respect to sustainability-oriented regional planning, goal functions, ecotargets and orientors have been identified [13]. Cost-benefit analysis is a well-known tool to assess which solution is best (most efficient), but it is difficult to monetarize non-financial costs and benefits, in particular the external costs of measures that are related to their negative impact on the environment. Multi-criteria analysis provides a flexible framework to include many different criteria and to weigh them against each other, thus deriving an "optimal" solution. In multi-criteria analysis, a number of criteria are developed in each of the categories under consideration. These criteria can be qualitative or quantitative. They will be quantitative if numerical indicator values were determined, e.g. by mathematical models, which allow to quantify to which degree a certain criterion is fulfilled. A large variety of mathematical techniques are available to derive the overall score of a certain intervention scenario based on the fulfillment of the different criteria, and the simplest one (weighted summation) is used most often [12]. If various techniques are used in one evaluation, the stakeholders may gain valuable insights [4]. Equity analysis, which determines whether different parts of society are affected differently by the intervention of interest, can complement multi-criteria analysis or can be handled as a part of it [14].

Generally, scenario development is an iterative process. Quantitative analysis (step 5), for example, might show some inconsistencies in the storylines, which will therefore be modified after step 5 has been performed for the first time. Besides, after a first discussion of the qualitative-quantitative scenarios, stakeholders might want to enhance either the qualitative or quantitative scenarios by other aspects or system components. In a regional planning process, steps 3 to 7 are likely to be repeated. After a first computation and evaluation of the system indicators, in particular the indicator definition can be refined, and new interventions may become interesting.

2.1. PARTICIPATION

Farrell [10] recommended to view environmental assessments as social processes or communication processes rather than as an end product, a document. This recommendation is also applicable to scenarios, which are often part of integrated environmental assessments. In the context of regional planning, scenario development should be regarded mainly as a tool to support the strategic thinking and decision-making of the stakeholders. Then the success of a scenario process should be judged based on what participants have learnt during the process. Certainly, the learning does depend on the quality of the developed scenarios themselves: their consistency, their meaningful indicators, the quality of the involved models, etc. This success criterion can be referred to as the credibility of scenarios. It is recommended to openly discuss the uncertainties of the involved quantitative assessments to thus increase the (long-term) credibility. The other two important success criteria are the saliency and the legitimacy of scenarios. The former relates to the relevance for the addressees of the scenarios (e.g. the stakeholders), the latter to the political acceptability, the satisfaction of the participants or addressees with the scenario process and the "ownership" of the developed scenarios. There are trade-offs among the three criteria, and depending on the stage of the scenario process or the dominance of societal vs. scientific interest, it might be useful to put the emphasis on one of the three criteria.

The question is how to organize participation in the scenario analysis process. Certainly, the type and number of participants will strongly influence the process and its results. In general, it can be recommended to involve stakeholders from the very beginning of the scenario process. However, it might be useful to involve different people at different stages of the scenario process, e.g. more technical experts in the beginning and more policy-makers towards the end.

2.2. SCALE ISSUES IN SCENARIO DEVELOPMENT

Scale issues in scenario development include all issues related to the spatial, temporal and institutional scales of the system for which scenarios are developed. They arise in all stages of the scenario analysis process, including the definition of the addressees of the scenarios, the formulation of the storylines, the quantification of the driving forces and the computation of indicators by models. Scale issues concern both the saliency and the credibility of the scenarios. It is recommended to explicitly address scale issues in any scenario analysis process.

As a first consideration, how can the spatial scale of the analysis be matched with the spatial sphere of influence or interest of the stakeholders? A correspondence is a prerequisite for saliency. With respect to the system definition and the development of storylines, it is important to consider cross-scale impacts. For example, a certain region (the chosen spatial unit for the scenarios) is subject to global-scale developments like climate change or change in world market prices that cannot be influenced by stakeholders inside the region but which have a strong impact on the region. Figure 1 illustrates a scheme which can be used to identify and lay out cross-scale impacts when developing local-scale scenarios. In the WAVES program, we designed the two regional-scale reference scenarios such that they fit to two global-scale scenarios that were developed by the IPCC [15]. In addition, we elaborated scenarios for each of eight scenario regions in a consistent manner, thus considering explicitly developments at smaller scales [17].

In most scenarios, the spatial units which are distinguished in the storylines and for the quantification of the driving forces are necessarily larger than the units of the models that are used to compute the system indicators. Thus, downscaling is required, and this might lead to unreasonable assumptions for the individual computational units. The challenge is to make scale implications and restrictions transparent to the addressees of the scenarios. In conclusion, the explicit consideration of multiple scales in the scenario analysis process leads to richer and more plausible environmental scenarios.

3. Applications in the Mediterranean region

The described qualitative-quantitative scenario approach has the potential to be applied to many problems related to the environmental future in the Mediterranean region. For example, what type of agricultural products can be grown in an economically and ecologically sustainable manner, e.g. without degrading water quality? How can freshwater be allocated among the irrigation, domestic and industrial water use sectors in an optimal manner? How can a region develop in which overgrazing has almost destroyed vegetation and soils? Scenarios can be derived for the whole Mediterranean region (e.g. to assess water quality issues of the Mediterranean Sea), individual countries, river basins or small sub-national regions. Depending on the region of interest and the problem field, the qualitative and quantitative parts can be flexibly chosen. In case of a reasonable amount of quantitative knowledge and tools, the share of the quantitative part of the scenario process may become much more important than in cases of low data availability and a small budget for the scenario process.

Unfortunately, I cannot illustrate the proposed scenario methodology with an example of integrated qualitative-quantitative scenarios for the Mediterranean. An interesting project that combines qualitative and quantitative approaches and will lead to land use change scenarios in the Mediterranean is the ongoing MEDACTION EU project (http://www.icis.unimaas.nl/medaction/).

Figure 1. Multi-scale relationship among drivers for community forest management

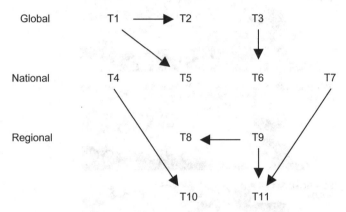

T1 Global demand for forest products
T2 UN convention support for community forest management
T3 GATT trade requirements
T4 World Bank loan conditions
T5 Currency exchange rates
T6 Support for mining
T7 Policy to promote oil palm plantations
T8 Regional tax base
T9 Designation of nearby national parks
T10   Local customary law
T11   Destructive use of forests by outsiders

Source: Wollenberg et al., 2000

In this chapter, I will present scenarios which show the potential impact of climate change on water resources and irrigation water requirements in the Mediterranean. They allow me to draw some general conclusions about the current possibilities for assessing the impact of climate change on freshwater issues. These climate change impact scenarios have been computed with the integrated global-scale model of water availability and water use WaterGAP 2. WaterGAP 2 is a global water resources and water use model which can be used to assess water scarcity problems [1]. With a spatial resolution of 0.5° by 0.5° (55 km by 55 km at the equator), the hydrological model WGHM of WaterGAP computes runoff and river discharge [7]. Fig. 2 (top) shows the long-term average water resources in the river basins of the Mediterranean region for the climate normal 1961-90, as computed by WGHM. WGHM was used to assess the impact of climate change on water resources. The studied climate change is based on the qualitative-quantitative global IPCC scenarios that show possible development paths between 1990 and 2100 [15]. The IPCC developed four global scenarios, which encompass both storylines as well as quantitative estimates of the development of population, Gross Domestic Product GDP, energy consumption and greenhouse gas (plus sulphate) emissions, distinguishing four world regions.

Figure 2. Long-term average water resources in the river basins of the Mediterranean region [mm/yr], computed by WaterGAP 2 for the climate normal 1961-90 (top), and percent change of water resources due to climate change between 1961-90 and the 2020s (bottom).

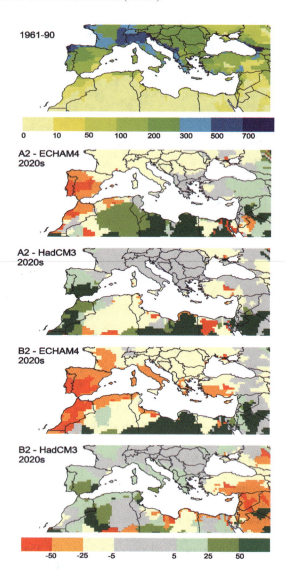

The storylines describe divergent futures along the lines of more or less globalization and more or less emphasis on the environment (Fig. 3). The differences among the storylines cover a wide range of key characteristics, such as technology, governance, and behavioral patterns.

Nakicenovic and Swart summarize the four storylines as follows (in their section 4.2.1):

- "The A1 storyline and scenario family describes a future world of very rapid economic growth, low population growth, and the rapid introduction of new and more efficient technologies. Major underlying themes are convergence among regions, capacity building, and increased cultural and social interactions, with a substantial reduction in regional differences in per capita income. (...)
- The A2 storyline and scenario family describes a very heterogeneous world. The underlying theme is self-reliance and preservation of local identities. Fertility patterns across regions converge very slowly, which results in high population growth. Economic development is primarily regionally oriented and per capita economic growth and technological change are more fragmented and slower than in other storylines.
- The B1 storyline and scenario family describes a convergent world with the same low population growth as in the A1 storyline, but with rapid changes in economic structures toward a service and information economy, with reductions in material intensity, and the introduction of clean and resource-efficient technologies. The emphasis is on global solutions to economic, social, and environmental sustainability, including improved equity, but without additional climate initiatives.
- The B2 storyline and scenario family describes a world in which the emphasis is on local solutions to economic, social, and environmental sustainability. It is a world with moderate population growth, intermediate levels of economic development, and less rapid and more diverse technological change than in the B1 and A1 storylines. While the scenario is also oriented toward environmental protection and social equity, it focuses on local and regional levels."

Figure 3. The IPCC global scenarios

	oriented mainly towards economic growth	oriented mainly towards the environment and social innovation
globalized world	A1	B1
regionalized world	A2	B2

Source: Nakicenovic and Swart, 2000 [15]

In the following analysis, only the greenhouse gas (plus sulphate) emission scenarios A2 and B2 are considered. In the A2 scenario, considerably more greenhouse gases are emitted after 1990 than in the B2 scenario, as the latter assumes a more environmentally conscious development of society than A2 (Fig. 4). The temperature and precipitation changes resulting from the two different emission scenarios were taken from two state-of-the-art global climate models, the HadCM3 model [11] and the ECHAM4/OPYC3 model [16]. The spatial resolution of the HadCM3 model is 2.5° latitude by 3.75° longitude, while the spatial resolution of the ECHAM4 model is approximately 2.8° by 2.8°. The changes in simulated long-term average monthly precipitation and temperature between the 2020s (2020-2029) and the climate normal 1961-90 were first interpolated to the 0.5° grid and then applied as input to WGHM (by scaling the observed monthly precipitation and temperature time series from 1961-90). Fig. 2 (bottom) shows the computed impact of climate change on water resources in the river basins of the Mediterranean region.

Obviously, the computed impacts depend more on the applied climate model than on the assumed emission scenario. This is also true for the 2070s when the emissions differ

Figure 4. The SRES greenhouse gas emission scenarios of IPCC A1, A2, B1 and B2, as well as the older IPCC IS92a scenario, which has been the basis of most published climate change modeling results until approximately the year 2001.

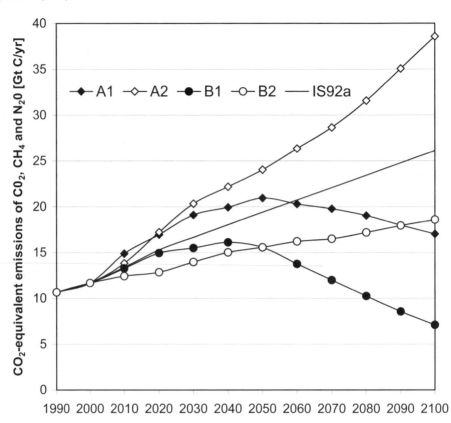

between A2 and B2 even more than in the 2020s (not shown). With the ECHAM4 climate scenarios, water resources are reduced by 5% to more than 50% until the 2020s in almost all the drainage basins of the Mediterranean. With the HadCM3 climate scenarios, resources either increase or remain constant, except for Turkey and some parts of North Africa (Fig. 2). The discrepancy between the impacts as based on the two climate models results from the very different precipitation pattern that the two models produce. For the emission scenario B2, for example, ECHAM4 computes a strong reduction of precipitation for most of the Mediterranean between the climate normal 1961-90 and the 2020s, in particular the eastern part of the Iberian Peninsula, while HadCM3 results in an increase of precipitation in the western Mediterranean, Southern Italy and Greece. Global climate models still have a low capability to simulate precipitation, both historic and future distributions. While different climate models agree quite well with respect to computed temperature changes, they differ strongly with respect to precipitation changes. Please note that on the scale of river basins, the changes in water resources can be larger for B2 than for A2, even though the emission changes are much smaller (compare the Iberian Peninsula, ECHAM4 model, in Fig. 2) such that we cannot simply conclude that a reduction in greenhouse gas emissions will lead to less changes in water resources. The changes until the 2070s are generally larger than until the 2020s.

Like water resources, irrigation water requirements depend on precipitation and temperature changes, but the impact of seasonal changes is different, and temperature is more important than in the case of water resources. Thus, the uncertainty in simulating precipitation plays out differently. Fig. 5 (top) shows the net irrigation requirement in each 0.5 degree cell of the Mediterranean region, as computed by the Global Irrigation Model GIM which is a module of WaterGAP 2 [9]. Due to data restrictions, GIM distinguishes only two crops, rice and other crops. It takes into account the area equipped for irrigation and the climate, and also models cropping patterns and the growing periods. The potential impact of climate change on net irrigation requirements is shown in Fig. 5 (bottom), but only for the emission scenario B2. The two GCM climate scenarios were implemented as in the case of the water resources computations.

When looking at the climate induced changes in irrigation requirements, consider that percentage changes are generally high when the base number, here the irrigation requirement per unit irrigated area, is small. An example is the northern part of Italy with its low requirement per unit irrigated area, where a strong percentage increase of irrigation water requirements occurs if there are only a few days with a decreased precipitation. Different from the water resources scenarios, irrigation requirements increase almost everywhere, which is caused by the fact that irrigation water requirements are more sensitive to temperature than runoff (renewable water resources).

Another difference between the water resources and the irrigation requirement scenarios is that the irrigation requirement scenarios do not differ as much between the ECHAM4 and HadCM3 climate scenarios, except for the Balkans. In the case of the Iberian Peninsula, the rather similar pattern of changes (as compared to the strongly discrepant water resources changes, (Fig. 2), is explained mainly by the fact that the increase in precipitation as computed by HadCM3, which leads to a strong increase in water resources, occurs mainly outside the growing period. As an example, in a grid cell in Central Spain, annual precipitation increases from 428 mm to 443 mm in the case of HadCM3 and decreases to 319 mm in the case of ECHAM4. The precipitation during the 150-day growing period starting in April, however, decreases for both climate scenarios, from 134 mm to 120 mm in the case of HadCM 3 and to 96 mm in the case of ECHAM4. This results in an increase of the net irrigation requirement per unit irrigated area from 776 mm to 803 mm and 823 mm.

Thus, net irrigation requirements increase by 3% and 6% while annual precipitation increases by 4% and decreases by 25%, respectively. Irrigation requirements are sensitive only to precipitation changes in the growing period, such that shifts in the seasonal distribution of precipitation lead to different impact on water resources and irrigation requirements. Computed temperature changes during the growing period are rather similar, plus 0.9°C for HadCM3 and plus 1.0°C for ECHAM4. Another reason for the similarity of irrigation requirement changes between climate scenarios might be that GIM does not model the impact of climate change on a crop with a fixed growing period but simulates the shift of the optimal growing period that will occur due to climate change. Cropping patterns and growing periods are modeled based on precipitation and temperature [9]; the growing period during which optimal temperature and precipitation

Figure 5. Net irrigation water requirements in the Mediterranean region [mm/yr averaged over 0.5° cell area], computed by WaterGAP 2 for the climate normal 1961-90 and the irrigated areas of 1995 (top), and percent change of net irrigation requirements due to climate change between 1961-90 and the 2020s (bottom).

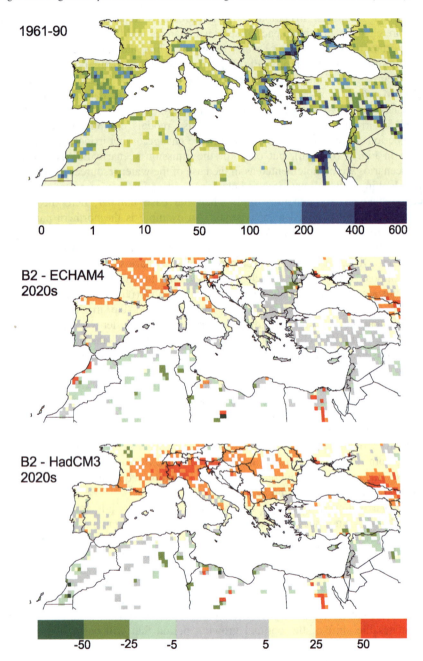

conditions prevail is determined, e.g. such that precipitation is high during the first 50 days of the growing period. In many cells, the two climate scenarios lead to a different shift of growing seasons towards the respective wetter and warmer period, which makes the irrigation requirements in two climate scenarios more similar [8].

4. Conclusions

The development of qualitative-quantitative scenarios is a state-of-the-art approach for assessing possible futures and thus for supporting sustainability-oriented regional planning. It is flexible because it can be applied to almost any type of problem, and with varying fractions of quantitative vs. qualitative analysis. If there is little data, little modeling experience and restricted funds, the fraction of the quantitative modeling analysis will be lower than in the case of good data, existing models and extensive funds. Still, the development of the then dominant qualitative part of the scenario will help to transfer knowledge between scientists and policy makers and will provide both scientists and policy makers with a clearer idea of possible futures and the options to achieve sustainable development. In a scenario analysis which explicitly considers scale issues, regional planners will learn about the feedback between global (or other coarser) scale developments and the development in their region.

From the scenario analysis of the impact of climate change on water resources and irrigation water requirements in the Mediterranean we conclude that
- due to the low capability of global climate models to realistically simulate precipitation, in a scenario analysis of climate change impacts for which precipitation change is a relevant driver (water resources, water use, agricultural productivity, erosion,...) it is advisable to apply the results from at least two different climate models
- the differences between the precipitation change patterns (and thus the derived water resources changes) as computed by two global climate models is often larger than the differences between the results of one model for two different greenhouse gas emission scenarios. Therefore, it might not be absolutely necessary to be consistent in a scenario analysis in which climate change impacts are compared to the impacts of the changes in other driving forces. It appears appropriate, for example, to use one emission scenario as interpreted by two climate models to compute two water resources scenarios and combine these scenarios with a water use scenario that is based on assumptions about the economic and demographic development that are not consistent with the respective assumptions for the emission scenario.

References

1. Alcamo, J., Döll, P., Henrichs, T., Kaspar, F., Lehner, B., Rösch, T., and Siebert, S. (2003) Development and testing of the WaterGAP 2 global model of water use and availability. *Hydrological Sciences Journal* 48, 3, 317-338.
2. Alcamo, J. (2001) Scenarios as Tools for International Environmental Assessment. Environmental issue report 24, European Environment Agency, Copenhagen.

3. Alcamo, J., Henrichs, T., and Rösch, T. (2000) World Water in 2025. Global modeling and scenario analysis for the World Commission on Water for the 21st Century, Kassel World Water Series 2, Center for Environmental Systems Research, University of Kassel, Kassel. http://www.usf.uni-kassel.de/usf/archiv/dokumente.en.htm

4. Bell, M.L., Hobbs, B.F., Elliott, E.M., Ellis, H., and Robinson, Z. (2001) An evaluation of multi-criteria methods in integrated assessment of climate policy, *Journal of Multicriteria Decision Analysis 10*, 229-256, DOI 10.1002/mcda.305.

5. Cosgrove, W.J., and Rijsberman, F.R. (2000) World Water Vision: Making Water Everybody's Business. Earthscan.

6. Döll, P., and Krol, M. (2003) Integrated scenarios of regional development in two semi-arid states of Northeastern Brazil, *Integrated Assessment 3*, 4, 308-320

7. Döll, P., Kaspar, F., and Lehner, B. (2003) A global hydrological model for deriving water availability indicators: model tuning and validation, *Journal of Hydrology 270, 1-2*, 105-134.

8. Döll, P. (2002) Impact of climate change and variability on irrigation requirements: a global perspective. *Climatic Change 54*, 269-293.

9. Döll, P., and Siebert, S. (2002) Global modeling of irrigation water requirements. Water Resources Research 38, 4, 8.1-8.10, DOI 10.1029/2001WR000355.

10. Farrell, A., Van Deveer, S.D., and Jäger, J. (2001) Environmental assessments: four under-appreciated elements of design. *Global Environmental Change 11*, 311-333.

11. Gordon, C., Cooper, C., Senior, C.A., Banks, H., Gregory, J.M., Johns, T.C., Mitchell, J.F.B., and Wood, R.A. (2000) The simulation of SST, sea ice extents and ocean heat transports in a version of the Hadley Centre coupled model without flux adjustments. *Climate Dynamics 16 2/3*, 147-168.

12. Janssen, R. (2001) On the use of multi-criteria analysis in environmental impact assessment in The Netherlands, *Journal of Multicriteria Decision Analysis 10*, 101-109. DOI 10.1002/mcda.293.

13. Müller, F., and Leupelt, M. (eds.) (1998) *Eco Targets, Goal Functions, and Orientors*, Springer-Verlag, Berlin.

14. Munda, G. (1995) *Multicriteria Evaluation in a Fuzzy Environment: Theory and Applications in Ecological Economics*. Physica-Verlag, Berlin.

15. Nakicenovic, N., and Swart, R. (eds.) 2000 Emission Scenarios. IPCC Special Report on Emission Scenarios, Cambridge University Press. can be downloaded from www.grida.no/climate/ipcc/emission/index.htm

16. Röckner, E., Arpe, K., Bengtsson, L., Christoph, M., Claussen, M., Dümenil, L., Esch, M., Giorgetta, M., Schlese, U., and Schulzweida, U. (1996) The atmospheric general circulation model ECHAM-4: Model description and simulation of present day climate, MPI-Report No. 218, MPI Hamburg.

17. Rotmans, J., van Asselt, M., Anastasi, Ch., Greeuw, S.C.H., Mellors, J., Peters, S., Rothman, D., and Rijkens, N. (2000) Visions for a sustainable Europe, *Futures 32*, 809-831.

18. Schwartz, P. (1998) The Art of the Long View, Wiley.

19. United Nations Environment Programme (UNEP) (2002) Global Environmental Outlook 3, Nairobi. (http://www.grid.unep.ch/geo/geo3/index.htm)

20. Wollenberg, E., Edmunds, D., and Buck, L. (2000) Using scenarios to make decisions about the future: anticipatory learning for the adaptive co-management of community forests, *Landscape and Urban Planning 47*, 65-77.

Chapter 4

Climate change and the occurrence of extremes: some implications for the Mediterranean Basin

J. P. Palutikof and T. Holt
Climatic Research Unit, University of East Anglia, Norwich NR4 7TJ, UK[1]

Intuitively we would expect climate change associated with increased emissions of greenhouse gases (so-called global warming) to be linked to substantial impacts in the Mediterranean. This region is vulnerable to present-day fluctuations in the weather, especially with respect to water supply for competing activities such as agriculture and tourism. The climate models used in the Third Assessment Report of IPCC Working Group I indicate substantial drying in the Mediterranean associated with future climate change, which can only exacerbate the current situation.

The potential impacts of climate change on the Mediterranean are described. We look first at the historical climate record, and the large-scale atmospheric characteristics, such as the North Atlantic Oscillation, which are related to the fluctuations in that record. Then, we look at future patterns of change in the Mediterranean as predicted by climate models. We examine techniques for predicting climate changes at the local scale. Scenarios are presented of the future change in temperature and precipitation for selected Mediterranean catchments, and analysed with respect to changes in the occurrence of extremes, such as high rainfall events, droughts and heat waves.

1. Climate change contexts

Working Group I of the Inter-governmental Panel on Climate Change (IPCC) stated in the Second Assessment Report that 'the balance of evidence suggests a discernible human influence on global climate' [5], and was sufficiently confident by the time of the Third Assessment Report to conclude that 'there is new and stronger evidence that most of the warming observed over the last 50 years is attributable to human activities' [6]. So-called 'global warming' is related to increased concentrations of greenhouse gases (carbon dioxide, methane, nitrous oxide etc.) due to emissions from industrial activities, transport and the intensification and spread of agriculture, including forest clearance. The term 'global warming' is something of a misnomer since although mean global

[1] This work was supported by the European Union under Contracts ENV4-CT95-0121 (MEDALUS), EVK2-CT-2001-00118 (MICE) and EVK2-CT2001-00132 (PRUDENCE). The Hadley Centre climate model data were supplied by the Climate Impacts LINK project (funded by the UK Department of the Environment, Food and Rural Affairs).

A. Marquina (ed.), Environmental Challenges in the Mediterranean 2000-2050, 61–73.
© 2004 *Kluwer Academic Publishers. Printed in the Netherlands.*

temperatures are increasing (see Figure 1), by around 0.6°C in the twentieth century, this warming trend is not constant in space or in time [9]. Some areas have experienced much steeper rates of warming, for example Northern Hemisphere continental areas in the period 1976–2000. Some periods have experienced widespread cooling, for example the years 1946–1975 over much of the Northern Hemisphere. Moreover, the effects of anthropogenic warming are not manifest simply in increased temperatures. All aspects of climate are likely to be affected, including variables such as rainfall and wind speeds. Changes may be expected not only in the mean climatology, but also at the extremes, so that the occurrence of phenomena such as droughts, floods and wind storm can be expected to change as a result of anthropogenic warming. Changes in these extremes generally have greater impacts on human activities than changes in the mean climate. It is clear, therefore, that climate change cannot be seen in simplistic terms as a gradual and homogeneous warming of our climate.

Figure 1. The combined global land and marine surface temperature record from 1856 to 2002, together with a smoothed filter to show the general trend

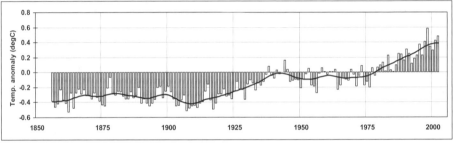

Source: Climatic Research Unit

Against this background of potentially complex changes in climate, the implications for regions which are already vulnerable to fluctuations in climate are potentially severe. The Mediterranean is such a region. Pressures on water resources throughout the region already exist, and are potential sources of conflict both within and between nation states. These pressures come not only from the traditional usages such as public water supply and agriculture, but also from activities such as tourism. The Mediterranean is the biggest tourism region in the world. The traditional Mediterranean climate is defined as one in which winter rainfall is at least three times the summer rainfall [11], and indeed over much of the Mediterranean Basin studied in this book, summer rainfall is practically zero. This means that at the times of the year when demands on water resources are high because of the influx of tourists, natural replenishment of aquifers and water courses is not taking place. This situation is not expected to improve in the future: according to the European Environment Agency [1] the number of tourists in the Mediterranean countries is expected to increase from 260 million in 1990 to 440-655 million in 2025.

In this chapter, the goal is to describe and discuss the potential impacts of climate change on the Mediterranean Basin. We look first at the historical climate record, and the large-scale atmospheric characteristics, such as the North Atlantic Oscillation, which are related to the fluctuations in that record. Then, we look at future patterns of change in the Mediterranean as predicted by climate models. Finally, we examine scenarios of climate change in two catchments in the Mediterranean, and analyze these scenarios with respect to the predicted changes in extremes.

2. Observed climate variability

A number of recent studies have looked at trends in climate variables globally, and over Europe, using station data. Frich et al. [2] mapped changes in the occurrence of temperature and rainfall extremes. Comparing two multi-decadal averages, they found that in the most recent period, there was a relative increase in the duration of heat waves, and a relative decrease in the number of frost days, throughout most of the Mediterranean Basin, although these changes were not statistically significant. For high precipitation extremes, there was no consistency in the trend in the Mediterranean region, and the changes that did occur were almost exclusively non-significant. Using station records throughout Europe, the European Climate Assessment (ECA) has looked at trends in temperature and precipitation between 1976 and 1999 [17]. We can use their results to look at trends along the northern shore of the Mediterranean. Just taking a few examples from their results, for mean temperature in the winter half year (October-March), stations in the west (Portugal to Italy) show significantly increasing temperatures, at a rate greater than 0.3°C/decade. However, in the eastern half of the basin (mainly Greece and Turkey), the trends are not significant and, in some cases, are slightly negative. This translates at the extremes into a significantly increasing number of winter warm days, but little change in the number of winter cold-spell days, except at a handful of Greek and Turkish stations which show a significant increase. With respect to mean precipitation, few southern European stations show significant trends. There is, however, a developing tendency towards fewer wet days and more precipitation per wet day.

In order to get some idea of detailed recent trends over the region, we have taken data from the NCEP/NCAR reanalysis project [10]. This project uses a state-of-the-art analysis/forecast system to assimilate data from 1948 to the present, with the goal of creating homogeneous gridded data sets of meteorological variables. To take the example of precipitation, we extracted data for three regions over the Mediterranean for the period 1979-98. These are the Eastern Mediterranean (25-45°E by 30-46°N), the Northern Mediterranean (36-46°N by 10°W-40°E), and Spain & Portugal. Standardized anomalies for the NCEP/NCAR reanalysis gridboxes within these domains were calculated (i.e., the average, over all gridboxes, of the individual gridbox differences from the 1979-98 mean divided by the gridbox standard deviation). We show the results for the hydrological year (October-September) in Figure 3. We can see that there is considerable interannual variability, but little evidence for trend in these statistics.

A number of authors [19], have looked at the relationship between the North Atlantic Oscillation, or NAO, and climate over Europe. The NAO is the normalized pressure difference between Iceland and the Azores (or sometimes Gibraltar). It is a measure of the amount of activity in the North Atlantic and, particularly in the winter half-year, is related to European climate. If the NAO is strongly positive, then the large-scale flow over the North Atlantic is predominately westward, bringing warm and wet maritime conditions to north-western Europe and drier weather to the western Mediterranean. A negative NAO indicates stagnant conditions, or blocking, when drier continental conditions prevail over northern Europe and the Mediterranean is wet. We can measure the degree of relationship by looking at the correlation between Mediterranean rainfall and the NAO. The correlation between the NAO and station rainfall in winter (December, January, February) over the western Mediterranean is -0.64, and even in the eastern Mediterranean it is as high as -0.45 [13].

Figure 2. Standardized anomalies of precipitation over three regions of the Mediterranean Basin

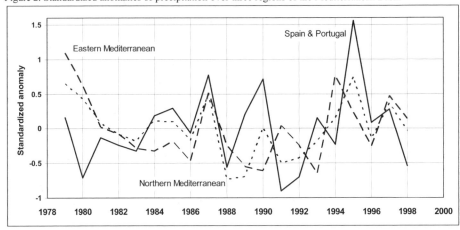

Source: NCEP/NCAR reanalyses (Kalnay et al.). [10]

3. Future patterns of climate over the Mediterranean

At the global scale, temperatures are expected to increase by 1.4°C - 5.8°C between 1990 and 2100 [7]. Land areas should warm more rapidly than the global averages, particularly in the winter season and at high latitudes. Warming below the global average is predicted for south and southeast Asia in summer, and southern South America in winter. In response to anthropogenic climate change, global precipitation is expected to increase, although the effects of this may be masked by large natural interannual variability. In the second half of the 21st century, it is expected that precipitation will have increased over northern mid- to high latitudes and Antarctica in winter. What are the implications of these anticipated changes for the Mediterranean?

Information on the future evolution of climate change in response to global warming is provided by three-dimensional climate models. These are complex, computer-based representations of the behaviour of the atmosphere, and the fluxes and feedbacks between it and the land, ocean and ice-covered surface. They generate information on changes in the climate over time at all points in a three-dimensional grid in response to a prescribed forcing. They may operate at lower resolution over the whole globe, in which case they are called General Circulation Models, or GCMs, or at higher resolution over a single region (Regional Climate Models or RCMs). The most commonly used forcings are the SRES emissions scenarios of IPCC [12]. In particular, the higher-emissions

SRES A2 scenario, describing a heterogeneous world with high population growth rates, and the lower-emissions SRES B2 scenario, describing a world of local solutions to economic, social, and environmental sustainability, are the most commonly modeled.

It is important to bear in mind that the output from a climate model forced with one of these emissions scenarios represents just one of a whole range of possible futures. For this reason, modeling institutions generally carry out more than one simulation with the same model and the same emissions scenario, but with a different starting condition. These 'ensemble members' each represent, with equal probability, a possible future climate.

In the 21st century, climate models suggest warming greater than the global mean warming over the Mediterranean Basin [3]. In the summer season, this warming is in excess of 40% above the global average warming, and is considered by some to be an artifact of model parameterization rather than a characterization of a probable future. The model consensus with respect to precipitation in the Mediterranean is that there will be little change in winter, and drying in the summer. For those model simulations forced with the SRES A2 scenario for the future, a large decrease (less than -20%) is predicted.

Changes in the occurrence of extremes may have greater implications for human activities, and be more noticeable, than changes in the mean climate. Three current research projects funded by the European Union under Framework Programme 5 are analyzing the occurrence of climate extremes, and how this may change in response to anthropogenic climate change, over Europe. The first is the PRUDENCE project (Prediction of Regional scenarios and Uncertainties for Defining EuropeaN Climate change risks and Effects), which is using high-resolution regional climate models to study the climate of 2070-2100 relative to that of 1961-90, and the impacts of the predicted changes (http://www.dmi.dk/f+u/klima/prudence/). Second is the STARDEX project (Statistical and Regional Dynamical Downscaling of Extremes for European regions), which aims to provide high resolution scenarios of changes in the frequency and intensity of extreme events (http://www.cru.uea.ac.uk/projects/stardex/). Third is the MICE project (Modelling the Impact of Climate Extremes), which uses information from climate models to explore future changes in extreme events in response to global warming, and their impacts (http://www.cru.uea.ac.uk/projects/mice/).

Figures 3 and 4 show some examples of the information which is emerging from these projects, and are taken from the MICE project. Indices of extremes have been calculated, based on the output from the Hadley Centre regional climate model, HadRM3H. In both figures, the results for one ensemble member forced with the A2 scenario are shown in the upper map and those for an ensemble member forced with the B2 scenario in the lower map.

Figure 3 shows, for the Mediterranean Basin, the difference between the two periods, 2070-2100 minus 1961-90, in the mean number of degree-days per year (using a threshold of 27°C). The degree-day measure used here is an accumulation of the excess of temperature above 27°C. One degree day is counted for each degree above 27°C, for each day in the record. Thus, if the temperature is 30°C on Day 1, and 33°C on Day 2, the number of degree days for the two days is 9. In terms of the impacts, this is a measure of human comfort and the occurrence of heat stress. The greatest changes take place in the A2 scenario and, in both maps, over the land. Particularly large changes are seen in the A2 scenario over North Africa (an increase of more than 1200 degree days/year) and over southern Spain and south-east Turkey (an increase of more than 900 degree days/year).

Figure 3. Differences (2070-2100 minus 1961-90) between the mean annual number of degree days above a threshold of 27°C, calculated from daily maximum temperatures.

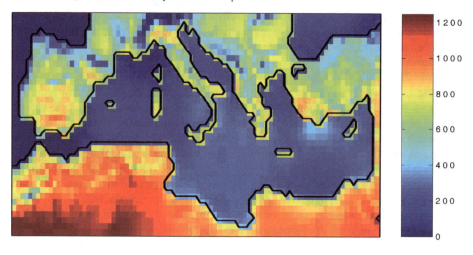

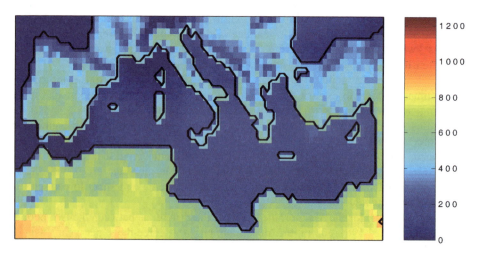

Source: HadRM3 model simulations for the SRES A2 scenario (above) and the SRES B2 scenario (below)

Figure 4. Differences (in number of days) between the lengths of the average annual maximum dry spell in 2070-2100 and in 1961-90.

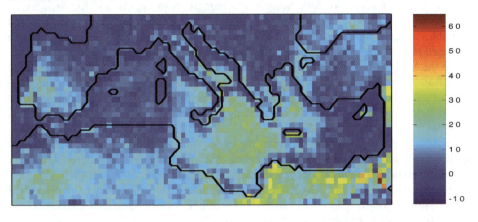

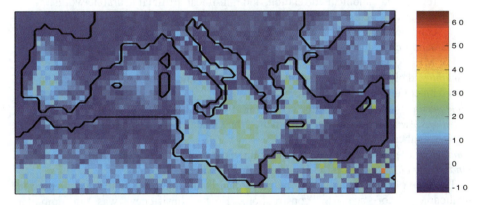

Figure 4 shows the change between the two periods in the average length of the longest dry period in each year, and is a measure of drought occurrence. The greatest increases are seen over the south-east of the study region, but over the land areas north of the Mediterranean Basin we also see large increases over south-western Spain, south-eastern Italy and western Turkey.

4. Creation of high-resolution scenarios of change

So far, we have looked at changes in climate across the whole of the Mediterranean Basin, with respect to the mean climate and the change in extremes occurrence. However, for impact analysis, it may be necessary to have much higher resolution information, both in space and in time, to provide input to impact models. Using data directly from climate models as a basis for impact analyses is not, at the present time, generally a satisfactory solution. First, a GCM generates output with a typical spatial resolution of 2.5° latitude by 3.75° longitude (the resolution, for example, of the UK Hadley Centre Unified Model). Second, the models, both GCMs and RCMs, may be unable to satisfactorily reproduce the present-day characteristics of meteorological variables at the required scales. In order to obtain the necessary information, at the required resolutions, downscaling procedures are commonly used.

In the MEDALUS project (Mediterranean Desertification And Land Use), funded by the European Union, statistical downscaling procedures were used to produce temperature and rainfall scenarios [14]. These scenarios then formed the basis of input to catchment models which predict the change in runoff due to changes in climate and land use.

Many reviews of statistical downscaling exist [20, 21]. Descriptions of the different techniques used for statistical downscaling are primarily methods-based [18, 20], according to the following major categories.

i. Analogue methods, in which the GCM-simulated circulation pattern is associated with the value of the local variable occurring in a set of historical observations on the day with the most similar circulation pattern;

ii. Deterministic transfer functions developed using techniques such as linear regression, canonical correlation, and non-linear artificial neural networks;

iii. Weather-type classifications and airflow indices, generally combined with a conditional (upon weather type/index) weather generator; and,

iv. Stochastic weather generators such as WGEN [16].

For MEDALUS, the scenarios had the following characteristics:

Variables: daily maximum (i.e. hottest on each day) and minimum (i.e., coldest on each day) temperatures and precipitation;

Time periods: three ten-year time series, for 1970-79, 2030-39 and 2090-99;

Locations: two Mediterranean catchments, the Guadalentin in South-east Spain and the Agri in southern Italy;

Number of sites: six sites in the Guadalentin, eleven sites in the Agri; and,

Underlying climate model: the Hadley Centre GCM HadCM3 [8].

The type of impact analysis to be performed is an important factor in designing the methodology for scenario construction. Here, consistency was an important consideration. First, the temperature (or rainfall) characteristics at any one site in a catchment on a particular day needed to be consistent with the temperature (or rainfall) at every other site on that day. An exceptionally warm day at one site should not be exceptionally cold at another. Second, the relationship between temperature and rainfall on a particular day at a particular site should be consistent. We would not, for example, normally associate wet conditions with exceptionally low night-time (minimum) temperatures, because the cloud cover prevents loss of heat by long-wave radiation.

To generate such scenarios, we used the following procedures. The daily rainfall scenarios are constructed in a two-stage procedure. First, for a single site, a reference rain day scenario is built using a circulation-typing approach (Method iii above) combined with a conditional first-order Markov chain to describe wet day/dry day probabilities (Method iv above). Then, the multisite scenarios are constructed by sampling from a benchmark file of multisite observations classified by season, weather type, and whether the day is wet or dry at the reference station (again, a Method iv approach). The temperature downscaling uses deterministic transfer functions (Method ii above) to predict daily maximum or minimum temperature from free atmosphere variables such as the 500 hPa geopotential height.

Consistency is maintained between the temperature and rainfall scenarios in two ways. First, the sea level pressure data used to define the circulation weather types which underpin the rainfall scenarios are also used to construct predictor variables for the temperature transfer functions. Second, separate temperature transfer functions are developed for wet and dry days. Between-site consistency is maintained in the rainfall scenarios by using the benchmark file of scenario days. For temperature, which is much more spatially-conservative than precipitation, it was not felt necessary to take any steps to maintain intersite consistency in the scenarios.

There are various problems associated with scenarios such as these which are based on statistical downscaling. For example, when comparing the 1970-79 scenarios with contemporary observations, the rainfall scenarios are always too dry (a characteristic found by many authors) and the temperature scenarios are too cool. However, and particularly in the case of temperature, the scenarios are a great improvement over raw GCM data. A further problem exists when we consider the differences which exist between the future and present-day scenarios. Inevitably, and this is true for all scenarios generated by statistical downscaling procedures, the changes which can occur between the present day and the future are constrained by the choice of scenario construction method. Taking the example of the rainfall scenarios here, because we use observed data as the basic building block of the scenarios, we constrain the changes in rainfall which can occur. Only the distribution and number of rain days can change, not the amount of rain per rain day.

Despite these drawbacks, statistical downscaling is a powerful tool to generate scenarios of future climate change at the high spatial and temporal resolutions required for impact analysis. These scenarios are simply one visualization of a whole range of plausible futures, and this must always be born in mind in their interpretation. To demonstrate the potential application of these scenarios, we will look at the changes in the occurrence of extremes in the temperature and rainfall scenario time series.

Measures of temperature extremes, and the changes in the two future decades 2030-39 and 2090-99, are shown in Table 1 for two sites in the Agri Basin. This is one of the two Mediterranean catchments for which downscaling was performed in the MEDALUS project, and is located in southern Italy. For this example, we selected two sites. Nova Siri Scalo is a low-lying (2 m above sea level), hot site where the potential for increased heat stress due to climate change may be of concern. For this site, we look in the daily maximum temperature scenarios at the change in the number of hot days ($\geq 35°C$) and in the number of degree-days above a threshold of 35°C. The scenario for 1970-79 comes very close to simulating correctly the observed occurrence of hot days, unlike the raw GCM data. Between 1970-79 and 2090-99, while the downscaled mean summer maximum temperature increases by 4.6°C (not shown in the table), the average number of hot days rises by 53 per year, representing a fourfold increase on the 1970-79 figure. In the decade 2090-99, the number of very hot days is equivalent to two months with maximum daytime temperatures continually at or above 35°C. Such a result has a clear significance for human health and comfort. Moliterno is a high-altitude (879 m) site where the behaviour of low temperatures is of interest. Again, the 1970-79 downscaled scenario measures of cold extremes closely

Table 1. Behaviour of temperature extremes in the Agri, expressed as annual averages. NoHD = number of hot days $\geq 35^\circ$C; HDG = degree days[1] above a threshold of 35°C; NoFD = number of cold days $\leq 0^\circ$C; FDG = degree days[1] below a threshold of 0°C. Hot-day measures are calculated from maximum temperatures, cold-day measures from minimum temperatures. Observations are for 1965-74 at Nova Siri Scalo, and for 1963-72 at Moliterno (selected to give no missing values).

	Observations	GCM data 1970-79	Downscaled scenarios for: 1970-79	2030-39	2090-99
Hot-day measures at Nova Siri Scalo:					
NoHD	16.8	0	15.2	18.1	68.2
HDG	33.5	0	31.2	32.1	233.4
Cold-day measures at Moliterno:					
NoFD	35	0.2	36.3	18.3	6.1
FDG	81.2	0.3	98.8	35.8	12.4

[1] Degree days are an accumulated measure of the excess (or deficit) of temperature above (or below) a threshold. One degree day is counted for each degree below 0°C (for FDG) or above 35°C (for HDG), for each day in the record. Thus, if the temperature is 37°C on Day 1, and 39°C on Day 2, HDG for the two days is 6.

approach the observed, whilst those based on raw GCM data are substantial underestimates. The number of low-temperature days falls from 36 per year in the 1970-79 scenario, to only 6 per year in 2090-99. Underpinning this change is an increase in mean winter TMIN from 1.7°C in 1970-79 to 6.6°C in 2090-99 (not shown).

Table 2 presents an analysis of rainfall extremes based on the scenarios. We look at two sites: Missanello in the Agri, and Lorca in the Guadalentin. Missanello is a relatively wet site, mean annual rainfall 804 mm, whereas Lorca is dry, with a mean annual rainfall of 234 mm. For rainfall, the scenario construction method is such that daily rainfall amounts in the scenarios cannot exceed observed values. Only the distribution and the relative proportion of wet and dry days can change in response to changes in weather-type occurrence. Under this constraint, and to explore the potential of the rainfall scenarios, extreme-value analyses based on the Generalized Extreme Value (GEV) distribution [4] were carried out. This distribution is a suitable tool for analysing the behaviour of annual extremes (maxima or minima) of a variable. For each scenario year, the number of days in the longest continuous spell of dry days, and the number of days in the longest continuous spell of wet days, was extracted. Thus, the sample size is just 10, the bare minimum for a GEV-based analysis. A Type 1 [4] distribution was assumed [15].

The results of fitting a GEV Type 1 distribution to the two data sets of maximum wet- and dry-day length, and calculating the return-period extremes, X_T, for T = 10, 20 and 50 years, are shown in Table 2. The results from the wet Agri site are more convincing. Whereas the 10-year return period dry spell extreme, X_{10}, at Missanello for the 1970-79 scenario is 86% of the observed length, at Lorca the figure is only 65%. For wet spells, X_{10} in the 1970-79 downscaled scenario at Missanello is 66% of observed, compared with 51% for Lorca.

The changes in return-period spell length between the scenarios are not a direct reflection of the change in mean rainfall. Taking the example of the relatively well-modelled dry-spell length at Missanello, X_T is substantially shorter in 2090-99 than in 1970-79 for all return periods, by around ten days.

This reduction in dry-spell length is despite the fact that mean rainfall at Missanello decreases monotonically between the 1970-79 and 2090-99 scenarios (not shown).

Table 2. GEV Type 1 (Gumbel) statistics and return-period extremes of annual maximum dry-spell and wet-spell length, in days. Observation period is 1951-70 at Missanello, and 1958-87 at Lorca.

	DRY SPELLS				WET SPELLS			
	Obs.	Scenarios			Obs.	Scenarios		
		1970-79	2030-39	2090-99		1970-79	2030-29	2090-99
MISSANELLO								
Longest run	73	53	54	47	16	7	10	8
GEV Mode	30.7	27.8	26.4	26.7	6.4	5.5	6.1	5.2
GEV Dispersion	11.7	9.6	9.7	6.2	2.1	0.9	1.3	1.0
Return period extremes (days)								
10-year	57.2	49.3	48.2	40.6	11.2	7.4	8.9	7.5
20-year	65.6	56.2	55.2	45.1	12.7	8.1	9.8	8.3
50-year	76.6	65.1	64.2	50.8	14.6	8.9	11.0	9.2
LORCA								
Longest run	156	86	63	84	6	4	5	5
GEV Mode	59.4	15.1	10.7	12.3	3.6	0.6	0.8	0.9
GEV Dispersion	23.1	25.3	18.8	23.0	0.9	1.0	1.3	1.6
Return period extremes (days)								
10-year	111.4	71.9	52.9	64.1	5.7	2.9	3.6	4.5
20-year	128.0	90.1	66.5	80.7	6.3	3.7	4.5	5.6
50-year	149.6	113.6	83.9	102.1	7.2	4.6	5.7	7.0

This emphasises the need to analyse the behaviour of extremes in their own right, rather than making the assumption that they will follow linearly the behaviour of the mean.

5. Conclusions

We have looked at past trends in climate over the Mediterranean region, as demonstrated by the record of surface observations, and at future trends as predicted by climate models. We have concentrated on changes in the occurrence of extremes, because these have potentially greater impacts on human activities and the environment than do changes in the mean climate.

High resolution scenarios of possible climate futures have been constructed for two catchments in the Mediterranean, the Guadalentin in Spain and the Agri in Italy. Analysis of these scenarios indicates the potential for a damaging increase in the occurrence of extremes in response to global warming. In particualr, a very large increase in the occurrence of days with high temperatures is suggested. In the scenario for the decade 2090-99, the number of very hot days (temperatures in excess of 35°C) is equivalent to two months with maximum daytime temperatures continually at or above 35°C. Such changes have serious implications for human health and comfort, and are likely to require substantial adaptation to be managed successfully.

References

1. European Environment Agency (2001) *Indicator Fact Sheet Signals 2001 – Chapter Tourism YIR01TO08 Tourism Arrivals.*
2. Frich, P., Alexander, L.V., Della-Marta, P., Gleason, B., Haylock, M., Tank, A.M.G.K., and Peterson, T. (2002) Observed coherent changes in climatic extremes during the second half of the twentieth century. *Climate Research 19*, 193-212.
3. Giorgi, F., Hewitson, B. and others (2001) Regional climate information – evaluation and projections. Chapter 10 in *Climate Change 2001: The Scientific Basis.* Contribution of Working Group I to the Third Assessment Report of the Intergovernmental Panel on Climate Change (eds. J.T. Houghton and others), Cambridge University Press, Cambridge, 583-638.
4. Gumbel, E.J. (1958) *Statistics of Extremes*. Columbia University Press, New York.
5. Houghton, J.T., Meira Filho, L.G., Callander, B.A.,. Harris, N., Kattenberg, A. and Maskell, K., eds. (1996) *Climate Change 1995: The Science of Climate Change.* Contribution of Working Group I to the Second Assessment Report of the Intergovernmental Panel on Climate Change, Cambridge University Press, Cambridge.
6. Houghton, J.T., Ding, Y., Griggs, D.J., Noguer, M., van der Linden, P.J., Dai, X., Maskell, K., and Johnson, C.A., eds. (2001) *Climate Change 2001: The Scientific Basis.* Contribution of Working Group I to the Third Assessment Report of the Intergovernmental Panel on Climate Change, Cambridge University Press, Cambridge
7. IPCC (2001) Summary for policymakers. A report of Working Group I of the Intergovernmental Panel on Climate Change. In *Climate Change 2001: The Scientific Basis.* Contribution of Working Group I to the Third Assessment Report of the Intergovernmental Panel on Climate Change (eds. J.T. Houghton and others), Cambridge University Press, Cambridge, 1-20. (Can be downloaded from http://www.ipcc.ch/pub/spm22-01.pdf).
8. Johns, T.C., Carnell, R.E., Crossley, J.F., Gregory, J.M., Mitchell, J.F.B., Senior, C.A., Tett, S.F.B., and Wood, R.A. (1997) The second Hadley Centre coupled ocean-atmosphere GCM: model description, spinup and validation. *Climate Dynamics 13*, 103-134.
9. Jones, P.D. and Moberg, A. (2003) Hemispheric and large-scale surface air temperature variations: an extensive revision and an update to 2001. *Journal of Climate 16*, 206-223.
10. Kalnay, E., Kanamitsu, M., Kistler, R., Collins, W., Deaven, D., Gandin, L., Iredell, M., Saha, S., White, G., Woollen, J., Zhu, Y., Chelliah, M., Ebisuzaki, W., Higgins, W., Janowiak, J., Mo, K.C., Ropelewski, C., Wang, J., Leetmaa, A., Reynolds, R., Jenne, R. and Joseph, D. (1996) The NCEP/NCAR 40-year reanalysis project. *Bulletin of the American Meteorological Society 77*, 437-471.
11. Köppen, W. (1936) Das geographische System der Klimate, in Köppen, W. and Geiger, R. (eds) *Handbuch der Klimatologie 3* , Gebrüder Borntraeger, Berlin.
12. Nakićenović, N. and Swart, R., eds. (2000) *Emission Scenarios.* IPCC Special Report on Emission Scenarios. Cambridge University Press, Cambridge. (Can be downloaded from http://www.grida.no/climate/ipcc/emission/index.htm).
13. Palutikof, J.P. (2003) Analysis of Mediterranean climate data: measured and modeled, in Bolle, H.-J. (ed) *Mediterranean Climate: Variability and Trends* Springer, Berlin, 125-132.
14. Palutikof, J.P., Goodess, C.M., Watkins, S.J., and Holt, T. (2002) Generating rainfall and temperature scenarios at multiple sites: examples from the Mediterranean. *Journal of Climate 15*, 3529–3548.
15. Palutikof, J.P., Brabson, B.B., Lister, D.H. and Adcock, S.T. (1998) A review of methods to calculate extreme wind speeds. *Meteorological Applications 6*, 119-132.
16. Richardson, C.W. (1981) Stochastic simulation of daily precipitation, temperature and solar radiation. *Water Resources Research 17*, 182-190.
17. Tank, A.M.G.K., Wijngaard, J.B., Konnen, G.P., Bohm, R., Demaree, G., Gocheva, A., Mileta, M., Pashiardis, S., Hejkrlik, L., Kern-Hansen, C., Heino, R., Bessemoulin, P., Muller-Westermeier, G., Tzanakou, M., Szalai, S., Palsdottir, T., Fitzgerald, D., Rubin, S., Capaldo, M., Maugeri, M., Leitass, A., Bukantis, A., Aberfeld, R., Van Engelen, A.F.V., Forland, E., Mietus, M., Coelho, F., Mares, C., Razuvaev, V., Nieplova, E., Cegnar, T., Lopez, J.A., Dahlstrom, B., Moberg, A., Kirchhofer, W., Ceylan, A., Pachaliuk, O., Alexander, L.V. and Petrovic, P. (2002) Daily dataset of 20th-century surface air temperature and precipitation series for the European Climate Assessment. *International Journal of Climatology 22*, 1441-1453.
18. Trigo, R.M. and Palutikof, J.P. (1999) Simulation of daily maximum and minimum temperature over Portugal: a neural network approach. *Climate Research 13*, 45-59.

19. Trigo, R.M., Osborn, T.J., and Corte-Real, J.M. (2002) The North Atlantic Oscillation influence on Europe: climate impacts and associated physical mechanisms. *Climate Research 20*, 9-17.
20. Wilby, R.L., Wigley, T.M.L., Conway, D., Jones, P.D., Hewitson, B.C., Main, J., and Wilks, D.S. (1998) Statistical downscaling of general circulation model output: a comparison of methods. *Water Resources Research 34*, 2995-3008.
21. Wilks, D.S., and Wilby, R.L. (1999) The weather generation game: a review of stochastic weather models. *Progress in Physical Geography 23*, 329-357

Chapter 5

Regional IPCC Projections untill 2100 in the Mediterranean Area

M. de Castro, C. Gallardo and S. Calabria
Facultad de Ciencias del Medio Ambiente
Universidad de Castilla-La Mancha, Avda. Carlos III s/n. 45071 Toledo, Spain

Over the last forty years much evidence has been accumulated as to how human activities can ultimately affect the global climate of our planet. However the extreme complexity of the Earth's climate system makes it very complicated to accurately analyse the importance of the induced climate alterations and the impact that they in turn could eventually have on the different ecosystems in the planet. One thing which is now a certainty is that the problem has a global dimension, and as a consequence, all the countries of the world must be involved in its solution. The Intergovernmental Panel on Climate Change, known by its acronym IPCC, is the key international body which has been established to assess the global climate change problem. It was founded in 1988 and at its first meeting held in November of that year in Geneva it decided to ask the scientific community for a report in which scientific facts on the global warming were established. In this manner politicians could be provided with a solid scientific base from which the requirements for action were developed. The IPCC is made up of three working groups (WG), each formed by hundreds of scientists, which deal respectively with the science of climate change itself (WG I), the impacts of climate change (WGII), and response strategies (WG III). The first IPCC assessment report from these three working groups was launched in May 1990. This was the key input to the international negotiations leading the way to prepare the agenda for the United Conference on Environment and Development (UNCED) in Rio de Janeiro in 1992, where more than 160 countries signed the Framework Convention on Climate Change. Since this date the IPCC working groups have released diverse technical reports and two additional assessment reports, the last of which was published in the year 2001 [5]. Some of the results obtained by the IPCC WG1 on climate change projections through the 21st century in the Mediterranean region are presented in this article. With these future climate scenarios it would be possible to estimate some of the impacts of global climate change in the region and to elaborate response strategies focused on adaptation and mitigation of its socio-economic consequences.

1. Global Climate Modelling

To understand and analyse in detail the likely climate change through this century (and beyond) so-called Global Climate Models (or GCMs) have been developed. A climate model is a simplified mathematical representation of the Earth's climate system, commonly defined as an interactive system consisting of five major components: the

A. Marquina (ed.), Environmental Challenges in the Mediterranean 2000-2050, 75–90.
© 2004 *Kluwer Academic Publishers. Printed in the Netherlands.*

atmosphere, the hydrosphere, the cryosphere, the land surface and the biosphere. These components are all linked by fluxes of mass, heat and momentum, and the numerous physical, chemical and biological interaction processes that take place between these components occur on a wide range of space and time scales, making the Earth's climate system extremely complex.

The most comprehensive current GCMs include some representation of the five major components of the climate system, along with the processes that go on within and between them. Global climate models in which the atmosphere and ocean components have been coupled together are known as Atmosphere-Ocean General Circulation Models (hereafter AOGCMs).

The AOGCMs are based upon physical laws describing the dynamics of atmosphere and ocean, expressed by mathematical equations. Since these equations are non-linear, they need to be solved numerically by means of well-established discretizing mathematical techniques, which use a three-dimensional grid over the globe. In the atmospheric module, for example, the time-spatial differential equations set that describe the large-scale evolution of momentum, heat and moisture are numerically solved. A similar differential equation system is solved for the ocean component. Currently, the spatial resolution of the atmospheric part of a typical AOGCM is about 250 km in the horizontal and some 10 to 30 levels in the vertical, and the ocean sub-model has a horizontal resolution of 125 to 250 km and a resolution of 200 to 400 m in the vertical. Processes taking place on spatial and temporal scales smaller than the model resolution, for example individual clouds in atmosphere models or mesoscale eddies in ocean models, are included through a parametric representation (called model parametrisations) in terms of the resolved basic quantities of the model. The AOGCMs are combined with mathematical representations of other components of the climate system, sometimes based on empirical relations, such as the land surface and the cryosphere. The most recent models also include representations of aerosol processes and the carbon cycle.

The model equation system is typically solved for every half hour of a model integration period. That is, at each integration time-step the model calculates updated values for all the physical variables that describe the state of the diverse climate system components in every three-dimensional grid covering the entire planet. This calculation process entails millions of simple mathematical operations every time-step, until the integration period (typically a few centuries) is completed. This means that climate simulations require the largest, most capable computers available.

AOGCMs are used to quantify the climate response to present and future human activities. The first step is to evaluate the quality of the models, which is essential in order to gain confidence in their ability to reproduce the main processes being carried out in the climate system. This is done by means of a systematic comparison between simulation results of current climate and observation data. For this, AOGCMs are run for extended simulation periods, typically many decades, under present conditions, that is with observed greenhouse gases (GHGs hereafter) concentrations and sulphate burden in the atmosphere. Models may also be evaluated by running them under different paleoclimatic conditions (for example the past Ice Ages). Once the quality of the model is well established, it is ready to provide time-dependent projections of the future climate change.

Table 1: The Emissions Scenarios of the IPCC-SRES

A1. The A1 storyline and scenario family describe a future world of very rapid economic growth, global population that peaks in mid-century and declines thereafter, and the rapid introduction of new and more efficient technologies. Major underlying themes are convergence among regions, capacity building and increased cultural and social interactions, with a substantial reduction in regional differences in per capita income. The A1 scenario family develops into three groups that describe alternative directions of technological change in the energy system. The three A1 groups are distinguished by their technological emphasis: fossil intensive (A1FI), non-fossil energy sources (A1T), or a balance across all sources (A1B) (where balanced is defined as not relying too heavily on one particular energy source, on the assumption that similar improvement rates apply to all energy supply and end use technologies).

A2. The A2 storyline and scenario family describe a very heterogeneous world. The underlying theme is self-reliance and preservation of local identities. Fertility patterns across regions converge very slowly, which results in continuously increasing population. Economic development is primarily regionally oriented and per capita economic growth and technological change are more fragmented and slower than in other storylines.

B1. The B1 storyline and scenario family describe a convergent world with the same global population, that peaks in mid-century and declines thereafter, as in the A1 storyline, but with rapid change in economic structures toward a service and information economy, with reductions in material intensity and the introduction of clean and resource-efficient technologies. The emphasis is on global solutions to economic, social and environmental sustainability, including improved equity, but without additional climate initiatives.

B2. The B2 storyline and scenario family describe a world in which the emphasis is on local solutions to economic, social and environmental sustainability. It is a world with continuously increasing global population, at a rate lower than A2, intermediate levels of economic development, and less rapid and more diverse technological change than in the B1 and A1 storylines. While the scenario is also oriented towards environmental protection and social equity, it focuses on local and regional levels.

In these future climate simulations the model is forced with time-dependent GHGs and aerosol concentrations, which are in turn derived from the so-called emission scenarios. These have been developed since 1996 by a group of experts within the IPCC, under the consideration of diverse hypothesis about future socio-economic and demographic developments [8]. Such hypotheses are based on coherent assumptions concerning future evolutions of the growth of the world population, energy intensity and efficiency, economic growth and other considerations. The result of this is an ensemble of forty emission scenarios, known by the acronym SRES (after Special Report on Emission Scenarios), covering a wide range of the main demographic, economic and technological driving forces of future GHGs and sulphur emissions. Each of these scenarios represents a specific quantification of one of the four narrative SRES storylines. All the scenarios based on the same storyline constitute a scenario "family". In the box above a brief description of the main characteristics of the four SRES storylines and scenario families is given. It is directly quoted from the last IPCC report [5].

In contrast to the two AOGCMs considered in the first IPCC assessment report, the third report includes output from about twenty AOGCMs worldwide. Roughly half of them had performed detail scenario experiments with time evolutions of GHGs and sulphate aerosols for the 20th and 21st century climate. Even though there are some differences in the climate simulated by each of these models, the comparison of the 20th century climate simulations with observations provided more confidence in the abilities of such models to simulate future climate changes and reduces the uncertainty in the model projections.

The IPCC Data Distribution Centre (DDC) is dedicated to collect in results from all of the above-mentioned AOGCM experiments. Such a collection allows a multi-model ensemble approach to the synthesis of projected climate change on a global scale. From the evaluation of the results included in the collection of 20th century climate simulations, it is clear that no one model can be chosen as "best" and it is better to use results from a range of models. For example, Lambert and Boer [7] show that for the ensemble of model simulations of the current climate, the multi-model means of temperature, pressure, and precipitation are generally closer to the observed distributions than are the results of any particular model. The multi-model ensemble mean represents those features of projected global climate change that are common to models as a group, and so it could be considered as more reliable.

Finally, it should be taken into account that a range of uncertainties affects all projections of climate change made with AOGCMs, despite the completeness of some of them. Therefore it is necessary to quantify uncertainty in so far as it is possible. Uncertainty in projected climate change arises from three main sources: uncertainty in forcing scenarios, uncertainty in model responses to given forcing scenarios, and uncertainty due to missing or misrepresented physical processes in models [5]. To tackle the first, a range of future emission scenarios instead of only one is used in the climate change simulations. In fact, since model simulations of the future climate depend on assumptions about emissions of GHGs and sulphate matter, the term "climate projections" should be used instead of "predictions". The second uncertainty source is somewhat circumvented by using the ensemble standard deviation and the range of indications of uncertainty in model results, though a clear characterisation is not yet established. And the third can be limited by requiring current climate model simulations to reproduce the so-far observed climate change.

2. Climate Projections in the Mediterranean Region from AOGCMs

As the SRES emissions scenarios were not approved by the IPCC until March 2000, not all of them have been used yet for performing future climate simulations with the more comprehensive current AOGCMs. In fact, two scenario groups (A2 and B2) have been used by most of the modelling climate centres, even though none of the SRES marker scenarios can be considered to be more likely than the others. Besides, results from only a few of the AOGCM are lodged at the IPCC Data Distribution Centre for public use. All of these AOGCM have been evaluated against current climate data, and in general the main large-scale climate features are correctly captured. The results of different climate models, however, may differ appreciably, in particular for small spatial and time scales. And in some cases such inter-model differences exceed the range of individual projected climate changes, which, however, do not invalidate the results. Instead valuable information can be extracted from this, because it allows an objective evaluation of the reliability involved in every individual climate model projection, as will be commented below. Due to such inter-model differences, it has been considered more convenient to show firstly the most relevant climate projections obtained from one AOGCM. And after this, a simple comparison will be made between the results of such a model and those offered by the ensemble composed from the output of several models.

The AOGCM HadCM3 was chosen as its global warming quantification because the end of the 21st century falls in the middle of the model ensemble range in both emission scenarios. The HadCM3 model was developed by and used at the Hadley Centre for Climate Prediction and Research in the United Kingdom. This choice should not be taken as an indication of the superiority of the HadCM3 model relative to others, although its quality is backed in the main by a good agreement between current climate simulation results and observations.

HadCM3 is a comprehensive coupled atmosphere-ocean general circulation model developed in 1998 at the Hadley Centre. A complete description of such a model can be found in Gordon et al. [4] and Pope et al. [9], therefore none of its technical details are here mentioned, except that it has a horizontal resolution of 2.5° of latitude by 3.75° of longitude (equivalent to about 300 x 300 km in the middle latitudes belt) and 19 levels from the surface up to the top of the atmosphere.

For climate experiments the HadCM3 model is initiated directly from an observed ocean state at rest, with a suitable atmospheric and sea ice state. During the simulation the atmosphere and ocean exchange information once per simulated day, and fluxes of heat and water are conserved exactly in the transfer between their different grids. The HadCM3 model run procedure for obtaining climate change 21st century projections is as follows: First the model is executed, with the mentioned initial conditions, over a period of 1000 years in which all concentrations of GHGs and aerosols are representative of the pre-industrial era. This long period is necessary for achieving an adequate coupling between atmosphere and oceans. The final state after this spin-up represents climate conditions before the year 1860. After that year the model simulation continues with the GHGs and aerosol concentrations increasing at a rate derived from available observations up to the year 1990. Beyond this year, the model is forced by evolving GHG and aerosol concentrations as deduced from any of the SRES scenarios. The results obtained for the period 1960-1990 can be compared to observational data for model evaluation. The climate change scenarios are obtained from the differences between 1960-90 and any other 30-year period through the 21st century. For the following climate change analysis three periods were chosen: 2010-2040, 2040-2070 and 2070-2100.

Winter season air surface temperature changes relative to the 1960-90 period for the A2 and B2 SRES scenarios are shown in figure 1a. A progressive warming throughout the 21st century over the entire Mediterranean region in both scenarios is evident. But the intensity and rate of such warming depend on the zone within the Mediterranean region as well as on the SRES scenario. In southern and northern zones winter temperatures increase more rapidly than in the eastern and western ones up to the last third of the century. In the first third of the century the warming rate in the western zone is similar in A2 and B2 scenarios, but in the northern and eastern zones the rate is somewhat higher in B2 than A2. In the second third the warming rate is similar in both scenarios. In fact, the spatial distribution of Mediterranean winter temperature changes in the 2040-2070 period relative to 1960-1990 is fairly similar in both SRES scenarios. And in the last third the warming accelerates with a higher rate in A2 than in B2 throughout all the Mediterranean zones. Thus, the warming is more intense in A2, reaching maximum values higher than 5°C in the Iberian Peninsula and northeast Africa, while B2 warming keeps below 4°C all over the Mediterranean region.

Similar features can be observed in the summer period (figure 1b), though in general the warming is more intense and rapid in this season than in winter. Furthermore the warming rate is somewhat higher in A2 than in the B2 scenario through all the century. But the most relevant result is that the summer warming in the European countries by the end of the century exceeds 6°C in the A2 scenario (highest local increase near 8°C) and exceeds 5°C in the B2 scenario in the western and northeastern zones. Such a huge warming means that many of the natural vegetation species in this area will be seriously threatened, and even more if seasonal precipitation changes in summer (commented on below) were to occur.

Figure 1a. Average winter season temperature changes for A2 and B2 SRES emissions scenarios in three periods through 21ᵗʰ century relative to 1960-1990. Isolines indicate the ratio (in %) between future and current interannual variability.

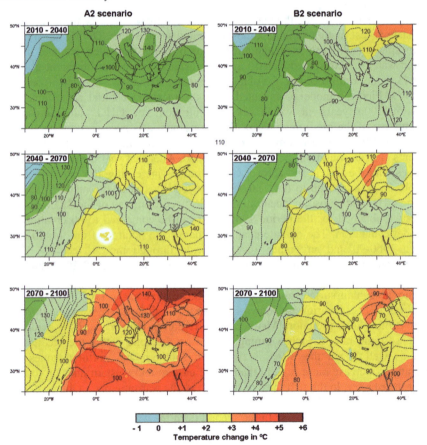

Concerning maximum and minimum temperatures, changes in winter are quite similar to those of the daily mean temperatures (not shown). But in the summer season, maximum temperatures increase more markedly than daily mean temperatures, reaching values well over 6°C through most of the Mediterranean region in the A2 scenario and the same value in the European part in the B2 scenario. Only on the coastal strips along the Mediterranean such impressive warming is moderated, keeping it under 4°C and 3°C in the A2 and B2 scenarios respectively. Obviously this implies that summer minimum temperatures increase less than daily mean temperatures, but in this case the higher minimum temperature changes correspond to B2 instead of to A2.

The simulated warming distributions for spring and autumn seasons (not shown) exhibit, in general, similar features to those mentioned above for winter and summer. But the temperature increasing values for these "transition" seasons are intermediate between winter and summer. The autumn warming relative to the 1960-90 period reaches a maximum of around 6°C in most of the Mediterranean zones by the end of the century in the A2 scenario, while it does not reach 5°C in the B2 scenario. On the other hand, spring temperature increases are in general 1°C less than in autumn.

Figure 1b. Average summer season temperature changes for A2 and B2 SRES emissions scenarios in three periods through 21th century relative to 1960-1990. Isolines indicate the ratio (in %) between future and current interannual variability.

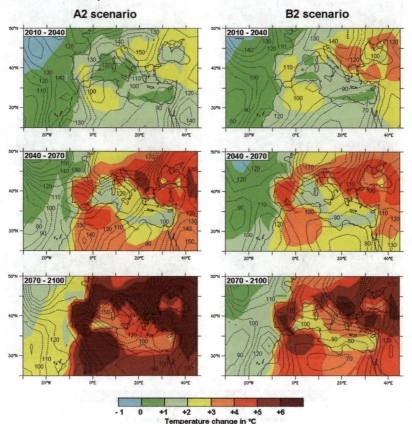

Seasonal precipitation changes exhibit, in general, a large spatial variability when they are expressed as a percentage change. But this can be considered, at least in part, as an "artefact" because in several zones of the Mediterranean region the absolute precipitation amounts are very small, specially in the southern half and generally in the summer season. For this reason it is preferable to analyze precipitation changes as expressed in absolute change values (mm/day). Obviously this kind of analysis has the inconvenience of not allowing a clear view of the relative importance of a seasonal precipitation decrease or increase in any particular zone. In other words, the same absolute change will have more relative importance in a dry zone than in a rainy area. But, on the other hand, absolute changes allow an easier quantification of future alterations of water availability in the diverse zones and seasons. Therefore, it was decided to show the results in terms of absolute changes of seasonal precipitation amounts relative to the 1960-1990 climate simulated by the AOGCM.

Figure 2a. Average winter seasonal precipitation changes (in mm/day) for A2 and B2 SRES emission scenarios in three periods through 21st century relative to 1960-1990. Isolines indicate the ratio (in %) between future and current interannual variability.

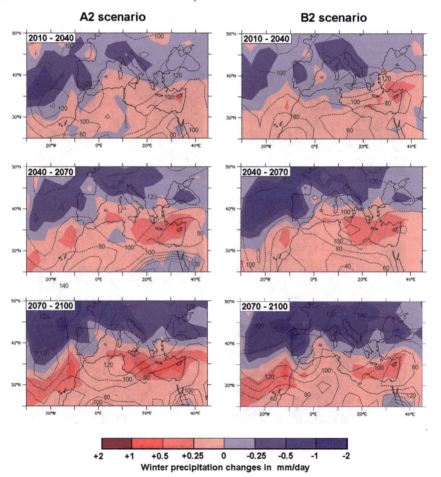

In contrast to simulated temperature changes throughout this century, which always have a positive sign, precipitation changes have diverse signs depending on the season and Mediterranean zones considered. Thus, it can be seen in figure 2a that the winter precipitation amount increases in the western and northern Mediterranean zones both in the A2 and B2 scenarios, while there is a decrease in the southern and eastern areas. This could be interpreted in terms of a southern displacement of the Atlantic storm tracks in winter months, which would give place to a greater moist convection activity when it is combined with the simulated surface air warming and greater latent heating in the moister atmosphere.

This hypothesis seems to be backed by Ulrich and Christoph [10], who found an increase in upper air storm track winter activity over the East Atlantic and Western Europe with rising greenhouse gas forcing, or by Knippertz et al. [6] who obtained an increase in the number of deep low pressure systems in the North Atlantic by using a AOGCM with increased CO_2 and sulphates. Nevertheless such as interpretation presents some uncertainty because so far there have been few studies examining changes in extra-tropical cyclones in a future climate.

However future precipitation changes simulated for the other seasons show more uniform behaviour all over the Mediterranean region. For example, in figure 2b a clear decrease in the entire region is seen in the summer season. Something similar has been obtained for autumn and spring (not shown), though in late autumn a small increase is obtained in western Mediterranean, but only until the last third of the century. In general, as is seen for the summer in figure 2b, the simulated precipitation decrease is less intense in B2 than in the A2 scenario. But the most important feature is that such a marked and long-lasting simulated precipitation decrease in most of the region, together with the foreseeable high warming through the century, would probably induce future irreversible impacts in several Mediterranean ecosystems and severe water storage shortages in many zones, among other climate threats.

An aspect of future climate projections as important as the change of average values (temperature, precipitation or whatever) relative to current climate is the possible alteration of inter-annual or intra-annual climate variability. For a correct analysis on this issue it would be necessary to examine output results from model ensemble experiments, just as was commented before, for a better assessment of average climate change. But in this case we were not able to perform such an analysis because only monthly values of climate variables from a single model experiment (HadCM3) were available. Instead of this, a simple statistical treatment was applied to these values with the purpose of obtaining an indication of changes of inter-annual variability in future climate projections relative to control experiment. This merely consists of percentage ratios between standard deviations of monthly values (temperature and precipitation) through each of the 30-years periods of the future climate (2010-2040, 2040-2070, 2070-2100) and that of the current climate (1960-1990). Thus, a ratio lower (*higher*) than 100 means that the projected future climate variability will be smaller (*greater*) than that of the current climate. The results obtained from such a simple statistical method are represented by isolines in figures 1 and 2.

Figure 1a shows that in the first third of this century the monthly winter temperature anomalies in scenario A2 will be more frequent than at present in the northern half of the Mediterranean and less frequent in the southern part. But in the next two thirds the

Figure 2b. Average summer seasonal precipitation changes (in mm/day) for A2 and B2 SRES emissions scenarios in three periods through 21st century relative to 1960-1990. Isolines indicate the ratio (in %) between future and current interannual variability.

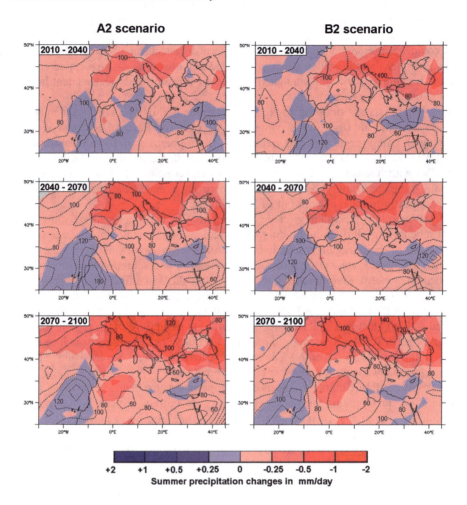

variability will be higher than the present climate all over the region, though the maximum values still appear in the northern part. In scenario B2 the monthly winter temperature variability is in general lower than in A2. In the summer (figure 1b) future temperature variability in scenario A2 will be higher than at present in most of the Mediterranean region, except in northeast Africa. The variability increases even more in the northern part through the century, specially during its last half. This result coincides with those obtained by Buishand and Beersma [1]. But, again in scenario B2 thermal variability is in general lower than in A2, in such a way that in most of the region such variability is smaller than in the current climate. This would mean, for example, that in the B2 scenario the frequency of relative cold air spills, that occasionally give relief during the suffocating summer days in the southern Mediterranean countries, would be less than in the present climate.

In the winter season (figure 2a) the frequency of monthly precipitation anomalies appears somewhat higher than in the present climate all over the Mediterranean through the first third of the century. But after this period variability becomes lower than at present in the southern half of the region. This behaviour appears in both scenarios, though in B2 the variability remains lower than at present through all the century in the southern part of the Mediterranean. Beside the significant dryness projected for this century, monthly precipitation anomalies in summer would be progressively smaller than at present all over the Mediterranean region through the century (figure 2b). This occurs in both scenarios, though in B2 the increasing rate is lower than in A2. As a pattern of general behaviour, it can be seen that increases in mean precipitation are likely to be associated with increases in inter-annual variability, and precipitation variability is likely to decrease in areas of reduced mean precipitation.

After this analysis of seasonal climate projections obtained from one AOGCM, figures 3 and 4 show a comparison with results derived from an ensemble average of several global models, including the HadCM3. For this comparison, only projected annual average climate changes for the 2070-2100 period relative to the current climate will be considered. The future annual warming distributions in the Mediterranean region are compared in figure 3. In general greater annual temperature changes are obtained in the individual model projections, but spatial distributions are quite similar to the model ensemble, as well as the differences between the A2 and B2 scenarios.

Figure 3. Annual average temperature changes for 2070-2100 period relative to 1960-1990, simulated by the HadCM3 model and by a model ensemble (taken from IPCC [5]), for A2 and B2 emissions scenarios.

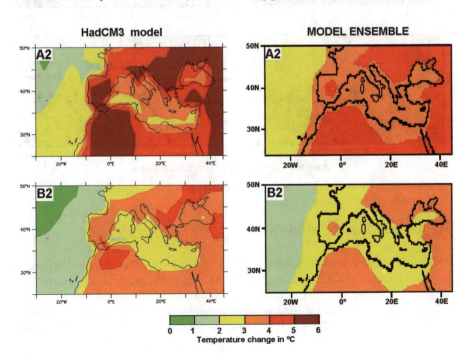

Greater differences between the single and ensemble model appear in annual precipitation changes, as figure 4 shows. On this occasion relative annual precipitation changes with respect to current climate values are exhibited. More drastic annual precipitation decreases result in the single model than in the ensemble simulations, and also the spatial distributions of changes are not as similar as those in the temperature. The magnitude of regional precipitation change varies considerably among models with the typical range being around 0 to 50%, where the direction of change is strongly indicated (all models coincide), and around -30 to +30% where it is not. This is an illustrative indication of the lack of reliability of the precipitation changes simulated by current AOGCMs in comparison to temperature. Nevertheless some similarities are still noted in the figure, for example the greater decrease in the southern Mediterranean countries and the more relevant dryness in A2 with respect to the B2 scenario.

Figure 4. Percentage of annual precipitation changes for 2070-2100 period relative to 1960-1990 , simulated by the HadCM3 model and by a model ensemble (taken from IPCC [5]), for A2 and B2 emissions scenarios.

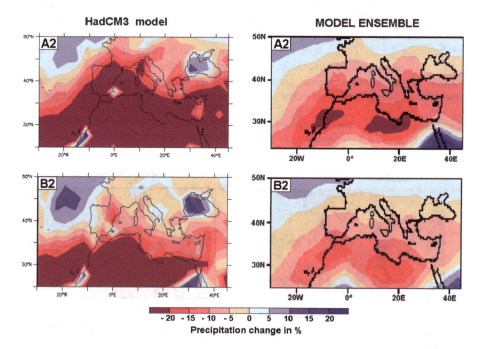

3. Regional Climate Models

As the IPCC report states, though there is an amount of qualitative large-scale agreement between AOGCMs, temperature and especially precipitation distributions show strong discrepancies when regional scale variations are considered. In a broad sense, this lack of reliability is thought to be mainly due to the poor spatial resolution and bulk parametrisation of physical processes in current AOGCMs. The coarse spatial resolution provokes a double effect: (a) It prevents a realistic reproduction of atmospheric processes with a size smaller than or similar to the model grid, and (b) it distorts the continent-ocean lines and orographic features, which generally leads to incorrect simulations. And most of the physical parametrisations in the AOGCMs have been developed for and tuned to its coarse resolution, thus resulting inadequate for reproducing smaller scale processes, that in some cases decisively contribute to some regional climate regimes. For example, it is well known that the climate in the Mediterranean region is determined partly by larger-scale features of the atmospheric circulation, partly by interactions between such large-scale flow and orography and sea-land contrasts (Mediterranean cyclogenesis) and partly by more local effects (sea-breezes or moist-convection forced by orography). And current AOGCMs are unable to reproduce such regional-scale features of the Mediterranean climate.

Introducing more detailed physical parametrisations or increasing the spatial resolution in AOGCMs would imply a considerable growth of computer time. For this reason a realistic analysis of regional scale consequences of global climate change with AOGCMs does not seem possible within the next 5-10 years. Therefore, it is necessary to consider other techniques to model sub-regional climates from current AOGCMs simulations. One of such techniques, which is regarded as the most promising [5], entails the use of so-called Regional Climate Models (RCMs).

The RCMs are essentially similar to the atmospheric module of any AOGCM, with some improved physical parametrisations, but applied to a limited area of the globe with much higher resolution. The regional models are embedded or nested in a global model. Initial conditions, time-dependent lateral meteorological conditions and surface boundary conditions derived from an AOGCM (or analysis of observations) are used to drive the RCMs. These driving data can also include GHGs and aerosol forcing. To date, the RCMs have been applied only in one-way nesting mode, i.e., with no feedback from the RCM simulation to the driving AOGCM. Therefore, the method consists of using the AOGCMs to simulate the response of the global circulation to large-scale forcings and the RCMs to (a) account for sub-GCM grid scale forcings (e.g., complex topographical features and land cover inhomogeneity) in a physically-based way; and (b) enhance the simulation of atmospheric circulations and climatic variables at fine spatial scales [5].

The first results with a RCM were obtained by Dickinson et al. [2]. Later studies have shown that this procedure is able to describe the effects of forcings with a size corresponding to a sub-grid in a global model on the circulations and processes determining the climate of a given region. The nested regional modelling technique is now used in a wide range of applications, from palaeoclimate to anthropogenic climate change studies. They can provide high resolution (up to 20 km or less) and multi-decadal simulations and are capable of describing climate feedback mechanisms acting at the regional scale. A number of widely used limited area modelling systems have been adapted to, or developed for, climate applications. More recently, RCMs have begun to couple atmospheric models with other climate process models, such as hydrology, ocean, sea-ice, chemistry/aerosol and land-biosphere models.

However, a nested RCM cannot correct large-scale errors generated by the AOGCM in which it is embedded. For example, if AOGCM does not properly simulate the location of polar front jet and storm tracks at mid-latitudes, the nested RCM cannot correct such an error, leading to unrealistic regional-scale results. Therefore it is necessary to choose a carefully validated AOGCM which realistically represents the large-scale circulations affecting the region of interest. It is also important that the RCM includes adequate physical representations to simulate the climate of a given region. There exists a great variety of parametrisations able to simulate satisfactorily convective precipitation processes, air-soil exchanges and radiative effects of clouds, GHGs and aerosols. Finally, the choice of grid size has to represent a compromise between the size of the most important mesoscale features influencing regional climate and computing power available. Current RCMs cannot thus simulate such long time periods as AOGCMs, because RCM need more computer time than global models, and since the integration time-step has to be in accordance with the smaller grid size. For example a RCM with 50 x 50 km grid mesh, 100 x 80 grids and 30 vertical levels needs approximately four times more CPU time than a current AOGCM with 2.5° x 2.5° grid-size and 17 layers. For this reason, the usual approach consists of simulating time-slice periods of several decades (e.g. 30-years) with a RCM.

The PRUDENCE project (*Prediction of Regional scenarios and Uncertainties for Defining European Climate change risks and Effects*) is a European-scale investigation with the following objectives:

a) to address and reduce the above-mentioned deficiencies in projections;
b) to quantify our confidence and the uncertainties in predictions of future climate and its impacts, using a variety of climate models and impact models and expert judgment on their performance;
c) to interpret these results in relation to European policies for adapting to or mitigating climate change.

PRUDENCE will provide a series of high-resolution climate change scenarios for 2071-2100 for Europe and the Mediterranean basin, characterising the variability and level of confidence in these scenarios as a function of uncertainties in model formulation, natural/internal climate variability, and alternative scenarios of future atmospheric composition. The project will provide a quantitative assessment of the risks arising from changes in regional weather and climate in different parts of Europe and the Mediterranean, by estimating future changes in extreme events such as flooding and windstorms and by providing a robust estimation of the likelihood and magnitude of such changes. The project will also examine the uncertainties of potential impacts induced by the range of climate scenarios developed from the climate modelling results. This will provide useful information for climate modellers on the levels of accuracy in climate scenarios required by impact analysts. Furthermore, a better appreciation of the uncertainty range in calculations of future impacts from climate change may offer new insights into the scope for adaptation and mitigation responses to climate change. In order to facilitate this exchange of new information, the PRUDENCE work plan places emphasis on the wide dissemination of results and preparation of a non-technical project summary aimed at policy makers and other interested parties. The partners of PRUDENCE consist of the major research centres in Europe working on climate and related topics. More information on PRUDENCE project can be found in http://www.dmi.dk/f+u/klima/prudence/index.htm.

4. Conclusions

Global climate modelling results throughout 21th century and focused on the Mediterranean region corresponding to two IPCC emission scenarios (A2 and B2 SRES) have been shown. The main results corresponding to temperature and precipitation seasonal changes can be summarized as follows:

a) The highest warming simulated in both emission scenarios corresponds to the summer, reaching values above 6°C in A2 and above 5°C in B2 throughout most of the inland Mediterranean areas for the last third of the century. In the winter season the projected temperature rise is in general 1-2°C less intense than in summer.

b) The maximum warming in the winter season is projected for the southern inland Mediterranean regions, while in the summer it is generally localised in the European countries. Despite the used global model coarse resolution, it is clearly observed that the warming will be less intense in coastal areas.

c) Projected seasonal precipitation changes have diverse signs depending on the season and Mediterranean zones considered. The amount of winter precipitation is projected to increase in the western and northern Mediterranean zones both in the A2 and B2 scenarios, while there will be a decrease in the southern and eastern areas. However precipitation changes projected for the other seasons show an uniform decrease all over the Mediterranean region, it being more remarkable in the summer season. In general, the simulated precipitation decreases are less intense in B2 than in the A2 scenario.

d) By means of a simple statistical treatment, some indications of projected changes of inter-annual variability in the future climate relative to the present are obtained. For the second half of the century temperature variability will be generally higher than the present climate in most of the Mediterranean regions, this being more remarkable in scenario A2. Projected changes for precipitation inter-annual variability are spatially less uniform than for temperature, showing that variability is likely to be reduced (enlarged) in areas where a decrease (increase) of seasonal precipitation is predicted .

e) A comparison of these single-model results with those derived from an ensemble average of several global models shows that spatial distributions of projected warming are quite similar, though less intense in general. But this is not the case at all for precipitation changes. This fact is an illustrative indication of the well-known lower reliability of the precipitation changes simulated by current AOGCMs relative to temperature ones.

f) Because part of this uncertainty is thought to be related to the coarse spatial resolution and bulk physical parameterisations used in current AOGCMs, a promising, but transitive, downscaling technique to address this issue entails the use of Regional Climate Models, until a sufficient computer power increase allows AOGCMs to circumvent such limitations.

References

1.Buishand, T.A., and Beersma, J.J. (1996) Statistical tests for comparison of daily variability in observed and simulated climates. *J. Climate 9*, 2538-2550.
2.Dickinson R.E., Errico, R.M., Giorgi, F., and Bates, B.T. (1989) A regional climate model for western United States. *Clim. Change, 15*, 383–422.
3.Giorgi, F. and Francisco, R. (2000) Evaluating uncertainties in the prediction of regional climate change. *Geophysics, Research Letter 27*, 1295-1298.
4.Gordon, C., Cooper, C., Senior, C.A., Banks, H.T., Gregory, J.M., Johns, T.C., Mitchell, J.F.B., and Wood, R.A. (2000) The simulation of SST, sea ice extents and ocean heat transports in a version of the Hadley Centre coupled model without flux adjustments. *Climate Dynamics 16*, 147-168.
5.IPCC (2001) *Climate Change 2001: The Scientific Basis*. In Houghton, J. T., Ding, Y., Griggs, D.J., Noguer, M., Van der Linden, P.J., and Xiaosu, D. (eds.), Cambridge University Press, Cambridge, 1994. (http://www.ipcc.ch)
6.Knippertz, P., Ulbrich, U., and Speth, P. (2000) Changing cyclones and surface wind speeds over the North Atlantic and Europe in a transient GHG experiment. *Climate Research 15*, 109-122
7.Lambert, S.J., and Boer, G.J. (2001) CMIP1 evaluation and intercomparison of coupled climate models. *Climate Dynamics 17, 2/3*, 83-106.
8.Nakicenovic, N., Alcamo, J., Davis, G., de Vries, B., Fenhann, J., Gaffin, S., Gregory, K., Grübler, A., Jung, T.Y., Kram, T., La Rovere, E.L., Michaelis, L., Mori, S., Morita, T., Pepper, W., Pitcher, H., Price, L., Raihi, K., Roehrl, A., Rogner, H.-H, Sankovski, A., Schlesinger, M., Shukla, P., Smith, S., Swart, R., van Rooijen, S., Victor, N., and Dadi, Z. (2000) *IPCC Special Report on Emissions Scenarios*, Cambridge University Press, Cambridge, 599. (http://www.ipcc.ch)
9.Pope, V.D., Gallani, M., Rowntree, P.R., and Stratton, R.A. (2000) The impact of new physical parametrisations in the Hadley Centre climate model - HadAM3. *Climate Dynamics 16*, 123-146.
10.Ulbrich, U., and Christoph, M. (1999) A shift of the NAO and increasing storm track activity over Europe due to anthropogenic greenhouse gas forcing. *Climate Dynamics 15*, 551-559.

Section III

Desertification Impacts

Chapter 6

Desertification in the Maghreb: A Case Study of an Algerian High-Plain Steppe

H. Slimani[1] and A. Aidoud[2]
[1] *Laboratoire d'Ecologie et Environnement, Univ. Sci. Technol. H. Boumédiene, El Alia, 16111 Algiers*
[2] *UMR , UMR 6553 (Bat 14A), Dynamique des Communautés, Univ. Rennes 1, Campus de Beaulieu, 35042Rennes cedex, France*

Natural and semi-natural ecosystems of the Maghreb have been subject, for several centuries, to grazing and cultivation as dominant land-uses. Living in a Mediterranean climate which is generally arid, plant, animal and human populations developed diverse adaptive strategies to cope with such rough conditions. Forests and rangelands have been undermined by erratic droughts that seem to be becoming more frequent. During the last two or three decades, vegetation and soil resources, in particular those of arid lands, have been marked by a rapid and severe degradation. The major changes observed were the regression of perennial plants (ligneous and grasses) and sand encroachment that are recognised to be relevant indicators of desertification in North Africa and the Middle East rangelands. A brief overall view of the Maghreb is given and illustrated by an example from the steppe rangelands dominated by alfa-grass (*Stipa tenacissima* L.) as a case study. The basic results of long term monitoring of what was originally a typical and dense steppe of the arid High Plains of Algeria have shown that alfa steppes have nearly disappeared from pediment surfaces since the 1980s. Causes are multiple but overgrazing remains in the present case, the main and direct origin of degradation because of diverse social and economic reasons, human population and need increases and increases in sheep grazing pressure. Irreversible changes were assessed using a grazing gradient approach. The gradient was represented by a transect starting from a long-term ungrazed exclosure, representing the pre-existing system and ending at free grazed rangelands. The monitored alfa stand showed, at the beginning of the study, a high homogeneity of vegetation and soil features. Vegetation pattern was a mosaic with individual tussocks and interstitial bare soil. The long term monitoring showed significant changes of the dominant species phytomass, plant composition and diversity and soil characteristics. The results, showing a regressive trend of the principal attributes of the vital ecosystem, could be generalized to all steppe habitats that had been originally dominated by alfa-grass. The aim of this work is to improve knowledge about extent, causes and mechanisms of desertification in order to direct restoration and rehabilitation actions.

1. The Maghreb Landscape

The natural landscape of the Maghreb has been changing rapidly over the last few decades because of increasing anthropogenic pressure. These changes are described as desertification. Since the Rio earth summit in 1992, desertification is defined as 'land degradation in arid, semi-arid and dry sub-humid areas resulting from various factors,

A. Marquina (ed.), Environmental Challenges in the Mediterranean 2000-2050, 93–108.
© 2004 *Kluwer Academic Publishers. Printed in the Netherlands.*

including climatic variations and human activities' [45]. According to this definition, in the Maghrebian countries, desertification results mainly from the conjunction of land over-use and drought.

During the last decades, forests, matorrals and steppes have been marked by severe degradation. However, the most affected ecosystems are those of the arid region (*s.l.*) *i.e.* arid, semi-arid and dry sub-humid regions that represent 93% of the total land area (excepting the hyper-arid region of the Sahara desert).

Recent dynamics of arid steppe rangelands may illustrate desertification processes. Maghrebian steppes have been subjected, for several centuries, to grazing as the dominant land use. Being under arid condition, these rangelands are naturally undermined by erratic droughts. Despite episodic droughts becoming more frequent, the important human population growth is the main social indicator of increasing population needs that may explain the over-exploitation of natural resources. At the same time, grazing pressure has rapidly increased, and in most of the Maghrebian steppes, overgrazing has become the main and direct cause of desertification. The degradation of steppe rangelands was strongly correlated to socio-economic indicators such as human and grazing animal population growth, nature and surface of cereal cultivation, sedentarisation rate of nomadic population.

An alfa-grass (*Stipa tenacissima* L.) dominated steppe is taken as a case study. Alfa-grass was the most representative species of the steppe vegetation in the North-West of Maghreb. Since the mid-1980's, a large part of the Algerian high-plain alfa-grass steppe has almost entirely disappeared [3,4]. Our purpose is to present the basic results of long term monitoring of an originally pure alfa stand. Ecological processes are depicted from a vegetation and soil point of view. Changes were assessed using a grazing gradient approach [41]. The gradient was represented by a transect starting from a long-term ungrazed exclosure, representing the pre-existing system and ending at open grazed rangeland. The monitored alfa stand showed, at the beginning of the study, a very low spatial heterogeneity of vegetation and soil.

Our objectives are to point out: (1) the direct causes and agents of desertification and responses in the Maghreb; (2) the indirect or 'up-stream' causes and mechanisms, (3) for discussion, the current trend and our expectations and predictions for the future.

2. Desertification of the Maghreb: an overview

From an official viewpoint, the Arab Maghrebian Union (UMA as acronym) includes five countries: Algeria, Libya, Mauritania, Morocco and Tunisia (including the Western Sahara). 85% of the total UMA area is a desert (mean annual rainfall R<100 mm.yr^{-1}). According to the desertification definition, we are only considering countries that present threatened zones *i.e.* arid (*s.l.*) but non-desertic zones. In the context of the Maghreb, the most affected and/or threatened ecosystems are those of arid lands (*s.s.*) with between 100 and 400 mm of mean annual rainfall (table 1). These limits correspond respectively to 0.05 and 0.20 or the P/PET (mean annual rainfall to potential evapotranspiration) ratio [34]. Below the 100 mm isohyet, Saharian vegetation is classified as a steppe but mainly restricted to sandy soils and run-off zones on large pediments.

It is not easy to give a complete idea of the whole threatened zone because of the scarcity and scattering of data. According to the FAO database, table 1 shows the main features that could help to explain the current overall trend of desertification in the Maghreb.

Table 1. Climatic, social and economic features of the Maghreb

	1960's	1990's	% increase*
Total area (in 1000 km²):	6034		-
% of non-arid and non-desert area (P/ETP>0.20)	9 %		-
% of arid area (0.05<P/ETP<0.20)	10 %		-
% of desert area (P/ETP<0.05)	81 %		-
Population (in1000 persons)	29700	77830	162
Cereals – Area Harvested (in 1000 ha)	9095	9600	6
Cereals – Production (in 1000 metric tons)	6416	9620	50
Cereals – Yield (in 100kg.ha^{-1})	4.91	8.98	83
Meat Production (in 1000 metric tons)	387	1388	259
Food supply (cereals in kg per capita)	216	124	-43
Food supply (meat in kg per capita)	13	18	37

P/PET= mean annual rainfall to potential evapotranspiration ratio
1960's to * 1990's average used except for population (1961 to 2000 increase)
Source: FAO, agriculture, 2002.

One of the most efficient indicators of desertification is the natural vegetation cleared mainly for crops but there is a lack of reliable data for the whole region. Some examples are taken from national reports (Morocco, Algeria and Tunisia mainly). Algerian national statistics report a decrease of forested lands from 4.4 x10^6 ha to 3.9 x10^6 between 1980 and 1998 (including reforestation) that means an annual average of about - 0.6%. This value is - 0.5 for Tunisia and - 0.3 for Morocco (1992-95). Locally this value may be higher. For instance, forest and woodlands regressed by 31% (annually – 1.5% on average) between 1977 and 1996 in the north-western region of Morocco [37]. In arid lands, a large number of perennials plants, that are a protection against water and mainly wind erosion [33], disappeared, either cleared for cereal cultivation, overgrazed or used for fuel (see the case study on alfa-grass below). Especially in rangelands, the destruction of perennial plants is mainly due to a forage deficit but also to a change in sheep breeding practices.

Cereals are by far the most important food for the Maghrebian population in terms of calories and traditional diet. From 1960's to 1990's, while cereal production increased by 50 % (table 1), global population was 150% higher. The cereal supply dropped by – 43%. The evolution of meat production, which is of a higher value, reflects an increase in livestock. However, this evolution is markedly different from one country to another. While the overall increasing factor was 3.6, it was 2.84 in Morocco and 5.28 in Algeria where the sheep population increased from 3.8 million in 1963 to 18.7 million in 1993. Yet, the whole Maghrebian region has suffered an increasing frequency of dry years during the last two decades. This means that climatic events are hazardous for livestock. In Morocco, natural forage for livestock shows a deficit of nearly 4 billion FU representing 40 % of total fodder needs [27]. As a result, the trend is an increase of cropland at the expense of rangelands and permanent grazing, and increasing pressure no matter what the availability of rainfall and vegetation, and more pressure on degraded lands.

Between the 1960s and the 1990s the area of cereal harvested increased by 6%, showing a relative stability while the production of cereals increased by 50% and cereal yield by 83% (table 1). In fact the harvested area has little significance regarding desertification. Cereals are mainly cultivated in rain-fed conditions and large areas of arid lands are included. Usually, in these conditions, the crop expectancy does not exceed 25% and then the land surface ploughed every year might be 2 – 3 times greater than the area harvested. Furthermore, cereal cultivation is called 'itinerant' because of fallow practices and abandonment when soil is severely eroded.

As an indication of this trend, the most efficient factor is the rapid human population growth with a doubling period of 25 years. The population has multiplied by a factor of 2.62 between 1961 and 2000. By comparison, this factor was of 1.21 for Western Europe but 3.01 for the whole of Africa. As a result, the population grew faster than the economy.

In relation to the global climatic change, as reported by [32], the increase of CO_2 and temperature will worsen the current conditions. Their impact will be extremely important but not because of their direct effects which would be negligible in comparison with human ones.

3. Desertification in Algeria: alfa steppe as a case study

Pure stands of alfa-grass, the previously so-called "alfa seas", covered large parts of north-western Africa between 100-500 mm isohiets [30]. As reported by Whal [49], the alfa steppe occupied about 70 % of the high-plains in Algeria during the late 19th century. The alfa steppes we are considering in this paper are those developed on sub-horizontal pediments commonly named 'plains' or 'plateaux' that are mainly used as open common rangelands for extensive sheep grazing. Alfa-steppe is considered to be the last level of forest degradation i.e. 'steppisation' [34]. However, for the past few decades, halophylious communities (gypso-saline habitats) excepted, alfa-steppe represents the 'historic indigenous system' with regard to nearly all other steppe plant communities.

3.1. STUDY AREA (PRE-EXISTENT SYSTEM) AND METHODS

The study area (Rogassa) is located in the western high-plains at a 1090 m mean altitude (figure 1). The climate is Mediterranean arid with a mean annual precipitation of 250 mm.yr^{-1} (table 2). Mean annual rainfall recorded during the monitoring period (1975 to 1993) was 224 mm.yr^{-1} with 60% falling between September and February. Mean annual temperature is 15.4°C with mean maximum temperature M = 36.7°C in August and minimum m = -0.9°C in January. According to climatic classification [31], such a value is characteristic of cold winters with 30 – 60 days of frost and a rest period of one to three months for vegetation.

The plateau of Rogassa is a pediment characterised by a limestone crust at 15 to 30 cm depth.

Soil of the pre-existent system was shallow with a low organic matter content (table 3). It belongs to the group of 'brown calcareous' soils that were interpreted as relics of original forest soils [42]. Low organic matter content and depth reflect a high vulnerability towards natural or human-induced changes [8], which explains why it is so difficult to restore such soils when damaged [38].

The vegetation of the pre-existent system was a pure stand of alfa-grass as the dominant species. The total aboveground biomass was on average 8300 kg dry matter per hectare. Considering only green parts, it contributed on average 81 % of total cover and 96 % of total above-ground biomass. *Stipa tenacissima* L. (named Alfa in the Maghreb and Esparto-grass in Spain) is a perennial tussock grass.

Figure 1. Location of the study area.

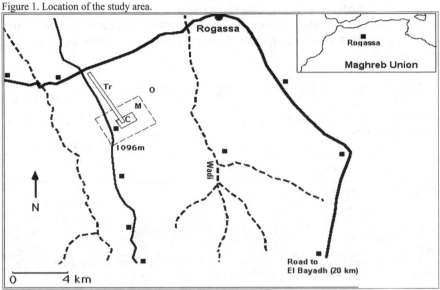

■ stations sampled within the pre-existent system (1976)
C: control exclosure (ungrazed), M: slightly grazed, O: open rangeland
Tr : transect (grazing gradient.

Table 2. Ecological characteristic of Rogassa Alfa steppe

Ecological characteristic	
Latitude	33°56' n,
LONGITUDE	00°51' e
Altitude	1095 m
Mean annual precipitation	250 m/yr
Annual rainfall recorded between 1975 and 1993	224 m/yr
(60% between September and February)	
Coefficient of variation of rainfall	24 %
Mean annual temperature	15.4°C
Mean maximum temperature in August (M)	36.7°C
Mean minimum temperature in January (m)	-0.9°C
P/PET	0.32

The permanent monitoring site (Rogassa plateau) had been chosen in 1975 as one of the best pure stands in North Africa with relatively homogenous ecological conditions, floristic composition and structure.

Vegetation was patchy with individual tussocks (alfa) and interstitial 'bare' soil. Such a structure is considered to be the common feature in arid lands and a response to aridity [39]. Therefore, under such conditions, the rate of soil surface erosion is closely linked to the density of the perennial matrix. Soil underneath the tussock forms a mound that ensures a set of important attributes depicted by [43]. In comparison with bare patches, water infiltration is higher, due to the stem-flow. Soil water availability is higher because of particular soil texture, structure and depth. Higher organic matter content indicates a higher trophic level. *S. tenacissima* was the main ecosystem 'matrix forming' species *sensu* Van Andel *et al.* [46]. Other species have a variable existence [23,46]. In arid ecosystems, these 'interstitial species' [24] are mainly ephemerals *sensu* Noy Meir [40] or 'arido-passive' [21] since their physiological function is interrupted during dry periods.

Table 3. Characteristics of cover and soil in Rogassa Alfa steppe (pre-existing system)

Global vegetation canopy:	50 ± 7	%
Perennial vegetation canopy (mainly *S. tenacissima*)	46 ± 9	%
Soil surface characteristics:		
Litter	17 ± 4	%
Gravel	16 ± 4	%
Sand	0.6	%
First soil layer (0 to 20-30 cm)		
Clay	13 ± 1	%
Silt	21 ± 3	%
Sand	66 ± 4	%
Organic matter	1.8 ± 0.5	%

Soil characteristics of the whole plateau of Rogassa are given by averaging 10 stations sampled in 1975-76 in the neighbourhood of the permanent site. Confidence intervals are given at p=0.05.

3.2. RESULTS

3.2.1. Ecosystem functioning (primary productivity): drought effect

Seasonal monitoring over 11 years, performed under conditions of controlled grazing, showed high fluctuations in plant community functioning (diversity and biomass) [2]. In Rogassa, annual net primary productivity (ANPP computed as the mean annual growth of the total above ground biomass) was 480 ± 140 kg dry matter. ha^{-1}.year^{-1} (p=0.05). This production is in close relation to rainfall (r = 0.963 p<0.001). For *S. tenacissima*, ANPP was 410 ± 110 kgDM. ha^{-1}.year^{-1} representing 85% of total ANPP. During the driest year (129 mm rainfall), total ANPP was only 59 kg DM.ha^{-1}.year^{-1}. The decrease in biomass affecting all the area between 1982 and 1984 may be explained by this drought. The maximum production was in 1977, 753 kg DM.ha^{-1}.year^{-1}. Inter-annual coefficient of variation of ANNP was 40% for perennials and 85% for ephemerals.

The ratio productivity/mean maximum biomass is 0.40 for *S. tenacissima*. This means that only 40% of the biomass is renewed yearly, which means a low biotic efficiency but a high durability. By comparison, this ratio was 84% for esparto (*Lygeum spartum*), another perennial Poaceae which is more productive but less drought-resistant.

Expressed in feed units (FU, energetic value equivalent to 1880 kcal for sheep), the mean productivity of the Rogassa rangeland was 140 ± 45 FU.ha^{-1} (p=0.05). A moderate stocking rate, considered to be the 'balance stocking rate', obtained from a maximum removal of 60% of annual production, should not exceed 0.25 sheep unit. ha^{-1}. However, this notion of "balance stocking rate" has no significance because of the extreme year-to-year variability of natural fodder resources. Measurements of productivity taken over the three main steppes of the Algerian high-plains showed a variability on average 2.5 times higher than the variability of rainfall [1]. In the Rogassa steppe, the minimum recorded was 20 FU. ha^{-1} for 129 mm.yr^{-1} and the maximum was 232 FU. ha^{-1} for 260 mm.yr^{-1}.

Between 1983 and 1985, the mean annual rainfall was only 150 mm.yr^{-1} (compared with a long-term mean of 250 mm/year). During the dry years 1983-1985, the fodder deficit amounted to 54% in relation to the so-called 'balance stocking rate'. A drought peak was included within a larger deficit period, which began in the early 70's. Steppe vegetation, and especially alfa, is fully adapted to drought as a natural factor of stress. However, in the recent conditions of use, such a drought leads to a critical situation in which the sheep population density greatly exceeds the rangeland carrying capacity. These were prerequisite conditions for overgrazing and degradation in the context of Maghrebian land-use legislation.

3.2.2. Ecosystem dynamics: overgrazing effects

The vegetation dynamics was studied over a period from 1976 to 1993 through the evolution of perennial (alfa) aboveground biomass in different situations of exploitation [7]. Along the transect (figure 1), the steppe community showed homogeneous density and composition in 1975. The mean aboveground biomass of alfa was 1460 ± 130 kgDM.ha^{-1} (green part). The regression of *S. tenacissima* biomass has been attested since 1984 by a progressive reduction of green limbs in open rangelands. Massive tussock decay had been observed and generalised between 1985 and 1987 particularly in the overgrazed rangeland (O zone). Only necromass remained within tussocks, and rhizomes were denuded *via* wind erosion and animal trampling. This denudation stage was considered to be an indicator of the limit of resilience capacity of *S. tenacissima* and thus a threshold of irreversibility [7]. In some cases, sand deposits might be a protection for tussock lower parts (rhizoms and renovation buds), and then re-growth of *S. tenacissima* could occur in limited patches. Even under the so-called 'moderate grazing' (M zone), 50% of tussocks degenerated in 1989 while it was 95% in the O zone.

In 1993, the transect was analysed in terms of biomass through a systematic sampling. The results are given in table 4. The global canopy varies from 53% in the protected area to 38% in the open rangeland, biomass from 2040 to 680 kgDM.ha^{-1} and forage production from 132 to 95 FU.ha^{-1}. If we consider the components (i.e. perennials *vs.* ephemerals), we note that the contribution of perennials decreased from the protected rangeland to the open one whatever the considered parameter. This variation was more significant for forage production which takes advantage of the increasing density of annuals (mainly weedy plants).

Table 4. Spatial variation along a grazing gradient

Zone	Global vegetation canopy %	Phytomass kgDM.ha⁻¹	Forage production FU.ha⁻¹.yr⁻¹
C	53 ± 11 (73)	2040 ± 750 (96)	132 ± 59 (82)
M	39 ± 19 (66)	1230 ± 770 (93)	85 ± 54 (55)
O	38 ± 11 (46)	680 ± 310 (84)	95 ± 43 (33)

C: control plot (exclosure); M: slightly grazed zone, O: heavily grazed zone
Interval of confidence is given at p=0.05
between brackets: contribution of perennials

Changes along the transect, show a significant variation of diversity (table 5) which is expressed by the plant species richness, the Shannon index and evenness. All parameters are lower in the protected area where the species composition remained nearly the same as with the pre-existing system. However, the difference between the M and O zones indicates that diversity may decrease when perturbation becomes higher. This confirms the Intermediate Disturbance Hypothesis, which predicts that diversity would be higher at an intermediate rate of disturbance [17]. In fact, it seems that the intermediate stage (M zone) corresponds to an enrichment of the pre-existing community due to an invasion by weed species promoted by the grazing animals (nitrogen addition) and psammophytic species by sand deposition.

Table 5. Spatial variation of diversity along a grazing gradient

Zone	Species richness	Species diversity*	Evenness	N
C	23.07 ± 1.88	1.30 ± 0.73	33.11 ± 17.01	7
M	29.75 ± 8.43	2.00 ± 0.45	47.08 ± 10.28	6
O	24.22 ± 3.02	1.39 ± 0.32	36.08 ± 7.09	25

C: control plot (exclosure); M: slightly grazed zone, O: heavily grazed zone
Interval of confidence is given at p=0.05, n = number of stations
* Shannon index and evenness were calculated according to Krebs [26]

Concerning soil surface and soil characteristics, the first indicator had been sand encroachment in the heavy grazed area (O zone). Sand deposit was limited to the grazed area. Variation of sand cover frequency was not correlated to the perennial canopy that was mainly represented by *S. tenacissima* and *Lygeum spartum* which became dominant in sandy soil. Ephemerals were relatively more abundant outside the protected area.

Table 6. Spatial variation of soil characteristics along a grazing gradient

	Organic matter (%)	Clay (%)	Total Silt (%)	Total Sand (%)
C	1,67 ± 0.31	13,15 ± 2,33	27,25± 1,75	59,60 ± 4,63
M	1,35 ± 0.35	9,38 ± 1,00	21,6± 1,88	69,02 ± 4,68
O	0,91 ± 0.10	5,15 ± 0,90	11,75 ± 0,78	83,10 ± 1,89

C: control plot (exclosure); M: slightly grazed zone, O heavily grazed zone
Interval of confidence is given at p=0.05

Meaningful changes of soil texture are indicated by an increase of sand content and a decrease of clay content (table 6). Significant difference means were shown between zones [44]. The decrease of both clay and organic matter is an efficient indicator of the degradation of the soil fertility since the two components are responsible for constituting the soil clay-organic complex which is responsible for absorbing nutriments.

The distribution of organic matter shows a decrease of about 50% between the open rangeland and the control exclosure. While in the last area the rate remains similar to those of the pre-existing system, the grazed area has kept relatively high values regarding most of the degraded conditions of the pre-existing systems.

To sum up, the results obtained over the grazing gradient show that:

(i) In the protected plot (C zone), global conditions had remained similar to those of the pristine system, attested by composition, biomass and soil organic matter content.

(ii) In the grazed area (M zone), pre-existent soil has been covered by a shallow layer of sand. Moderate grazing and sand deposit favoured weed species leading to

(iii) high diversity and improved plant productivity at the community scale. Some new field observations suggested that C zone is now shifting to M zone conditions because of a significant increase of *Lygeum spartum* [4].

(iv) In the overgrazed rangeland (O zone), the destruction of *Stipa tenacissima* has been followed by soil erosion and the destruction of the whole pre-existent system. New conditions have been created with sandy conditions that have favored the encroachment of pre-existing species such as *Lygeum spartum* and *Plantago albicans* and the introduction of other weed species. Extreme changes are indicated locally by the typical psammophytic grass *Stipagrostis pungens* and by the installation of woody perennial plants of low palatability such as *Noaea mucronata* and *Atractylis serratuloides*.

According to these results, the role of overgrazing as the main and direct cause of this fast degradation of alfa is confirmed [7,44].

The results from the monitoring of such a simplified system has indicated a general pathway of directional changes but could also generate a hypothesis on mechanisms in order to separate trends and fluctuations [10]. All observations made on steppe rangelands [5] had revealed that sand spreading was the common feature of the recent degradation. An efficient indicator during the transition stage of degradation (early warning indicator), is the reduction of perennial biomass. However natural fluctuations must be taken into account.

In steppe rangelands, perennial species are capable of persisting after episodic disturbances caused by the synergetic action of both drought and grazing. These xerophytic species are considered to be 'adapted to disturbance' because of a set of characteristics [46]. In steppe rangelands, extensive grazing has been the main occupation and land use of the steppe for several centuries. A nomadic way of life has been a social adaptation coping with arid environment. Such a traditional practice has allowed relative ecological balance to be maintained as well as overcoming other climatic crises.

A comparison of the distribution of *S. tenacissima* steppe in the western high-plains assessed during the 1970s with previous descriptions suggested that most of the communities had remained relatively unchanged for a long time (several decades). Such 'alternative steady states' also termed 'suspended stages of succession' by Laycock [28] attest to the slowness of change before the 1980s. On the other hand, between 1978 and 1990, successive sampling [5] in dense steppe over nearly 500,000 ha, the mean frequency of *S. tenacissima* declined from 30 to 3% and the green biomass from 1750 to 75 kgDM.ha while sparse stands that had been covering approximately 700,000 ha, had entirely disappeared. The fact is that the high-plains pure stand of *S. tenacissima* had practically disappeared and the four million hectares of alfa steppe still reported in 1992 by official publications and scientists must be revised [7]. Claims that grazing was the main cause of such a decline were difficult for mainly subjective reasons: (i) the high total aboveground biomass of alfa pure stands, (ii) alfa was known to be a nearly unpalatable plant with a low feed value [6,20,35] and (iii) harvesting alfa for the paper-making industry was another well-known use that can affect this plant. This activity began in 1962. Even though the limbs were harvested manually, it was certainly detrimental to the reproduction of the plant. This use of alfa markedly declined on reaching the 70's (figure 2): in other words, it did not account much for the degradation during the 80's.

Figure 2. Evolution of amounts of alfa harvested in Algeria from 1963 to 1999

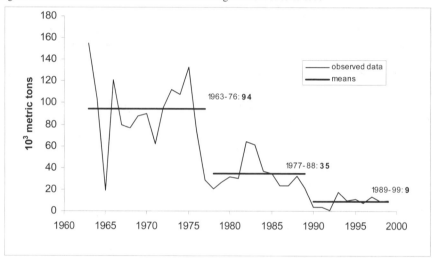

Source: national statistics

3.2.3. Socio-economic 'up-stream' causes

From a socio-economic point of view, in the late 80's, according to the Rogassa local statistics, the actual stocking rate on rangelands (0.5 to 0.75 sheep unit. ha-') increased beyond the carrying capacity. While the fodder deficit was 50 to 70% for a medium year, if resilience conditions were admitted the deficit would exceed 80% during dry years. In fact, in 1993, the Rogassa rangeland had lost 40% of its fodder production compared with the pre-existent system. Such a decrease may seem rather low compared with the huge decrease in global biomass. This low decrease is explained by a relative gain in the annual species that degradation favours in the first stage [4].

The first economic indicator of desertification is the increased number of sheep, which has prevailed in the steppes since the 1960s (figure 3). The evolution of livestock over more than a century, in relation to the rainfall, shows considerable fluctuations, due to various ecological or socio-economic reasons [14,25]. However, the main factor of this variability was the rainfall, which is directly correlated to natural fodder availability. For instance, the important drought during the 40's had been catastrophic for livestock.

Since the late 60's, the steady increase of the sheep population has not seemed to depend any longer on annual rainfall, as the steppe has suffered the longest dry period of the century. Recurrent droughts seem to have lost their function as a natural livestock regulator.

Complex socio-economic factors of different levels are involved in the disruption, which has affected the pastoral steppe. In the western steppe of Algeria, human population, as a main social indicator, increased by 2.4 between 1966 and 1993. The government has made huge efforts to improve rangeland management and to combat degradation and desertification [11] but the results generally did not show the benefits of this endeavour. The decisions made at the beginning of the 80's in conjunction with an exceptional drought led to increased grazing pressure on rangelands by removing most of the protection status. Namely, all breeding co-operatives -some of which dated back to the 60's - were dissolved; some alfa-steppes, which normally belonged to the forestry estate, were changed to pasture during dry periods. The local ban (specific to western high-plains) on cereal cultivation apart from depressive soils (daya) was lifted.

Figure 3. Comparative evolutions of sheep population and rainfall in Algeria

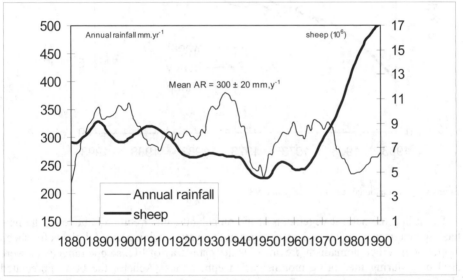

Rainfall is given for El Bayadh, the closest station to Rogassa site. The graph represents the moving average (n=10) of annual rainfall.

The already chronic fodder deficit worsened. As additional feeding, breeders used concentrate fodder (mainly barley), state-subsidised and mostly imported in order at first to help breeders during drought periods. But this practice came into widespread use. The cultivation of barley has been promoted mainly for speculative reasons [13]. Production of barley evolved closely related to sheep numbers whereas hard wheat, the main feeding resource of the rural population, tended to stagnate and even to decrease (figure 4). Like flock feed concentrates, the overall staple foods were subsidised. The price of meat, which did not fall into this category, has been constantly increasing and during the

80's, sheep breeding became a speculative niche covering a large part of the household economy. The new feeding practices explain how steppe vegetation, which was at first the main fodder resource, soon became itself a "supplement" and quoting [15], the whole steppe has turned into a real 'open-air sheepfold'.

Figure 4. Evolution of livestock and of barley and hard wheat in Algeria (trend, 1963-1993)

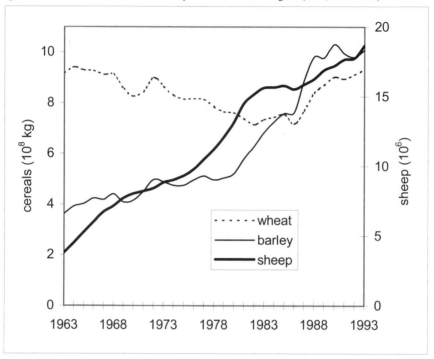

Source: Algerian national statistics organism ONS

Regarding the natural vegetation of the rangelands, this trend and above all the new feeding practices have led to deep changes in the way of using perennial plants for sheep (appetence). When nomadism prevailed as the main way of life, steppe rangelands were used only during the spring months, when ephemerals (Acheb), the best appreciated plant category, were abundant. Perennial vegetation was grazed occasionally, in case of drought and especially during the lambing and lactation periods. During dry periods, traditionally additional food (e.g. barley) was given to ewes. In the alfa rangelands, green limbs of alfa-grass were usually given to ewes mainly as necessary bulky food.

With the widespread use of feed concentrates (barley and corn) ensuring nearly 90% of the energetic needs of flocks, the bulky ration necessary for this massive supply of food, was provided by alfa. The use of alfa as bulky food had been probably favoured by the increasing price of straw [13]. In spite of its low palatability, alfa green limbs are being grazed non-stop, tussocks have been reduced to a dry tuft and irremediably lost when both wind and trampling occur.

In addition to this devastating effect on the alfa steppe, another indicator of change has been the near disappearance of the 'Hamra' (brown headed), typical sheep bred in this region. These, also named 'Béni Guil' or 'Deghma' in Eastern Morocco, were known for their hardiness *i.e.* their ability to make profitable use of natural food mainly grazed on the rangelands. Since hardiness is no longer a requirement because of the use of feed concentrates, the Hamra race was soon substituted by the "Berguia" (white) sheep breed. This hybrid coming from eastern Algeria (Ouled Djellal) of a bigger size and seemingly more appreciated on the market, has became widespread in the western steppe of Algeria since the beginning of the 90's.

These new practices widely explain the rapid destruction of *S. tenacissima* which proved to be the most sensitive perennial plant.

4. Discussion and conclusion

It is widely admitted that it is difficult to define the scope of desertification, as an environmental concept because of the multitude and complexity of the issues involved. The CCD (United Nations Convention to Combat Desertification) definition is generally used as a simplification but it does not answer many questions posed. For instance, this definition considers desertification as synonymous with 'degradation' which is often associated with managed land use. Danckwerts and Adams [18] consider that 'degradation' cannot be defined without an *a priori* definition of user objectives. In fact, user objectives are often defined in terms of productivity as the main indicator of resilience regardless of a number of vital attributes as assumed by Behnke and Kerven [12] for species diversity. It is an example of considerations that must necessarily be taken into account in the prevision of rehabilitation and sustainable management.

Desertification is now admitted as a global change and there is general agreement about the threat to arid zones, especially African ones. We have emphasised overgrazing as the first cause of desertification in Africa. This type of degradation corresponds to a set of step-wise changes [38]. Soil alteration is considered to be the last step of degradation and thus the indubitable threshold of irreversibility [22,36]. Early effects of grazing are biotic changes among which removal of perennial plants is one of the key processes [7,9,34,47] and could be used in an early warning system. However, changes in abiotic environment are more informative but slower and thus long-term diachronic studies in permanent plots are needed. Indeed, dynamic hypotheses have been often built on the basis of synchronic sampling. The synchronic approach is used as a generalisation of the 'vegetation series' concept [29] presuming that several 'alternative steady states' [50] of a 'successional mosaic' may coexist for a long time as contiguous systems derived from the same 'historic indigenous system' [9].

Theoretical models of change in arid lands have been based recently on non-equilibrium which introduce a large number of stochastic events in such a variable system of dynamics [19,38,48]. Changes recorded in the arid steppes of the Maghreb seem to proceed partly in this way because of the strong anthropic pressure. But little information exists currently to predict future changes in these systems. The decline of 50% in soil organic matter as occurred along the Rogassa grazing gradient during less than one decade seems to be catastrophic in comparison to examples given by Bullock and Le Houérou [16] in other parts of the world. However, the worst current values recorded at the Rogassa stand are not yet similar to those of a desert and may allow rehabilitation *i.e.* recovering an optimal part of the previous functioning.

Since irreversible degradation occurs with stepwise change, we have been witnesses, in the steppe rangelands of Algeria, to a major 'step' or 'transition' over about ten years. The new conditions may remain stable or 'suspended' for few decades until a new crisis. The marked decline of the biotic and abiotic vital attributes has been recorded; fortunately real 'desert conditions' have not been reached yet. We believe that such a global decline of the natural system integrity corresponds to the 'desertification concept' even if the change does not seem to coincide exactly in the 'etymological' sense.

The causes of desertification are multiple but some of them have a close association with local peculiarities: ecosystem features, local economic and social contexts. Similar causes that have occurred in the Rogassa steppe have led to other effects in the other steppe rangelands of Algeria. Since there is very little probability of the conjunction of certain causes, such as persistent drought and socio-economic conditions leading to particular practices regarding natural resources, the results cannot be easily generalised from one system to another. Restoration and rehabilitation methods and actions must be adapted to the local context as well. Since degradation occurs over a large environmental continuum from slightly altered to totally devastated systems, scientists are confronted with a lack of norms to assess the nature of change, intensity, scale and degree of reversibility.

The predictions about the possible consequences of a global climate change on arid ecosystems are mainly confined to hypotheses. On the whole, it would be seriously negative because of a probable increase of aridity. However the consequence of a possible climate change would be trivial in comparison with human mismanagement.

The majority of actions taken to combat desertification during the 30 past years have failed because of the difficulties of finding a balance between the production rate of natural resources and their exploitation by man. The current strategy relies on national and sub-regional, at the UMA scale, action programmes that are the main implementation strategies advocated by the United Nations Convention to Combat Desertification (CCD). Adopting a participative bottom-up approach, these programmes should emphasize the creation of an "enabling environment" by permitting all stakeholders, especially non-governmental organizations, to play a prominent role in reversing the current trend.

References

1. Aidoud, A. (1989) Les écosystèmes steppiques pâturés d'Algérie: Fonctionnement et dynamique, Thèse de Doctorat d'Etat. Univ. Sci. Technol. Houari Boumediène, Alger.
2. Aidoud, A. (1992) Les parcours à alfa (Stipa tenacissima L.) des Hautes Plaines algérienne : variations inter-annuelles et productivité, in Gaston, A., Kernick M., and Le Houérou, H.N. (eds.), *IVᵉ congrès international des terres de parcours*, CIRAD, Montpellier, 22-26.
3. Aidoud, A. (1993) Las poblaciones vegetales bajas en el Norte de Africa, in Pastor-López, A., and Seva-Roman, E. (eds.), *Restauración de la cubierta vegetal en ecosistemas mediterráneos*, Instituto de Cultura Juan Gil-Albert, Alicante, 53-80.
4. Aidoud, A. (1994) Pâturage et désertification des steppes arides en Algérie, cas de la steppe d'alfa (*Stipa tenacissima* L.), *Paralelo 37*, 33-42.

5. Aidoud, A., and Aidoud-Lounis, F. (1991) Les ressources végétales steppiques des Hautes Plaines algériennes: évaluation et regression, in Gaston, A., Kernick M. and Le Houérou, H.N. (eds.), *Proceeding of the fourth International Rangeland Congress*, CIRAD, Montpellier, pp. 307-309.

6. Aidoud, A., and Nedjraoui, D. (1992) The steppes of Alfa (Stipa tenacissima L.) and their utilisation by sheeps, in Thanos, C.A. (ed.), Plant-Animal interactions in Mediterranean type ecosystems. *Medecos VI*, 62-67.

7. Aidoud, A., and Touffet J. (1996) La régression de l'alfa (Stipa tenacissima L.), graminée pérenne, un indicateur de désertification des steppes algériennes, *Sécheresse* 7, 187-193.

8. Albaladejo, J., Martinez-Mena, M., Roldan, A., and Castillo V. (1998) soil degradation and desertification induced by vegetation removal in a semiarid environment, *Soil Use and Management 14*, 1-15.

9. Aronson, J., Floret, C., Le Floch, E., Ovalle, C. and Pontanier, R. (1993) Restoration and rehabilitation of degraded ecosystems in arid and semi-arid lands, I. A view from the south. Restoration, *Ecology 1,1*, 8-17.

10. Bakker, J.P., Olff, H., Willems, J.H., and Zobel, M. (1996) Why do we need permanent plots in the study of long-term vegetation dynamics, *Journal of Vegetation Science 7*, 147-156.

11. Bedrani, S. (1994) La place des zones steppiques dans la politique agricole algérienne, *Paralelo 37*, 16-52.

12. Behnke, R., and Kerven, C. (1994) Redesigning for risk: tracking and buffering environmental variability in Africa's rangelands, *Natural Resource Perspectives 1*, 1-8.

13. Benfrid, M. (1998) La commercialisation du bétail et de la viande rouge en Algérie, *Options Méditerranéennes*, CIHEAM, Montpellier, sér. A, 35, 163-174.

14. Boukhobza, M. (1982) *L'agro-pastoralisme traditionnel en Algérie: de l'ordre tribal au désordre colonial*, Off. Publ. Univ., Alger.

15. Boutonnet, J.P. (1992) Production de viande ovine en Algérie. Est-elle encore issue des parcours ? in *Actes du IVe Congrès International des Terres de parcours*, CIRAD, Montpellier, 906-908.

16. Bullock, P. B., and Le Houérou, H.N. (1996) Land degradation and desertification, in *Climate Change 1995, Impacts, Adaptations and Mitigation of climate change*, Contribution of Working group II to the 2d Assessment Report of Intergovernmental Panel on Climate Change, Cambridge Univ. Press, Cambridge, pp. 171-189.

17. Connell, J.H. (1978) Diversity in tropical rain forests and coral reefs, *Science 199*, 1302-1309.

18. Danckwerts, J.E. and Adams, K.M. (1991) Dynamics of rangeland ecosystems, in Gaston, A., Kernick, M., and Le Houérou, H.N. (eds.), *Proceeding of the fourth International Rangeland Congress*, CIRAD, Montpellier, 1066-1069.

19. Dodd, J.L. (1994) Desertification and degradation in sub-saharian Africa, *BioScience* 44 (1), 28-34.

20. El Hamrouni, A., and Sarson, M. (1974) Valeur alimentaire de certaines plantes spontanées ou introduites en Tunisie. Note de Recherche, *Institut National de la Recherche Forestière 2*, 1-17.

21. Evenari, M. (1985) The desert environment, in Evenari, M., Noy-Meir, I. and Goodall, D.W. (eds.), *Ecosystems of the world, 12A: Hot deserts and arid shrublands*, Elsevier, Amsterdam, 1-22.

22. Friedel, M.H. (1991) Range condition assessment and the concept of thresholds: A viewpoint, *J. Range Manage, 44, 5*, 422-426

23. Grubb, P.J. (1985) Problems posed by sparse and patchily distributed species in species-rich plant communities, in Diamond, J. and Case, T.J. (eds), *Community ecology*, Harper and Row, New York, pp. 207-225.

24. Huston, M.A. (1994) *Biological diversity: the coexistense of species on changing landscapes*, Cambridge University Press, Cambridge.

25. Jarre C. (1933) Le problème de l'élevage ovin en Algérie et la coopération, Thèse Doct., Univ. Alger.

26. Krebs, C.J. (1989) *Ecological Methodology*, Harper Collins Publisher, New York.

27. Laouina, A., Chaker, M., Nafaa, R., and Naciri R. (2001) Forest and steppe grazing lands in Morocco, degradation processes and impacts on runoff and erosion, in *Land use changes and cover in the Mediterranean region*, Ricamare International Workshop, Medenine, 24-37.

28. Laycock, W.A. (1991) Srable state concepts applied to range condition criteria used to evaluate North American rangelands, in Gaston, A., Kernick, M., and Le Houérou, H.N. (eds.), *Proceeding of the fourth International Rangeland Congress*, CIRAD, Montpellier, pp. 134-137.

29. Le Houérou, H.N. (1981) Long-term dynamics in arid-land vegetation and ecosystems of North Africa, in Goodall, D.W., and Perry, R.A. (eds.), *Arid-land ecosystems 2*, Cambridge University Press, Cambridge, pp. 357-384.

30. Le Houérou, H.N. (1969) La végétation de la Tunisie steppique, *Ann. Inst. Nat. Rech. Agron. 42*, 1-624.

31. Le Houérou, H.N. (1986) The desert and arid zones of northern Africa, in Evenari, M., Noy-Meir, I., and Goodall, D.W. (eds.), *Ecosystems of the world 12B: Hot deserts and arid shrublands*, Elsevier, Amsterdam, pp. 101-147.

32. Le Houérou, H.N. (1990) Global change : population, land-use and vegetation in the Mediterranean Basin by the mid-21st century, in Paepe, R., and *al.* (eds), *Greenhouse Effect, Sea Level and Drought*, Kluwer Academic Publishers, Dordrecht, pp. 301-367.

33. Le Houérou, H.N. (1992) An overview of vegetation and land degradation in world arid lands, in Dregne, H.E. (ed.), *Degradation and restoration of arid lands*, International Center for Arid and Semi-Arid Land Studies, Texas Technical University, Lubbock, pp. 127-163.

34. Le Houérou, H N. (1995) Bioclimatologie et biogéographie des steppes arides du nord de l'afrique. Diversité biologique, développement durable et désertisation, *Options méditerranéennes sér.* B, 1-396.

35. Le Houérou, H.N., and Ionesco, T. (1973) Appétibilité des espèces végétales de la Tunisie steppique, *Documents Travaux et Projets FAO/Tun./71/525*, Rome.

36. Mainguet, M. (1994) *Desertification: Natural background and human mismanagement*, 2d edition. Springer, Berlin.

37. Merzouk, A., Alami, M.M., Berkat, O., and Sabir, M. (2001) Effects of rangeland changes on water balances and water quality in Morocco: a Rif mountains case study, in *Land use changes and cover in the Mediterranean region*, Ricamare International Workshop, Medenine, pp. 13-23.

38. Milton, S.J., Dean, W.R.J., Du Plessis, M.A., and Siegfried, W.R. (1994) A conceptual model of arid rangeland degradation, the escalating cost of declining productivity, *Bioscience 44*, 70-76.

39. Montana, C. (1992) The colonisation of bare area in two-phases mosaics of an arid ecosystem, *Journal of Ecology 80*, 315-327.

40. Noy Meir, I. (1973) Desert ecosystems: Environment and producers, *Annual Review of Ecology and Systematic* 5, 195-214.

41. Pickup, G. (1992) Dynamics of rangeland ecosystems, in Gaston, A. Kernick M., and Le Houérou, H.N. (*eds.*), *Proceeding of the fourth International Rangeland Congress*, CIRAD, Montpellier, pp. 1066- 1069.

42. Pouget, M. (1980) Les relations sols-végétation dans les steppes Sud-Algéroises, Trav. Doc. ORSTOM, Th. Doct. Univ. Aix- Marseille.

43. Puigdefabregas, J., and Sanchez, G. (1996) Geomorphological implications of vegetation patchiness on semi-aride slopes, in Anderson, M.G. and Books, S.M. (eds), *Advances in Hillslope processes V. 2* Wiley Chichester, pp. 1028-1060.

44. Slimani, H. (1998) Effet du pâturage sur la végétation et le sol et désertification. Cas de la steppe à alfa (Stipa tenacissima L.), de Rogassa des Hautes Plaines occidentales algériennes, Th. de magister, Univ. Sci. Technol. Houari Boumediene, Alger.

45. UNEP (1992) *World Atlas of Desertification*. Edward Arnold, London, 69

46. Van Andel, J., Van Baalen, J., and Rozun N.A.M.G. (1991) Population ecology of plant species in disturbed forest and grassland habitats, in Rozema, J., and Verkleij, J.A.C. (*eds.*), *Ecological Responses to Enviromental stresses*, Kluwer Academic Publishers, pp. 136-148.

47. Verstraete, M.M., and Schwartz, S.A.(1991) Desertification and global change, in Henderson-Sellers, A., and Pitman, A.J. (eds.), *Vegetation and climate interactions in semi-arid regions*. Kluwer Academic Publishers, pp. 3-13.

48. Westoby, M., Walker B., and Noy-Meir, I. (1989) Opportunistic management for rangelands not at equilibrium, *Journal of Range Management* 17, 235-239.

49. Whal, M. (1897) *L'Algérie*, F. Alcan, Paris.

50. Wittaker, R.H., and Levin, S.A. 1 (977) The rôle of mosaic phenomena in natural communities, *Theoretical Population Biology 12*, 117-139.

Chapter 7

Prospects for Desertification Impacts for Egypt and Libya[1]

M. Nasr
Ain Shams University, Cairo, Egypt
Center for Development Research, Bonn University

New deserts are forming in some areas of the world [38]. This process is referred to as "desertification". For most people, the term conjures up an image, an emotive picture of inexorably shifting sands encroaching on valuable farmland [5]. But is the desert really expanding? An increasing number of scientists are now arguing that the image associated with "desertification" is a mirage. There is no general consensus regarding the definition, causes, or impact of desertification. Desertification has been defined in many different ways by researchers in different disciplines, which have included soil scientists, hydrologists, agronomists, veterinarians, economists and anthropologists. Most definitions of desertification, therefore, vary according to the judgment and expertise of the researchers involved. Whereas some researchers consider desertification to be a great danger to the sustainable development of arid and semi-arid areas, others doubt that the phenomenon occurs at all. These different opinions on desertification are mainly due to the lack of an overall concept, the dearth of information available at global and regional levels and the different objectives and interests of the countries in the north and south.

The problem of desertification in arid and semi-arid areas can be traced back for centuries. There has always been a correlation between long-term changes in climate and changes in human activities. As long as the population density of both men and livestock in a desertification-endangered area remained sufficiently low, the ecological consequences of human activities remained relatively insignificant or were concentrated within a very limited area. In regions where food security and poverty alleviation are priorities, as in the Middle East and North Africa (MENA) region, the primary emphasis regarding land is its availability, the diminution of land degradation, and efficient land and water management. The message by the FAO is to encourage countries in arid and semi-arid areas to identify reasons for land degradation. Few researcherS dispute that most MENA countries, particularly Egypt and Libya, have appropriate technologies to combat desertification, but that the technologies are not used efficiently enough due to insufficient knowledge of the socioeconomic contexts, incorrect identification of the causes of the arid land problems and ineffective management of natural resources, i.e. water. The study addresses two questions: What is desertification, i.e. describing the concept and how to monitor productivity impacts of desertification processes? What is the magnitude of the desertification problem in Egypt and Libya, both current and in the past? Policy actions and implications are discussed regarding what is being done in both countries to address desertification problems by both governmental and non-governmental efforts.

---

[1] This paper is drawn from the Discussion Paper published by the Center for Development Research of Bonn University by the same author in 1999.

A. Marquina (ed.), Environmental Challenges in the Mediterranean 2000-2050, 109–122.
© 2004 *Kluwer Academic Publishers. Printed in the Netherlands.*

1. Concepts and Techniques for Identifying Desertification Processes

1.1 WHAT IS DESERTIFICATION ?

The phenomenon of desertification is very old, but national and international awareness and the desire to control it are very recent. In the public mind, desertification is often associated with the idea of 'desert advance'. In the 1970s, there was great concern that the Sahara desert was advancing into the non-desert lands of West Africa.

The first definition of desertification agreed upon in 1977 at the United Nations Conference on Desertification (UNCOD) in Nairobi is as follows: "Desertification is the diminution or destruction of the biological potential of land, and can lead ultimately to desert-like conditions".

When the first attempts were made to carry out a quantitative assessment of desertification and to implement various practical recommendations of the Plan of Action to Combat Desertification (PACD), this definition was found to be inadequate and insufficiently operational. A new definition was required to distinguish between desertification and other problems of climatic change.

Based on special studies and extensive discussions at the UNEP Consultative Meeting on the Assessment of Desertification in Nairobi in 1990, the following new definition was suggested: "Desertification is land degradation in arid, semi-arid and dry sub-humid areas resulting from adverse human impact. Land in this concept includes soil and local water resources, land surface and vegetation or crops. Degradation implies reduction of resource potential by one or a combination of processes acting on the land".

In 1991 the UNEP Panel of Senior Consultants and the Governing Council of UNEP underscored "the need for further refinement of the definition, taking into consideration recent findings about influence of climate fluctuations and about the resilience of soils". The following authoritative definition of desertification was approved at the 1992 Earth Summit and adopted by the United Nations Convention to Combat Desertification (UNCCD): "Land degradation in arid, semi-arid and sub-humid areas resulting from various factors, including climatic variations and human activities" [32].

1.2 TOWARDS A DEFINITION OF DESERTIFICATION

As there is still much confusion on the definition, diagnosis, and measurement of deser-tification, a few questions will be raised:

Why not define desertification as the extension of desert margins into non-desert areas? In this case we would have to deal with at least 30 to 40 years of data to investigate shifts in desert margins.

Why not define all the forms of observed degradation as land degradation, rather than desertification? Why not say that all processes acting against sustainable land use are degradation processes? These suggestions call not for a redefinition of the problem, but simply for the use of a suitable terminology.

1.3 PROBLEMS AND INDICATORS RELATED TO PROCESSES OF LAND DEGRADATION

Problems posed by land degradation processes cannot be generalized across land-use zones. Land degradation differs in rangeland, rainfed, and irrigated areas [11]. In Libya, the main problems of land degradation are those of the rangeland areas such as overgrazing, shrub clearing, and soil erosion. *Salinisation* is the main land degradation problem in irrigated agriculture like in Egypt. It involves a number of interrelated processes occurring in the soil, for example water-logging, increasing salt content, and alkalinisation, in which some nutrients can no longer be absorbed due to the increasing pH-value of the soil. This problem is caused by the overuse of water through inadequate irrigation techniques, accompanied by inefficient drainage systems. Moreover, in Egypt, soil erosion is the principal problem in rainfed areas.

Soil erosion results from the uprooting of shrubs, which leads to the destruction of the soil structure and thus to accelerated erosion of the soil by wind and water; soil infiltration of rainwater decreases, and surface runoff increases. Examples of this type of land degradation problems are to be seen in both countries.

2. Status of Desertification in Egypt and Libya

2.1 LAND, WATER AND POPULATION

The land area of Egypt and Libya together amount to 277 million ha, which represents about 29% of the MENA region. The land area of Libya amounts to 177 millions ha, which is about 1.8 times the land area of Egypt (Table 1). Whereas 95% of the land in Egypt is irrigated, only 22% is irrigated in Libya. More than 90% of the land in both countries is considered as hyper-arid. Range-land represents 10% of the land area in Libya while it represents 3% in Egypt. The population size of Egypt (68 millions in 2002) is about 13 times that in Libya (5.4 millions). The population growth in both countries is one of the highest in the world averaging 2.8% in 2000. Urban population has doubled in recent years to 54% of the total population.

The water situation has been described as "precarious" in terms of both quality and quantity. Egypt and Libya consume more than 100% of their renewable water resources. Rainfall in both countries averages 300 mm per year. At least 400 mm per year is needed for successful agriculture. Rainfall ranges from a maximum of 250 mm/year in Egypt to a minimum of 5mm/year in the north of Libya. It is estimated that MENA groundwater reserves amounted to 15.3 km³ in 1987 [27]. Egypt, Libya and Sudan share groundwater basins. The present rate of groundwater withdrawal cannot be considered sustainable in the long run since it far outpaces the rate of replenishment and will fairly rapidly deplete the aquifers. Surface water in Egypt flows into Nile. The internal renewable water resources per inhabitant in both countries are among the lowest in the world, the average being 1016 m³/year per inhabitant, compared with about 7000 m³/year per inhabitant for the world total [12].

The structure of the water demand among individual water users is 1 m³ of drinking-quality water per year, 100 m³ of water per year for domestic purposes and a further 1000 m³ of water per year for food production, making a total of 1101 m³ per person per year. In both countries, the internal renewable water resources per inhabitant are 1000-1667 m³/year, and the countries are consequently considered as water-stressed countries. While 85% of water withdrawals are directed to agriculture in Egypt, only 31% are directed to agriculture in Libya, as the industry sector consumes a considerable amount of water.

Table 1. Estimated Land Use in Egypt and Libya (in 1000 hectares)

Country	Irrigated Area	Rainfed Area	Rangeland Area	Hyperarid Area	Total Dryland Area
Egypt	2486	10	2604	94900	100000
Libya	234	1659	17172	157655	176720
Total	2720	1669	19776	252555	276720
%	1%	0.6%	7%	91.4%	100%

Source: Calculated from UNEP 1996 and Dregne, H.E., & Chou, N.T. (1992) Global Desertification Dimensions and Costs, in *the Degradation and Restoration of Arid Lands*, Lubbock, Texas Technical. University [10].

2.2 LACK OF SOUND DESERTIFICATION STATISTICS

The information base on the current magnitude of and trends in desertification in the world, particularly in Egypt and Libya, is very poor. The data needed to classify land is available for only a very few areas, and for very few years [36]. For most of Africa, and particularly for the African part of the MENA region, very little is known about the extent of land degradation [33]. In Africa attempts to quantify degradation, even in small areas, have so far failed to come up with sound estimates [28].

There is no comprehensive assessment of the degradation of irrigated land, rainfed cropland, or rangeland in the region. There are no time-series data available on the development of desertification in the region. The only information available for the MENA region, are the estimates provided by Dregne and Chou in 1992 (Table 2). Many figures are derived from replies to a questionnaire sent by UNEP in 1982 to 100 countries. The answers probably mean very little [30]. "In Africa, governments were completing it in many cases at the height of a drought" [24]. "Experts even from sophisticated governments say they had great difficulty answering the questions. They had little of the data that they were asked for. There were no proper guidelines on how to answer critical questions about the degree of desertification of land" [15].

Table 2. Estimated Desertified Irrigated Land in Egypt and Libya (in 1000 ha)

Country	Irrigated Area	Degree of Desertification				Total incl. Moderate	Desertified Area %
		Slight	Moderate	Severe	Very Severe		
Egypt	2486	1735	700	50	1	751	30%
Libya	234	179	50	5	0	55	24%
Total	2720	1914	750	55	1	806	30%

Source: Calculated from UNEP 1996 and Dregne, H.E., & Chou, N.T. (1992) Global Desertification Dimensions and Costs, in *the Degradation and Restoration of Arid Lands*, Lubbock, Texas Tech. University[10].

Despite the deficient data, we used estimates by Dregne and Chou from 1992 to quantify the magnitude of desertification in Egypt and Libya as the only source of information available (tables 2, 3 and 4). For the purposes of comparison, we tried to obtain data from satellite remote sensing images and selected local research reports and studies for both countries. Most reports on desertification were based on Dregne and Chou [10] estimates, which were derived from UNDCPAC [41], reflecting conflicting definitions. One example of misleading statistics on desertification is the claim by UNDCPAC that 35% of the Earth's land area is threatened with desertification. This 35% is the area that is arid, at least half of which is very arid by UNDCPAC's own acknowledgment; this zone is not in danger, as about half of it is too arid for any form of agriculture [43]. This also applies to Egypt and Libya, most of which is hyperarid (more than 90%, table 1), consisting of pure desert, and far too arid for any form of farming. Another example is the misuse of terms such as "desert expansion" in some international debates despite of the assurance by national authorities that the most serious problems in semi-arid areas do not occur at desert margins (PACD 1977). Few people live in such areas, and the most destructive forms of land use, such as overcultivation and overgrazing, do not take place there [43]. An additional example is that in Egypt, Libya, Algeria, Saudi Arabia, and Morocco, the problem was seen as desert expansion. The countermeasure, therefore, was the planting of sand dunes, which is costly. Warren and Agnew [43] say that active sand dunes seldom threaten valuable land. Even some areas covered by active dunes have been shown to be stable. The same difficulty was encountered by the United States Soil Conservation Service, when it plotted damage to farmland in the drought and recession years of the 1930s. Many areas seen then as irreparably damaged [11] are producing record crops today.

A selected measure of the desertification process depends on the definition used. The definition itself is a type of diagnosis of the desertification process. The more precise a diagnosis is, the more effective the selected measure will be. Warren and Agnew (1988) say that if the diagnosis is falling productivity due to over-exploitation, over-grazing, or over-watering, then the measure is better management of land and water. If the problem is climate change in a certain area, permanent withdrawal is called for. If the problem is a near-complete devegetation, in the absence of climatic change, re-seeding or re-planting is suitable treatment. If the problem is the expansion of the desert margin, then some kind of holding-line might be an appropriate treatment. If the problem is drought for no more than two or three years, food aid may be adequate. Since desertification is a complex process, involving a mix of conflicting definitions, causes and effects, no single indicator alone can adequately reflect the interaction of its several components.

2.3 MAGNITUDE AND TRENDS IN DESERTIFICATION IN EGYPT AND LIBYA

The degree and type of land degradation varies between the two countries. The magnitude of the problem is shown in Tables 2, 3 and 4. According to Dregne's assessments [10] each hectare of desertified land has been categorized into one of four classes according to its degree of desertification (slight, moderate, severe, very severe). Slight, moderate, and severe desertification are usually reversible, but very severely degraded land cannot at present be rehabilitated cost-effectively. Land categorized as slightly desertified shows little or no degradation; (less than 10 percent loss in potential yield); moderately desertified land shows 10-25% degradation, severely desertified land 25%-50% degradation, and very severely desertified land more than 50% degradation. The main indicator of degradation in Egypt is salinity, combined with waterlogging. According to these estimates, most of the affected irrigated areas lie in Egypt.

These problems are due to bad irrigation management (over-irrigation) in the absence of adequate drainage systems. Of the total irrigated area affected in Egypt of 2,5 million ha, 1000 ha (0.4%) are very severely degraded, while the degree of degradation of the rest (99.6%) is reversible. Despite the doubts about the reliability of theses estimates, they show that the magnitude of the problem in Egypt is very limited. As for Libya, 100% of degraded irrigated areas are reversible. Moreover, these estimates say nothing about current trends or future developments. Estimated trends in the development of yield, area harvested, and production of grain during the 1975-2001 period show an increase accompanied by the reduced application of fertilizers. This result offers some relief from the alarm caused by UNDCPAC estimates and calls for more precise studies based on surveys based on remote sensing data.

Rainfed cropland represents only 0.1 % (almost 10,000 ha) of Egypt's total drylands, of which about 9000 ha (90%) are estimated by UNDCPAC to be slightly degraded. In Libya, about 580 thousand ha (35%) of total rainfed area is degraded, of which 65% is slightly degraded.

Table 3. Estimated Desertified Rainfed Cropland in Egypt and Libya (in 1000 ha)

Country	Total Rainfed Area	Degree of Desertification				Total incl. Moderate	% of Desertified Area
		Slight	Moderate	Severe	Very Severe		
Egypt	10	9	1	0	0	. 1	10%
Libya	1659	1079	540	40	0	580	35%

Source: Calculated from UNEP (1996) Dregne, H.E., & Chou, N.T. (1992) Global Desertification Dimensions and Costs, in *the Degradation and Restoration of Arid Lands*, Lubbock, Texas Tech. University [10].

According to the estimates analysed by Dregne and Chou [10], the degradation of rangeland is more extensive than that of irrigated or rainfed areas in both countries. It is estimated that more than 80% of the rangelands in Egypt and Libya are degraded. The principal cause of degradation is overgrazing, combined with the cutting of woody species for use as fuel.

Table 4. Estimated Desertified Rangeland in Egypt and Libya (in 1000 ha)

Country	Total Rangeland Area	Degree of Desertification				Total incl. Moderate	% of Desertified Area
		Slight	Moderate	Severe	Very Severe		
Egypt	2604	504	300	1800	0	2100	81%
Libya	17172	3472	1700	11800	200	13700	80%

Source: Calculated from UNEP (1996) and Dregne, H.E., & Chou, N.T. (1992) Global Desertification Dimensions and Costs, in *the Degradation and Restoration of Arid Lands*, Lubbock, Texas Tech. University [10].

3. Changes of Desertification in Egypt and Libya in the Last 19 Years

3.1 MONITORING INDICATORS OF DESERTIFICATION PROCESSES BY SATELLITE

There is a big discrepancy between the time frame of human monitoring activities and the time frame of desertification processes. Due to the lack of data and information on the real magnitude of desertification and its changes with time, it has been expedient to analyse satellite images with multispectral properties and to use thermal imaging techniques for vegetational cover.

The spatial resolution of satellites ranges from high-resolutions capable of imaging objects only five meters in size to low resolutions covering tens of kilometers [8]. The temporal resolution ranges between geo-stationary satellites and those that cover a certain area every 10-16 days. The spectral resolution ranges between long-wave infrared (IR) and ultrashort-wave (UV). With regard to the collection of environmental data, satellite remote sensing systems have the advantage that they can provide both regional and global data, use their unique sensing capabilities to monitor changes ranging in duration from half an hour to several weeks, and process the data collected for the purposes of comparison [9].

3.2 NOAA SATELLITES TO MONITOR DESERTIFICATION IN EGYPT AND LIBYA

The key element in controlling soil erosion in both countries is vegetation cover. This study used thermal images generated by satellite remote sensing systems over a 18-year period of time to represent the vegetation cover in Egypt and Libya. Hielkema, a Remote Sensing Specialist with FAO, considered the application of scales of up to 1:50,000 as acceptable. Below this scale, e.g. at 1:10,000 or at the farm level, satellite remote sensing would no longer be practicable. This is why the study used NOAA systems to monitor vegetation as an indicator of desertification processes in both countries. Because the NOAA AVHRR satellite was established in 1982, the study only covers the period from 1982 to 1999.

Since each satellite image consists of a very large number of digitized dots, a *Normalized Difference Vegetation Index* (NDVI) was computed as a measure of the amount and vigor of the vegetation on the surface. The magnitude of NDVI is related to the level of photosynthetic activity in the observed vegetation. In general, higher values of NDVI indicate greater vigor and amounts of vegetation. The NDVI is derived from data collected by the National Oceanic and Atmospheric Administration (NOAA) satellites, and processed by the department of Global Inventory Monitoring and Modeling Studies (GIMMS) at the National Aeronautics and Space Administration (NASA). The NOAA-AVHRR sensor collects data of 1.1 km resolution. The NDVI was calculated for the entire MENA region and for all 17 countries for 18 years (1982 – 1999) to investigate vegetational development year by year.

By monitoring the development of vegetation, the statistical analysis of satellite imagery based on computed values of NDVI for both countries over 17 years (Table 5) shows no significant changes. To monitor the vegetation changes two years were selected, one at the beginning of the time series and one at the end. Within the studied period, the most similar years in terms of the amount and seasonal pattern of rainfall are the years 1982 and 1997 for the MENA region, 1986 and 1997 for Egypt and Libya.

Table 5. NDVI Data of Egypt and Libya, 1982 - 1997

Years	**Egypt**			**Libya**		
	Mean	Standard deviation	Modified scale	Mean	Standard deviation	Modified scale
1982	195.8485	31.3910	0.4539	157.1915	24.4347	0.3051
1983	187.9674	28.4057	0.4235	148.7700	17.2491	0.2726
1984	192.5571	31.3610	0.4412	153.4415	21.2761	0.2906
1985	193.5897	33.0175	0.4452	154.3234	22.1119	0.2940
1986	200.3974	32.5293	0.4714	156.2907	26.3332	0.3016
1987	200.2746	32.3641	0.4709	154.0733	20.8941	0.2931
1988	195.4493	33.9882	0.4523	152.4889	23.7364	0.2870
1989	201.7054	37.7950	0.4764	153.7857	22.9233	0.2920
1990	203.979	33.6244	0.4852	158.8168	24.3231	0.3113
1991	200.8913	38.0776	0.4733	158.6771	23.5893	0.3108
1992	192.1594	35.0680	0.4397	153.7505	23.186	0.2918
1993	197.2241	36.1341	0.4592	155.6304	25.5154	0.2991
1994	191.2612	33.8337	0.4362	155.1895	24.3608	0.2974
1995	193.5000	34.4647	0.4448	159.9614	27.1759	0.3157
1996	199.3906	31.6135	0.4675	159.6012	23.8469	0.3143
1997	200.2008	32.3858	0.4706	156.2979	23.8669	0.3016

* Modified scale – 0.30 to + 0.67

Two NOAA images are compared below (figure 1):

- from 15 June 1997 (start of dry season) for the generation of country-specific and overall MENA vegetation maps with information on the density and distribution of vegetation; and
- from 26 February 1982 and 15 February 1997 (rainy season) to provide information on vegetation dynamics and land-degradation processes

The study compares the edges of deserts in Egypt and Libya in 1986 and 1997. Comparison indicated a regeneration of the vegetational cover. The satellite images of both countries show no alarming damage to vegetation. Areas with extensive vegetation were bright or brightened on the satellite image. Although a drought occurred between 1982 and 1997 and rainfall was below average, the edges of the deserts had not shifted in 17 years, and some desert areas had even become greener. On the contrary, the study estimates that the vegetational boundary has shifted into the deserts in both countries (see NOAA satellite maps of the MENA region and Egypt)[2].

---

[2] NOAA has generated 36 satellite maps of MENA countries that show changes in the distribution of vegetational cover between 1982 and 1997. These maps and calculations of NDVI are available at the ZEF at Bonn University.

The examination of the satellite images revealed no evidence substantiating a trend towards desertification in Egypt and Libya. In some places, regeneration of the vegetation could hardly be expected, given the continual destruction of woody plants as a result of high population pressure. But even the destruction of the shrub vegetation may have been positive for promoting the growth of grasses. In fact, the satellite images document the presence of more greenery in the desert, but do not provide any precise indication of the nature or quality of the vegetation.

Figure 1: NOAA images of the MENA Region. and Egypt for 1982 and 1997

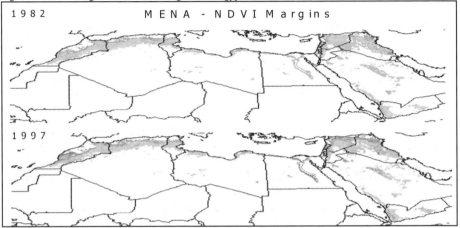

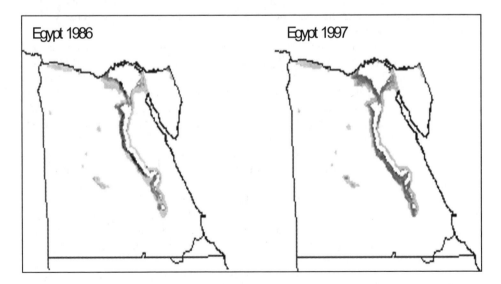

Land degradation may have occurred but there has been no change to more desert-like (less vegetated) conditions. Of course, some small areas have suffered from one or more types of land degradation, but only for a very short time because the soil's ability to regenerate itself has enabled Egypt and Libya to show a net increase in vegetational cover over the last 17 years. In arid and semi-arid areas, the climate does fluctuate wildly between years and decades. As a consequence, there is also a marked fluctuation in the condition of natural and cultivated vegetation. Indeed, there are some indicators of land-degradation processes in both countries, but it has demonstrated a remarkable regenerative ability. Trials in Egypt and Libya have clearly demonstrated that tremendous improvement in the rangelands can be attained in only a few years after the introduction of proper stocking or deferred rotational grazing (Arrar 1989).

Table 6. Comparative Development of Yields (t/ha) of Grain in Egypt and Libya, 1975- 2000.

	1975- 1980	1980- 1985	1985- 1990	1990- 1995	1995- 2000
Egypt	3.3	3.5	4.5	5.1	5.2
Libya	0.4	0.6	0.9	1.1	1.2

Source: FAO 2003

The destruction of vegetation has always proceeded from regions under human influence in response to the need for agricultural areas, roads, watering places, firewood, etc. For both countries as a whole, however, it may be said that human influences have been positive and tended to increase vegetational cover in coastal areas and near rivers, where most of the population is concentrated. The relatively high population growth in both countries has forced people to utilize unused lands, where water is relatively available to meet the increasing demand for food.

Human efforts involving the use of biotechnology, including genetic engineering and the nitrogen-fixing ability of leguminous plants, offer considerable promise of narrowing the differences in yield between farmers and extension stations, without degrading the natural resource base.

Human efforts have led to increasing time trends of grain yield, areas harvested and production of grain in Egypt and Libya. Nevertheless, as in all satellite studies, the observed increase in the region's vegetational cover in no way denies the occurrence of other types of land degradation and provides no measure of land productivity. To gain some indication of the magnitude of other land-degradation problems in both countries, other information relating to vegetation was consulted. This information included the amount of rainfall, crop yields, soil organic matter, use of fertilizers, and the number of animal grazing in each country over the same period of time.

Unfortunately, no information exists on the desert areas of both countries that could be compared with non-desert areas in each country. The results of table 6 indicate that grain yields have increased in Egypt and Libya between 1982 and 2000. Thus, the increase in vegetational cover and the increase in cultivated areas and grain yield, considered in conjunction with the reduced use of fertilizers, show that the region has not suffered significantly from the diminution or destruction of the biological potential of the land during the years 1982 through 1999.

This result agrees with the conclusions reached by FAO [12] studies that the MENA region is to be classified as mainly productive crop, pasture and forest land, and that desertification in the region is mostly moderate.

4. Policy Implications

4.1 IMPROVING THE DATA BASE

- The assessment of the current status of land degradation throughout both countries shows that there is a lack of hard, precise data that allows conclusions to be drawn regarding the extent and rate of desertification. This result calls for more precise studies based on surveys in combination with remote sensing data and aimed at determining the magnitude of the problem and the extent to which man is responsible.
- There are no reliably accurate estimates of the total economic loss resulting from land degradation in Egypt and Libya. Cross-country comparison, however, provides an idea of the differences between various land-use systems, and thus contributes to the investigation of the impact and causes of desertification processes and the development of possible countermeasures. At present, there is not even a rough estimate available of the off-site and other indirect economic losses resulting from desertification in both countries. There is a need for more extensive country-specific research that investigates the differences between various socio-economic situations and avoids generalization.

4.2 RESEARCH INTO ECONOMIC ALTERNATIVES AND OPTIONS

Assessments of exactly how much land turns into desert each year depend on definitions. It may be that spending money on better use of the desert in Egypt and Libya is much wiser than spending it on measures to combat desertification. This includes that spending on basic research for the better management of desert land and water resources may be more efficient for the cultivation of the desert. Investments in improving water-harvesting facilities and systems in both countries have proved to be economically viable and socially acceptable (e.g. in Egypt). Funds can be used to intervene in the life of individual villages to bring soil degradation to a halt. The current funding of anti-desertification measures should gradually shift from desertification control and rehabilitation to sustainable desert cultivation with the appropriate use of available water resources. Investments to cultivate the desert by increasing the amount of rainwater harvested can be an effective measure for combating desertification.

4.3 ON PRACTICAL ISSUES OF DESERT LAND USE

On the practical side of programmes and projects for the utilization of desert land and water resources utilization, a lot of experience has been accumulated in Egypt and Libya over time. Much of that knowledge resides in the local communities of both countries. Even though improvements in land and water use under desert conditions are very site-specific, the following guidelines can generally be made:

- that target groups be integrated and participate actively in any water-harvesting projects implemented;

- that only simple, small-scale projects be implemented;
- that the small-scale projects be consolidated into regional and national plans of action;
- that the decision-making processes of the national support services be decentralized to the sites concerned and
- that the role of international organizations be largely an advisory one, and less one of an implementing national plans of actions, as this should properly be the task of decentralized, national organizations.

5. Prospects for 2020 and 2050

Generally, in both countries, human efforts involving the use of biotechnology, including genetic engineering and the nitrogen-fixing ability of leguminous plants offer considerable promise of narrowing the differences in yield between farmers and extension stations, without degrading the natural resource base. Human efforts have led to increasing time trends of yields of cereals, cultivated areas and production of grain in both countries.

Based on the recorded time trends of the last 19 years and using the least square method, projections were made for selected indicators that are closely associated with the vegetation cover of threatened areas in both countries. The following table shows positive prospects of human efforts to control land degradation in both countries.

Table 7. Projections of some selected indicators of vegetation cover in Egypt and Libya in 2020 and 2050

Indicator	2001	2020	2050
Egypt			
* Total cultivated area (1,000 ha)	3,291	4,571	6,346
* Cultivated area of permanent crops (1,000 ha)	466	915	1,508
* Pasture area (1,000 ha)	00	00	00
* Forest area (1,000 ha)	34	37	39
* Area of cereals (1,000 ha)	2,762	3,930	5,594
* Crop yield of cereals (t/ ha)	7.28	10.60	15.71
* Number of animals (1,000 heads)	11,424	21,812	59,760
* Livestock density (head/ ha)	3.47	4.77	9.41
Libya			
* Total cultivated area (1,000 ha)	15,450	15,722	15,988
* Cultivated area of permanent crops (1,000 ha)	335	349	357
* Pasture area (1,000 ha)	13,300	13,300	13,300
* Forest area (1,000 ha)	840	850	869
* Area of cereals (1,000 ha)	320	295	190
* Crop yield of cereals (t/ ha)	0.623	0.743	0.756
* Number of animals (1,000 heads)	7,210	8,018	8,634
* Livestock density (head/ ha)	0.50	0.51	0.54

Source: Results of simple regression analysis of statistical time trends during the period, 1982- 2001

References

1. Abdulrahman, M., and Bamatraf, (1994) Water Harvesting and Conservation Systems in Yemen, in FAO (ed.) *Proceedings of the Expert Consultation about Water Harvesting for Improved Agricultural production*, FAO, Rome.
2. Aleryani, M.L., and Bamatraf, A. M. (1993) Water Resources in Kuhlan-Affar/Sharis Districts, Annex C, in FAO *Final Report: Dryland Resource Management in the Northern Highlands of Yemen*, REA/FAO/ICARDA Joint Project, Sanaa.
3. Arrar, A. (1989) Current Issues and Trends in Irrigation with Special Reference to Developing Countries, in FAO *Resource Conservation and Desertification Control in the Near East, Report of the International Training Course*, DSE, FAO, GTZ, UNESCWA, Amman.
4. Arar, A. (1993) *Optimization of water use in arid areas* Arab Consult, Amman.
5. Bazza, M. (1994) Operation and Management of Water Harvesting Techniques, in FAO (ed.) *Proceedings of the Expert Consultation about Water Harvesting for Improved Agricultural production*, Water Report, Rome.
6. Bill, R.U.D., and Fritsch, J. (1994) Grundlagen der Geo-Informationssysteme, *Hardware, Software und Daten*, Institute for Geography, Karlsruhe.
7. Biot, Y. (1995) *Rethinking Research on Land Degradation in Developing Countries*, World Bank Discussion Paper, Washington.
8. CIESIN (1996) *The Importance of Satellite Remote Sensing for Global Change Research*, CIESIN Thematic Guides, from Internet web site: www.ciesin.org/TG/RS-home.htm).
9. CIESIN (1996a) *The Use of Satellite Remote Sensing*, CIESIN Thematic Guides, from Internet web site: www.ciesin.org/TG/RS-home.htm)
10. Dregne, H.E., and Chou, N.T. (1992) Global Desertification Dimensions and Costs, in *the Degradation and Restoration of Arid Lands*, Texas Technical University, Lubbock.
11. DSE (German Foundation for International Development) (1989) *Resource Conservation and Desertification Control in the Near East*, DSE, Bonn.
12. FAO (1997) *Irrigation in the Near East Region in Figures*, FAO-Water Report No. 9, Rome.
13. FAO (1998) *Food for All*, FAO Report, Rome.
14. FAO (1998) *Statistics of Land and Water Use and Agricultural Production in the MENA Countries*, FAO, Rome.
15. Forse, B. (1989) *The Myth of the Marching Desert*, CIESIN Organization, Washington.
16. Frasier, G.W. (1994) Water Harvesting, Runoff Farming Systems for Agricultural production, in FAO (ed.): *Proceedings of the Expert Consultation about Water Harvesting for Improved Agricultural production*, FAO-Water Report No. 3, Rome.
17. Glantz, M.H., and Orlovsky, N.S. (1983) Desertification: A Review of the Concept in *Desertification Control Bulletin 9*, 15-22
18. GTZ/QRDP (1993) *Estimation of development costs for the area of Qasr Rural Development Project* QRDP-Internal report, Marsa Matrouh.
19. Hielkema, J.U. (1989) Introduction to Environmental-Satellite Remote-Sensing Techniques and Systems, in: DSE (1989) *Resource Conservation and Desertification Control in the Near East*, DSE, Bonn.
20. Hillel, D. (1982) *Negev, Land, Water, and Life in a Desert Environment*, Praeger Publishers, New York.
21. Joudeh, O, (1994) Integration of Water Harvesting in Agricultural Production, in FAO (ed.) *Proceedings of the Expert Consultation about Water Harvesting for Improved Agricultural production*, FAO-Water Report No. 3, Rome.
22. Nasr, M. (1998) *Agriculture: The Biggest User of Water*, International Water Conference, Bonn, October.
23. Nasr, M. (1995) The Technical, Social and Economic Aspects of Water Harvesting and Water Supply of Rainfed Desert Farming Systems, *National Agricultural Research Project (NARP) I-12*,
24. Nelson, R. (1989) *Dryland management: The desertification problem*, World Bank Technical Paper No. 116, Washington.
25. Olsson, L. (1985) *An Integrated Study of Desertification: Applications of Remote Sensing, GIS and Spatial Models in semi-arid Sudan*, Meddelanden fran Lunds Universites geografiska Institution, Avhadlingar.
26. Qaim, M. (1996) Modernization of Farming systems in the Northwest Coastal Zone of Egypt, Qasr Rural Development Project Area, M.Sc. thesis, Kiel University, Institute of Agricultural Policy.
27. Saad, K.F., and Shaheen, A. (1988) *Evaluation of Water Resources in the Arab World*, Arab Center for Dry Land studies, International Institute for Hydrological and environmental Engineering, Paris.

28. Siegert, K. (1994) *Water Resources Engineer, Water Resources, Development and Management Services*, FAO-Land and Water Development Division, Rome.

29. Siegert, K. (1994a) Water Harvesting for Improved Agricultural Production, in: FAO (ed.) *Proceedings of the Expert Consultation about Water Harvesting for Improved Agricultural production*, Water Report No. 3, Rome.

30. Swift, J., and Maaliki, A. (1984) *A cooperative Development Experiment Among Nomadic Herds*, Overseas Development Institute, London.

31. Taher, A. (1993) *Agricultural Development in the Northwest Zones of Egypt, Soils and Water Research Institute*, Agricultural Research Center, Cairo.

32. UNCCD (1999) *United Nations Convention to Combat Desertification*, Secretariat for the Convention to Combat Desertification, Bonn.

33. UNCCD (1998) *Desertification – Its causes and Consequences*, Pergamon, Oxford.

34. UNCCD (1998a) *Down to Earth*, UNCCD, Bonn.

35. UNCED (1992) *Status of Desertification and Implementation of the United Nations Plan of Action to Combat Desertification*, UNEP, Nairobi.

36. UNCOD (1997) *Plan of Action and Resolutions, United Nations Conference on Desertification*, UNCOD Nairobi.

37. UNEP (1982) *Assessment of Desertification, UNCED Part I*, UNEP, Nairobi.

38. UNEP (1984) *Assessment of Desertification, UNCED Part I*, UNEP, Nairobi.

39. UNEP (1996) *Status of Desertification and Implementation of the United Nations Plan of Action to Combat Desertification, UNCED Part I*, UNEP, Nairobi.

40. UNEP (1996a) *Status of Desertification and Implementation of the United Nations Plan of Action to Combat Desertification, Executive Summary*, UNEP, Nairobi.

41. UNDCPAC (1987) *Rolling Back the Desert*, UNEP, Nairobi.

42. Warren, A. (1984) Productivity, Variability and Sustainability as Criteria of Desertification, paper presented to the *International Symposium in the REC Programme on Climatology*, Held in Mytelene.

43. Warren, A., and Agnew, C. (1988) *An Assessment of Desertification and Land Degradation in Arid and semi-arid Areas*, University College, Ecology and Conservation Unit, London.

Chapter 8

Prospective Desertification Trends In The Negev - Implications For Urban And Regional Development

B.A. Portnov[1] *and U.N. Safriel*[2]
[1]*Department of Natural Resources and Environmental Management, Haifa University*
[2]*Department of Evolution, Systematics and Ecology, Hebrew University of Jerusalem*

Most Mediterranean countries contain a diversity of land types, of various degrees of aridity and soil fertility. In the southern Mediterranean, coastal areas are usually less dry and more fertile than inland regions, many of which are semi-arid and arid deserts; the southern regions of several northern Mediterranean countries are drylands, e.g. Spain, Italy and Greece [15]. In Israel, there is a north-south gradient of aridity: the north and centre of the country are dry-sub-humid, whereas the south is semi-arid, arid, and hyper-arid. With the surge of population increase, intensifying urbanisation in Israel out-competes agriculture and industry. As a result, agriculture and polluting industries are pushed from fertile land to the semi-arid northern Negev, or even to the arid and hyperarid, central and southern Negev, respectively. However, agriculture and heavy industries are nearly always the main causes of desertification: they cause land degradation due to soil erosion, soil and water salinisation, and loss of biodiversity. Population and urbanisation pressure is thus putting more drylands at risk of further desertification. These processes are bound to intensify in Israel, and probably also in many other Mediterranean countries. The first victim of desertification caused by agriculture is agriculture itself. As a result, dryland agriculture is often abandoned, leaving behind an ecosystem that ceases to provide environmental services, and is costly or even impossible to rehabilitate. Thus, by pushing agriculture from non-desert to desert, not only agriculture itself may be doomed, but also alternative uses of the desert, such as conservation of biodiversity, recreation, eco-tourism and tourism, are excluded. Thus, urbanisation trends in non-desert drylands and other lands, not only bring desertification to more arid drylands, but also deny a country's non-desert population the use of these drylands as a resource for recreation and for alternative income-generating livelihoods. Furthermore, due to a sustained accelerated population growth, urbanisation will eventually also reach the desert, but will then find there a further desertified, less hospitable environment than it would otherwise have done. In this paper, we propose an *alternative scenario* for sustainable development of a Mediterranean country having both desert and non-desert regions. Rather than concentrating now on further urbanizing the fertile and already heavy populated regions and as a result introducing much more agriculture and polluting industry into the desert, *we propose either full or partial cessation of the urbanisation process in the fertile, heavy populated areas, and, instead of introducing agriculture into deserts, introduce urbanisation into desert regions.*

The present paper attempts to answer the following questions:
- What are the main development alternatives available for the Negev?
- How can desirable types of development in the Negev be encouraged?

The paper starts with a brief overview of environmental conditions in the Negev. Then it outlines both current and prospective desertification trends in the region. Following this introduction, two alternative strategies of the Negev development – agricultural vs. urban development path - are introduced and discussed in some

A. Marquina (ed.), Environmental Challenges in the Mediterranean 2000-2050, 123–138.
© 2004 *Kluwer Academic Publishers. Printed in the Netherlands.*

detail. However, the process of desert urbanisation requires special design and planning features that will make it viable, and these features will be elaborated in this paper. The critical feature of this urbanisation is that it will be economic on land use, and thus the remaining large non-urbanized desert areas will have a non-desertified, functional environment, available for other environmentally-compatible uses, while agriculture will, as much as economically required, be confined to fertile lands.

1. Negev: Climate and Environment

Figure 1. Climatic zones and administrative districts of Israel

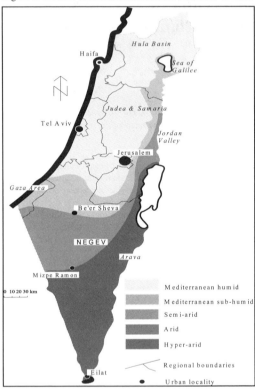

Source: Bitan, A. and Rubin, S. (1991) *Climatic Atlas of Israel for Physical and Environmental Planning and Design,* Ramot Publishing Co., Tel-Aviv, ICBS (1996-2001) *Statistical Abstracts of Israel No. 47- 52 (annual).* Israel Central Bureau of Statistics & Hemed Press, Jerusalem. [4]

Geographically, the Negev region extends over three out of five Israel's climatic zones – semi-arid, arid and hyperarid [3][1]

---

[1] Two other climatic zones, located outside the Negev, are: a) *Mediterranean dry sub-humid zone,* which stretches along the Mediterranean plain, and encompasses the valleys north of Haifa and the Galilee, and b) *Humid zone,* represented by a small enclave of high-elevated areas in northern Israel.

- *Semi-arid zone* lies north of Be'er Sheva, the major urban center of the region; this zone includes the northern Negev, the upper reaches of the Judean Desert, the northern Jordan Valley, the Hula Valley and the Sea of Galilee (Lake Kinneret) region. The mean annual precipitation (P) in this zone is 300-500 mm, while its mean annual evapotranspiration (EP) ranges from 1500 to 1700 mm. The Aridity Index (AI=P/EP) for the area thus reaches 0.20-0.35 (Figure 1).

- *Arid zone* is located south and east of the semi-arid zone, south of Be'er Sheva to Mitzpe Ramon, and east to the lower reaches of the Jordan Valley. The area is known as the central Negev (dry, in English). The mean annual precipitation in the region does not exceed 100-300 mm, while the mean annual evapotranspiration ranges from 1700 to 1800 mm (AI=0.05-0.20).

- *Hyperarid zone* lies south and east of the arid zone. It includes the central and southern Negev, from Mitzpe Ramon to Elat, the Dead Sea Basin and the Arava Valley. The mean annual precipitation in this zone does not exceed 30-90 mm, and the mean annual evapotranspiration ranges between 1800 to 2800 mm (AI<0.05).

The climate of the Negev is hot and dry during the summer, and cool during the winter. Summer daily average temperatures range between 17-33°C in July and August, but maximums of 42-43°C occur in May and June, during heat waves called *sharav* in Hebrew. Relative humidity during the hot hours of the day is usually 25-30%, and as low as 5-15% during the 'sharav' waves. Winter temperatures in the Negev range between 5-15°C, but temperatures near ground surface may drop below 0°C [5]. Rainfall in the Negev occurs only in winter, and varies annually. In the central Negev, for instance, the amount of rainfall varies annually from 20 to 160 mm [4].

2. Development Patterns

Until the establishment of the State of Israel in 1948, mainly nomadic Bedouin tribes inhabited the semi-arid and arid parts of the Negev. The Bedouins, whose number did not exceed at that time 15,000 people [10] were spread over an area of about 10,000 km^2. Originally, the Bedouins subsisted on sheep, goat and camel herding. However, in the middle 19th century, they also initiated patchwork farming in *wadis*[2] that had previously been used exclusively as rangeland, and grew cereals through dry farming, often as a cash crop [1, 11]. In rainy years they also cropped in the late winter and spring in the area of up to 2500 km^2, reducing it to 600 km^2 in dry years [18].

In the pre-state period, some 20 new agricultural communities (moshavim and kibbutzim) were established in the Negev. By 1948, their population grew to some 6,000 residents [10]. After the establishment of the State in 1948, an intensive urban development of the region was initiated. This development led to the establishment of a number of new settlements scattered across the area at varying distances from each other. In this fashion, seven new 'development towns' - Arad, Dimona, Mitzpe-Ramon, Netivot, Ofaqim, Sderot and Yeroham - were established in the 1950's and early 1960's, while the existing towns of Be'er Sheva and Elat received a significant growth impetus [20].

---

[2] *Wadi* is a dry valley with rainfall-dependent surface run-off.

The initial growth of these towns was sustained by the involuntary location of new immigrants and direct government investment in their economy [7]. Since the early 1970's, this policy has gradually been replaced by various incentives designed to encourage both inward migration and private investment; these incentives included such measures as government loan guarantees, tax exemptions, and the provision of public housing [13].

In the late 1960s, the first permanent urban settlement for the Negev Bedouins – *Tel Sheva* – was established. Then during the 1970s-1980s, six new Bedouin townships - *Rahat, Kuseifa, Aro'er, Lakiya, Houra* and *Segev Shalom* – were established. In these towns, the physical layout was set so as to accommodate the tribal structure, allowing for a tribal territoriality [2, 8].

Currently, there are 15 urban sites in the Negev. They occupy less than 0.6 per cent of its land area but concentrate more than 90 per cent of the region's population (Table 1). The rest of the region is used for military training areas and installations, nature reserves and national parks, agriculture, and industrial sites.

Table 1. Urban Localities in the Negev

Locality	Area under jurisdiction in dunams*	Urban area in dunams (estimate)	Population in 1983	Population in 1995	Annual population growth, %
Be'er Sheva	54,454	26,000	110,100	150,040	3.02
Ofaqim	9,545	4,898	12,665	20,740	5.31
Elat	58,103	7,677	18,725	32,510	6.13
Dimona	30,539	7,332	26,815	31,050	1.32
Yeroham	34,099	1,822	6,225	7,715	1.99
Lehavim	960	588	NA	2,490	NA
Metar	8,263	1,103	NA	4,670	NA
Mitzpe Ramon	645,001	1,005	2,945	4,255	3.71
Netivot	5,695	3,540	7,755	14,990	7.77
Omer	12,773	1,337	4,500	5,660	2.15
Arad	75,934	4,786	12,560	20,265	5.11
Rahat	8,851	5,207	9,970	22,050	10.10
Tel Sheva	4,762	1,474	2,485	6,240	12.59
Qiryat Gat	8,440	10,320	25,825	43,700	5.77
Sderot	4,302	3,943	9,060	16,695	7.02
Total:	961,721	81,032	249,630	383,070	4.45
Be'er Sheva sub-district	12,946,000	12,946,000	275,000	418,900	4.36
Urban areas, %	7.43	0.63	90.77	91.45	

*1 dunam = 1,000 m².
Source: ICBS (1996-2001) *Statistical Abstracts of Israel No. 47- 52 (annual).* Israel Central Bureau of Statistics & Hemed Press, Jerusalem, NCRD (1998) *Statistical Yearbook of the Negev.* Negev Center for Regional Development & Negev Development Authority, Be'er Sheva. [10, 16]

3. Desertification Processes in the Negev[3]

By the end of the 19th century, desertification in the Negev was most likely to result in soil salinisation in dry sub-humid areas, and a definite loss of natural vegetation and soil erosion in dry sub-humid and some semi-arid areas. In both sub-humid and arid parts of the region, ecological and hydrological processes were disrupted. This resulted in an overall decline in productivity of the regional ecosystem. It is not clear whether or not the southern arid parts of the Negev had been desertified by the end of the 19th century. However, historical studies of traditional pastoralism in the Negev indicated that the existing rangelands in the region seemed to be stable at a low-level equilibrium state [17, 24].

In the dry sub-humid areas there is soil salinisation due to irrigation in dry sub-humid valleys, and increasing impenetrability of dry sub-humid woodland and 'bush encroachment' leading to degraded range quality on the one hand, and woodland fires leading to soil erosion on the other. In the semi-arid areas, there are indications of sheet soil erosion on irrigated agricultural land, and of highly intensified gully erosion, both in regions of agricultural activity and of grazing activity. However, reliable quantitative assessments of the magnitude of these processes are not currently available.

4. Current Practices of Combating Desertification

The process of combating desertification in the Negev and other drylands of Israel includes four major components: a) control of scrubland grazing, b) afforestation programs, c) water transportation, and d) increasing the efficiency of water use in agriculture.

4.1. CONTROL OF SCRUBLAND GRAZING

To reduce overgrazing in the dry sub-humid areas, mainly by goats feeding on scrubland, the 'Black Goat Law' (The Law for Vegetation Protection [by Goat Damages]) was enacted in 1950. Prior to 1948, the number of goats in the region was estimated at 185,000. By 1950 this number was reduced to 71,000, so was the pressure on the natural scrubland. As a result, the latter rapidly developed into typical eastern Mediterranean woodlands. Subsequently, the number of goats kept in the Negev increased to 115,000 by 1973; it went down to 70,000 (1994) and then again increased to 74,000 in 1998 [10]. The overall positive effects of the reduced grazing pressure demonstrates the high resilience of the dry sub-humid Mediterranean woodland ecosystems, attributed to the long co-evolution of these systems with human-induced disturbances, including livestock grazing [17].

---

[3] This and the following section (on combating desertification in the Negev) are based, in part, on BIDR (2000).

4.2. AFFORESTATION PROGRAMS

The afforestation programs in Israel fall under three laws: 'The Forest Law', 'The City Building Directive', and 'The National Parks and Nature Reserves Law.' In 1961, the State of Israel contracted the Jewish National Fund (JNF), a national non-governmental organisation, to carry out afforestation activities in Israel. By 1993, JNF had planted over 200 million trees (around two-thirds of which were Aleppo pines), divided between 280 afforestation plots. By 1993 these plots covered 690 km^2 (or 5 per cent of the total area of the Negev). By 1999, the size of these plots increased to 911 km^2, or to 7 per cent [10].

The 22nd Country Master Plan is the National Master Plan for Forests and Afforestation for the coming 25 years. During this period the amount of afforested and managed woodlands in Israel will increase to 1,606 km^2 (15% of the country's dry sub-humid and semi-arid lands). JNF instituted a 'watchdog system' for identifying development plans that are incompatible with the Master Plan, so that timely counter-measures can be taken.

4.3. WATER TRANSPORTATION

Transforming rangeland and rain-fed cropland to irrigated croplands on a large scale requires the planning and execution of large-scale water transportation projects. The first project of this kind was a 66' diameter pipeline drawing water from the *Yarkon* River in the centre of Israel to the Negev; it covered a distance of some 130 km. The annual output of this pipe was about 100 million m^3 of water. The second large-scale project was the National Water Carrier completed by 1964. This carrier is a combination of underground pipelines, open canals, interim reservoirs and tunnels supplying about 400 million m^3 annually. Via this system, water from Lake Kinneret (the Sea of Galilee) in the north of the country, located about 220 m below sea level, is pumped to an elevation of about 152 m above sea level. From this height the water flows by gravitation to the coastal region, where it is pumped to the Negev.

These water transportation projects are accompanied by the exploitation of groundwater and of flash-floods; to date, 115 dams and reservoirs with a total capacity of 100 million m^3 have been constructed. In addition, Israel operates 30 desalination facilities, in which Israel's National Water Company, *Mekorot*, desalinates 9.8 million m^3 of water annually, much of it for domestic use, especially in the far south, which is out of reach of the National Water Carrier.

4.4. INCREASING THE EFFICIENCY OF WATER USE IN AGRICULTURE

Prior to 1948, crops in the Negev were irrigated by surface (flood and furrow) irrigation. However, the farmers quickly realised that economically viable agriculture is constrained by both the scarcity of water and the uncertainty of its supply. This recognition resulted in attempts to drill wells and draw underground water. However, the quantities obtained were quite small, and the salinity of water was often too high for agricultural use. Therefore, sprinkler and drip irrigation substituted surface flooding practices.

Water use efficiency, which is the ratio between the amount of water taken up by the plant and the total amount of water applied, increased considerably. (The efficiency of water use is about 45% in surface irrigation and 75% in sprinkler irrigation, in drip irrigation, it is about 95%). Outdoor computers that control drip irrigation and fertigation have improved the water use efficiency even further.

Protected agriculture, based on greenhouses, is another common solution. In a dryland greenhouse evapotranspiration is minimized, but cooling in summers and in some drylands warming in winter nights, are required. Technologies related to protected agriculture in Israel include synthetic fabrics, cooling and warming devices and mechanisms, drip irrigation and fertigation, growth substrates, and supplies of insect pollinators. Either fully-closed chambers or partly opened at prescribed seasons and times of the day, dryland greenhouses can be fertilized with CO_2 and be protected from insects, thus reducing the use of insecticides.

Agricultural production in the Negev greenhouses is very intensive, with very high water and soil/space-use efficiency. Hydro-geological surveys have revealed that the Negev and the Arava valley possess considerable reserves, mostly fossil, of saline underground water with a variable concentration of salts. Some of the cash crops, bred for salt-tolerance (especially tomatoes, melons and grapes), are of higher quality (sweeter and firmer fruit) when irrigated by brackish, rather than by low-salinity water. Their quality and out-of-season availability makes them ideal for export.

5. Prospects for Future Development

Figure 2. Factors affecting development of desert regions

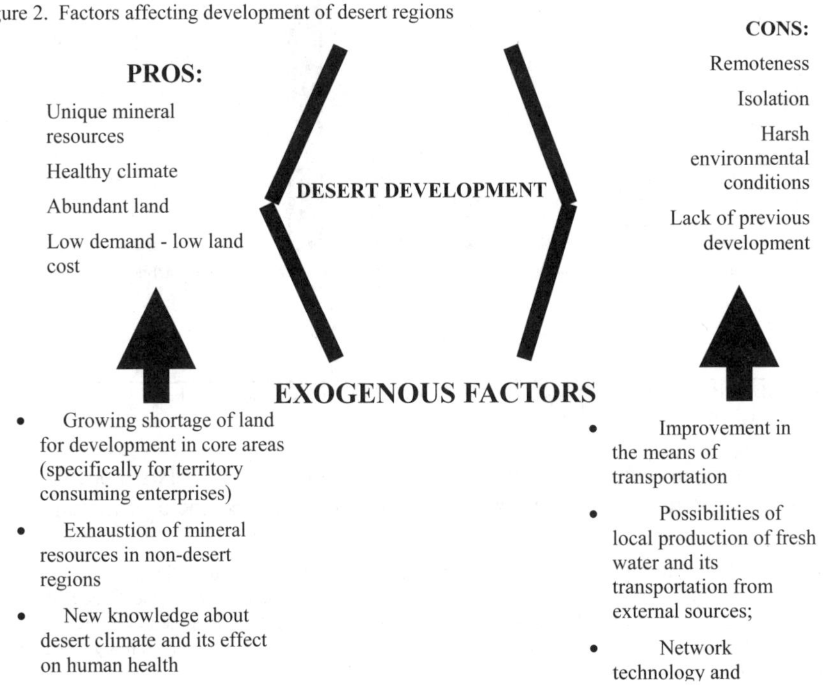

PROS:

Unique mineral resources

Healthy climate

Abundant land

Low demand - low land cost

DESERT DEVELOPMENT

CONS:

Remoteness

Isolation

Harsh environmental conditions

Lack of previous development

EXOGENOUS FACTORS

- Growing shortage of land for development in core areas (specifically for territory consuming enterprises)

- Exhaustion of mineral resources in non-desert regions

- New knowledge about desert climate and its effect on human health

- Improvement in the means of transportation

- Possibilities of local production of fresh water and its transportation from external sources;

- Network technology and

In the coming years, the development pressure on the Negev may intensify due to a number of factors. They include: the growing shortage of land for new development in central regions of the country; improvement in the means of transportation (both water and thoroughfare), and new knowledge about desert climate, which indicates its suitability for asthmatics, arthritics, and other health-risk groups (Figure 2).

The availability of land resources in the Negev is especially significant in the local context. Israel is a small country (22,145 km^2), whose population is concentrated predominantly along the Mediterranean coast, in close proximity to its two major cities - Tel Aviv and Haifa (Figure 1). The population of these cities and their immediate hinterland (the Tel Aviv, Central, Haifa districts) amounts to some 3.4 million residents, or nearly 55% of the country's population [10].

In the past decades, the most intensive population growth has occurred around these population centers. In the Tel Aviv district, for instance, the population density in 1948-2000 grew, on average, by some 95 persons per km^2 per year (Figure 3), from 1,800 residents per km^2 in 1948 to 6,750 residents per km^2 in 2000.

Figure 3. Annual increase in population density (persons per km^2) by district of Israel in 1948-1995.

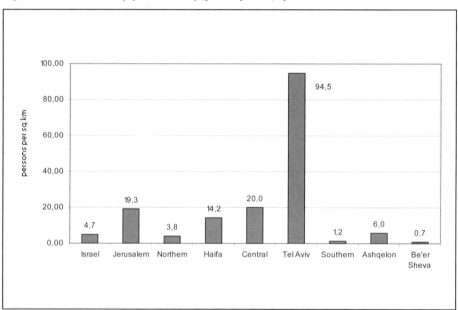

Source: ICBS (1996-2001) *Statistical Abstracts of Israel No. 47- 52 (annual).* Israel Central Bureau of Statistics & Hemed Press, Jerusalem[4]. [10]

Unsurprisingly, urbanisation in the central, sub-humid regions of Israel occurs at the expense of agricultural land. In past decades, such land has become a prime target for

---

[4] Between 1948 and 1995, the most substantial increase in population density occurred in the centrally located districts of the country – Tel Aviv, Central, Jerusalem and Haifa. Concurrently, in the Negev (the Be'er Sheva sub-district), the increase of population density was rather marginal – less than one person per km^2 per year. As land resources for new development in the central part of the country rapidly diminish the development pressure on the Negev should be expected to increase.

real estate development [12]. In turn, this process 'pushes' agriculture elsewhere, primarily to the northern border of the arid Negev region. In view of this trend, there may be two alternative strategies to the Negev development:

a) *Agricultural path:* Urbanisation in the 'core' regions of Israel continues at the current rate, leading to a step-by-step transfer of agriculture and other land extensive functions (such as industrial waste dumps, military installations and primary industries) towards the country's desert fringe;

b) *Urban path:* Restriction on further urban growth in the central, non-arid regions of the country and the concentration of such growth in the Negev.

In the following subsections, we shall consider each of these development alternatives in some detail.

5.1. AGRICULTURAL PATH

The recently adopted State Master Plan of Israel – '*Tama-35'* - introduces the concept of 'concentrated dispersion.' According to this concept further development should be dispersed at the national level and concentrated at the level of individual regions. In line with this concept, the plan puts emphasis on the continuing growth of four existing metropolitan centres – Tel Aviv, Jerusalem, Haifa and Be'er Sheva (see Figure 1). Three of these metropolitan areas (excluding Be'er Sheva) are located in the central part of the country. Therefore, their expansion will cause the further concentration of development in the central part of the country. In turn, it would accelerate the relocation of non-urban functions, currently surrounding these urban centres (specifically agriculture), into peripheral locations, such as the semi-arid and arid parts of the Negev. *Is the agricultural path a viable option for the Negev?* The answer to this question is rather negative. There are both ecological and economic considerations.

As Thomas and Middleton [15] note, agriculture-related activities (the removal of natural vegetation, overgrazing and other agricultural activities) are the major causes of desertification. According to FAO's estimates [6], agricultural activities are responsible for up to 30-35 percent of land degradation in the Middle East. Although in Israel agriculture is based on advanced technologies (such as sprinkler and drip irrigation), large-scale agriculture production in this arid region will most definitely increase the risks of soil salinisation and soil erosion.

The consumption of fresh water and economic capacity of agriculture are two other important factors. In Israel, agriculture consumes annually about 60 per cent of the national water supply, while it contributes to only 2.4 per cent of the national GDP. In contrast, the urban sector consumes annually 40 per cent of the national water supply, whereas it generates 97.6 per cent of the national GDP [14].

In most parts of the Negev, the economic capacity of agricultural production is lower than the national average. This is due to the relative remoteness of the region, poor soil quality and higher than average water consumption due to high average annual evapotranspiration. Thus, according to the results of the 1995 Agricultural Survey of Israel [9], agricultural production in the central Negev (Region 6-South) is the least efficient among the country's agricultural regions, both in terms of output per dunam and gross value added (Figure. 4).

One comment is important. The Eastern valleys and the Arava region (Region 4; Figure 4) represent a notable exemption from the general trend. Agricultural production in these regions is of high efficiency. There are a number of explanations for this phenomenon.

First, agricultural production in the Arava is highly intensive due to common use of greenhouses. Therefore, production per dunam is high, compared to agricultural production elsewhere. Second, the Arava region has a rich fossil reservoir of brackish water, but the risk of soil salinisation due to the use of this water for irrigation is reduced – in many greenhouses crops are not grown directly on the soil but on tables with artificial substrates. Lastly, winters in the region are relatively warm, and this creates a comparative advantage in seasonal production of exported cash crops.

Another drawback of introducing large-scale agricultural development to the Negev is its negative environmental consequences. The agriculture brought to the Negev is likely to be based on treated wastewater that will not have been desalinized, and hence it will pose a risk of salinizing the soil, which, in the long run, will reduce productivity.

Thus, agriculture that is already not economically viable, may become even less viable, and will also damage, perhaps irreversibly, the ability of the environment to provide ecological services.

Figure 4. Efficiency of agricultural production by agricultural region of Israel

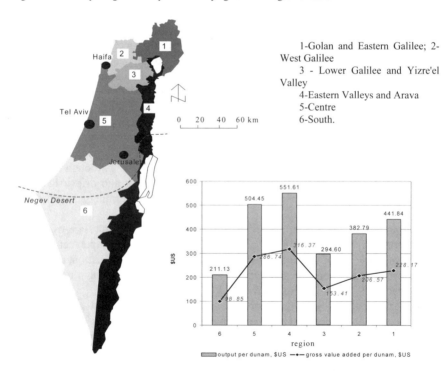

1-Golan and Eastern Galilee; 2-West Galilee
3 - Lower Galilee and Yizre'el Valley
4-Eastern Valleys and Arava
5-Centre
6-South.

Source: ICBS (1998) *1995 Agricultural Survey.* Israel Central Bureau of Statistics, Jerusalem. [9]

5.2. URBAN DISPERSION PATH

Compared to agricultural development, the adverse effect of urban development on the desert environment may be limited. The reason is rather simple: geographically, urban development is compact. As Table 1 shows, all the urban settlements in the Negev occupy less than 0.6 per cent of its land area, while housing more than 95 per cent of the region's population. The concentrated urban development may thus be least damaging for the desert ecosystems, and compared to the agricultural path, may thus have less severe desertification consequences. However, there are a number of preconditions, which may make the 'urban path' feasible. They are: a) achieving the critical mass of urban population; b) territorial contiguity of urban settlement; c) provision of unique urban functions and attractiveness to investors, and d) climate-responsive urban environment.

5.2.1. *Achieving the Critical Mass of Urban Population*
Upon reaching a certain population threshold, many urban places start to grow faster. Such a threshold may be termed the 'critical mass.' The main reason for the change in the pace of growth is simple. Residents of smaller localities are often denied access to social amenities that are concomitant with a larger settlement size. As the population of a community increases, it crosses the threshold for higher-level services, and offers more varied opportunities for employment, social services and leisure.

As Portnov and Erell [20] demonstrate, the 'critical mass' for peripheral desert localities is not fixed. It depends on the physical remoteness of such localities from the major population centers of a country, so that more remote localities should be larger in order to become sufficiently attractive for both migrants and private investors.

In Israel, the 'critical mass' (CM) of small peripheral localities (measured in thousands of residents) was found to relate to the town's remoteness as follows:

$$CM=3.2*IR$$

where IR is index of remoteness, estimated as the road distance from the closest major population of the country in km (ibid). According to this equation, a town, which is 100 km apart from one of the country's major population centers (Tel Aviv or Jerusalem), should have at least 320,000 residents (3.2*1000*100) to become sufficiently attractive to both investors and migrants.

5.2.2. *Territorial Contiguity of Urban Settlement*
Most often, peripheral desert towns are small and may not thus reach the 'critical mass,' which is essential for their sustained growth.[5] To achieve such a "critical mass', desert

---

[5] There are different definitions of sustainability. Their analysis is far beyond the scope of the present paper. Following Turner (1993) and Portnov and Pearlmutter (1999), we adopt the following simplified definition: *a non-diminishing growth that endures over a prolonged period of time.* In already densely populated 'core' areas, measures restricting further urban growth may be required. However, in sparsely populated, specifically peripheral, desert regions, future growth of urban settlement may become a desirable objective. Such a growth may create a sufficient variety of employment and cultural opportunities, making desert towns more attractive and desirable for both their present-day residents and newcomers.

towns may be grouped in clusters. 'Urban cluster' is a group of urban settlements located in close proximity to each other and connected by strong socio-economic and functional links [21]. Very often, both investors and migrants consider urban clusters as integrated functional units, making their location decision hierarchically: first, among clusters, and then among individual urban settlements in the 'preferred' cluster.

The sizes of urban clusters are not fixed. They depend on the actual patterns of inter-urban commuting. If commuting conditions are better (in terms of roads, availability of public transportation, etc.), urban clusters extend over a larger area. As in the case of individual towns, the minimal threshold of cluster efficiency ('Critical Mass') is a function of the cluster's geographic location: Urban clusters established in more geographically remote areas should be larger in order to sustain the growth of individual towns (Figure 5).

5.2.3. Provision of Unique Urban Functions and Attractiveness to Investors
Neither physical location nor opportunities for functional exchanges *alone* may guarantee the successful development of urban localities in peripheral desert areas.
A clearly identified and well-established *motivating force* is required to drive such growth.

Figure 5. Urban development in a desert region as clusters of towns

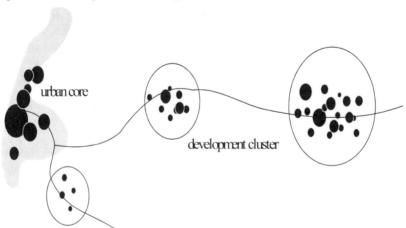

Source: Portnov, B.A. and Erell, E. (2001) Urban Clustering: The Benefits and Drawbacks of Location, Ashgate, Aldershot [20]

There are numerous functions and services that contribute to urban growth, but relatively few provide a sufficiently strong impetus to generate the wide range of regional multipliers essential for sustained growth.

Among such functions are universities and large hospitals, whose role in the formation of cities is profound. Unless a small peripheral town possesses a considerable comparative advantage such as a coastal location or other unique natural resources, the establishment of the above educational and medical facilities may enhance considerably its development potential, at least initially.

5.2.4. Climate-responsive Urban Environment

There are two essential advantages of peripheral desert cities to centrally located non-desert settlements. One of them is the relatively low land price, which may result in a relatively low housing costs.

This is a powerful advantage, specifically for a small country like Israel that is poor in land resources. This advantage may be especially important at the initial phase of urbanisation. Though land prices in a desert region may increase as the region develops, they will always be lower than those in the country's overpopulated central areas.

Another potential advantage of a desert city is the higher availability of open spaces outside the city perimeter. As a rule, a desert city does not have a developed agricultural hinterland, which often impedes a further expansion of non-desert cities.

However, the importance of this advantage may diminish as a desert city grows. There are two main reasons:

Figure 6. The proposed strategy of development for the Negev region

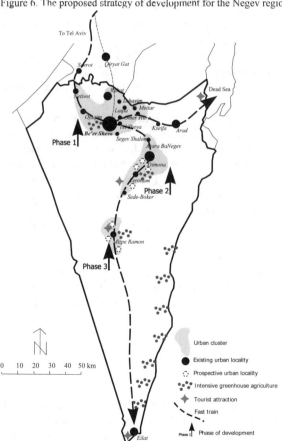

- *First*, functionally, the urban environment is not homogeneous. In an established city, residential areas are surrounded by different land uses - industry, cemeteries, transport terminals, etc. If residential areas expand beyond this ring, they may become alternated by non-residential uses. This patchwork pattern disrupts functional links between residential neighbourhoods; it may also increase the level of air pollution in residential areas.
- *Second*, the territory of a desert city may not expand indefinitely. If it happens, the functional integrity of such a city rapidly diminishes. For instance, the city centre may become inaccessible to many of the city's dwellers. In the harsh desert climate, a long walk, waiting for a bus, may become prohibitive. Therefore, desert cities should be small and compact in order to function efficiently. If a need for expansion nevertheless exists, clusters of small, functionally integrated towns may become a viable option (Figure 6).

Finally, we should mention a number of design features that may be especially beneficial for a desert city:

- *High density of development.* Narrow streets, small open spaces and high-density low-rise development of residential quarters provide effective shading and protection from blowing dust. Compact urban form also reduces the length of intra-city commuting, which is essential for residents of a desert city. As empirical studies [19], [22] show, people in extreme climates are more tolerable to high densities. If residents of climatically moderate cities prefer large open spaces, dwellers of climatically harsh regions often tend to smaller, more 'intimate' urban areas.
- *The use of water in open spaces.* In a dry desert climate, the use of small pools, fountains, sprinklers installed above pedestrian passageways may be very efficient. These devices may provide the efficient cooling of outdoor spaces. In contrast, the efficacy of such devices may be lesser in humid non-desert climates.
- *Enclosed courtyards.* Small courtyards attached to desert houses (if properly ventilated, sized and oriented) may provide cool air to indoor spaces and a place of retreat for residents, specifically during evening and morning hours. In contrast, in non-desert cities relatively large open spaces surrounding buildings may be more beneficial.
- *Building clusters.* In a desert city, buildings should be arranged in compact clusters with a clearly emphasized built perimeter. Such a perimeter protects internal spaces from cold winter winds and from sand storms during the summer. In contrast, in a non-desert city, building clusters may be more open towards the exterior, following the 'garden city' concept.

6. Conclusion

In the arid and hyper-arid Negev many farmers turn either to non-agricultural livelihood, or to aquaculture and greenhouse agriculture that are mostly detached from the soil. These trends reduce soil erosion and soil salinisation. On the other hand, in the northern, semi-arid Negev, more land comes under cropping, mainly horticulture, which is irrigated with treated urban wastewater. This wastewater is safe for plants and human health, but it is not desalinized, hence it creates a long-term threat of soil salinisation. At least in the short run, soil erosion in the Northern Negev due to irrigation, as well as soil salinisation, may increase. On the other hand, as pastoralism is giving way to alternative livelihoods practiced by the Bedouin population, the pressure on rangelands may be reduced and the risk of further desertification may partially be averted.

However, the pace of desertification in the Negev largely depends on development processes elsewhere, primarily in the central, non-desert regions of Israel. As land resources for new development in these regions rapidly diminish, the development pressure on the Negev's resources may intensify. If the current trend of urban concentration in the central part of the country (which occurs mainly at the expense of agricultural land) continues, agricultural activities will inevitably be forced towards the northern, semi-arid fringe of the Negev. This may result in the intensification of the desertification process in the region, due to agriculture-induced soil erosion and salinisation.

The strategy for the Negev development we advocate is based on urban development, possibly accompanied by compact pockets of intensive greenhouse agriculture (Figure 6).[6] Such a development may justify the often large investment required for the provision of drinking water, and compared to the agricultural path, will have a much lower impact on the dryland ecosystems' land resources, and hence may have less severe desertification consequences.

References

1. Abu-Rabia, A. (1994) *The Negev Bedouin and Livestock Rearing.* Berg, Oxford/Providence.
2. Ben-David, Y. (1993) *The Urbanization of the Nomadic Bedouin Population of the Negev 1967-1992.* Jerusalem Institute for Israeli Studies, Jerusalem (in Hebrew).
3. BIDR (2000) *The 1st National Report on the Implementation of the United Nations Convention to Combat Desertification.* Midreshet Ben-Gurion, J. Blaustein Institute for Desert Research (http://www.unccd.int).
4. Bitan, A., and Rubin, S. (1991) *Climatic Atlas of Israel for Physical and Environmental Planning and Design,* Ramot Publishing Co., Tel-Aviv.
5. Etzion, Y., Portnov, B.A., Erell E, Pearlmutter D. and I. Meir, (2001) An Open GIS Framework for Recording and Analyzing Post-occupancy Changes in Residential Buildings – A Climate-related Case Study,' *Building and Environment, 36, 10,* 1075-90.
6. FAO/AGL (2000) Terrastat – *Human-induced Land Degradation due to Agricultural Activities* http://www.fao.org/waicent/faoinfo/agricult/agl/agll/terrastat/default.asp.
7. Fialkoff, C., (1992) Israel's Housing Policy during a Period of Massive Immigration, in Golani, Y., Eldor, S., and Garon, M. (eds), *Planning and Housing in Israel in the Wake of Rapid Changes,* Ministries of the Interior and of Construction and Housing, Jerusalem, 169-177.
8. Gradus, Y. (1999) Planning in desert environment: three cases of responsive planning, in Portnov, B.A., and Hare, A.P. (eds.) *Desert regions: population, migration, and environment.* Springer, Heidelberg.
9. ICBS (1998) *1995 Agricultural Survey.* Israel Central Bureau of Statistics, Jerusalem.

---

[6] Although we are strongly convinced that the agricultural path *may not* become a viable option for *large-scale development* in the Negev, small pockets of intensive greenhouse-based agriculture, such as those proven efficient in the Arava region (see p. 12), may be economically feasible. Moreover, they will also provide an additional source of employment for the local urban population.

10. ICBS (1996-2001) *Statistical Abstracts of Israel No. 47- 52 (annual)*. Israel Central Bureau of Statistics & Hemed Press, Jerusalem.
11. Kressel, G.M., Ben-David, J., and Abu-Rabi'a, K. (1991) Changes in land usage by the Negev Bedouin since the Mid- 19th Century. *Nomadic Peoples 28*, 28-55.
12. Lipkis-Beck, G. (1997) Sitting on a gold mine. *Jerusalem Post,* January 22, 1997.
13. Lipshitz, G. (1997) Immigrants from the Former Soviet Union in the Israeli Housing Market: Spatial Aspects of Supply and Demand', *Urban Studies, 34, 3,* 471-488.
14. MFA (2001) Israel's Chronic Water Problem. Jerusalem, Israel Ministry of Foreign Affairs (http://www.israel-mfa.gov).
15. Middleton, N., and Thomas, D. (1997) *World Atlas of Desertification.* Arnold, London.
16. NCRD (1998) *Statistical Yearbook of the Negev.* Negev Center for Regional Development & Negev Development Authority, Be'er Sheva.
17. Pervolotsky, A. (1995) Conservation, reclamation and grazing in the Northern Negev: Contradictory or complimentary concepts. *Network Paper 38a,* Pastoral Development Network. London, ODI. Pervolotsky, A., and Zeligman, N. (1993) The myth of overgrazing in the Mediterranean scrub landscapes. *Ecology and Environment 1,* 9-17 (in Hebrew).
18. Porat, C. *From wasteland to inhabited land.* Yad Izhak Ben-Zvi Press, Jerusalem (in Hebrew).
19. Portnov, B.A. (1992) *Rational Use of Urban Land in Reconstruction Areas,* Stroyizdat, Krasnoyarsk (in Russian).
20. Portnov, B.A., and Erell, E. (1998b) Long-term Development Peculiarities of Peripheral Desert Settlements: The Case of Israel', *International Journal of Urban and Regional Research, 22, 2,* 216-32.
21. Portnov, B.A., and Erell, E. (2001) *Urban Clustering: The Benefits and Drawbacks of Location*, Ashgate, Aldershot.
22. Pressman, N. (1988) Developing climate-responsive winter cities, *Energy and Building, 11,* 11-22.
23. TAMA-2020 (2000) *State Master Plan #35*. Jerusalem. Ministry of Interior, Planning Department.
24. Zeligman, N., and Pervolotsky, A. (1994) Has intensive grazing by domestic livestock degraded Mediterranean Basin rangelands? in Arianoutsou, M., and Groves, R.H. (eds.) *Plant-Animal Interactions in the Mediterranean-Type Ecosystems*. Kluwer Academic Publishers, Dordrecht.

Chapter 9

The Prospects of the Impact of Desertification on Turkey, Lebanon, Syria and Iraq

K. Haktanir, A. Karaca, S. M. Omar
Department of Soil Science, Faculty of Agriculture,
University of Ankara   06110 Ankara, Turkey

The prospective study area is 1.413.232 sq km and geographic coordinates are 30-42 latitude, 26-48 longitude, and includes Turkey, Lebanon, Syria and, Iraq (Mesopotamia). The relevant area is in arid, semi-arid and sub-humid areas (Black sea region).

Desertification is a global problem and most of Turkey, Lebanon, Syria and Iraq are affected by it. Climatic limitations, extensive clearance of natural vegetation, excessive grazing, agricultural expansion into the natural habitats, unsuitable irrigation techniques and soil salinisation have contributed to land degradation, reduced water supplies and limited agricultural production in these areas. It is estimated that nearly 61 per cent of agricultural land in Iraq is threatened by salinity problems.

Soil erosion is affected by climatic properties such as unequal and limited rainfall, temperature and wind. Most of the region is expected to continue as very hot deserts under climate change scenarios. Climate change may put stresses on those areas. With increased evaporation, the problem may become even worse under global warming. The Mesopotamia Basin would suffer drastic desertification since the area only receives between 150-300 mm of rainfall annually but experiences 1500-2500 mm of evaporation per year. The region also experiences extreme temperatures and windstorms. Erosion has a major influence on desertification. About 73 per cent of cultivated land and 68 per cent of prime agricultural land is prone to erosion in Turkey.

Population growth and agricultural mismanagement have also exacerbated desertification problems in those areas. Population growth will have increased 50 per cent by 2025 in Iraq, Syria and Turkey (Tigris and Euphrates Basin).

Deforestation, loss of biodiversity, environmental and socio-economic factors and human activities are the main factors triggering desertification processes on vulnerable land in those areas. Forests cover about 26 % of the landmass in Turkey, 8% in Lebanon, 3% in Syria and 0% in Iraq, but over half the forest area in Turkey is degraded and unproductive.

1. Desertification and vulnerability

Good arable land is lost by desertification through wind and water erosion, overgrazing, unsustainable farming practices and urbanisation. The world loses over 6 million hectares of its land per year by desertification. This land differs from natural deserts because its results from human activity, and they can be controlled if the causes are recognized early enough. As only 70 sq. meters is the minimum area of arable land required to feed a person, even a medium-rate increase in population growth will give

A. Marquina (ed.), Environmental Challenges in the Mediterranean 2000-2050, 139–154.
© 2004 *Kluwer Academic Publishers. Printed in the Netherlands.*

rise to land scarcity for about four billion people by the year 2050. Land degradation has been a dominant problem throughout the past decade. Most land is either desertified or vulnerable to desertification.

Desertification in Turkey is mainly caused by improper land-use, excessive grazing, fuel wood and plant collection. Turkey has 77 million ha of surface area, and of this 20 million ha and 31 million ha are located in arid and semi-arid climatic regions respectively. In addition, more than 75% of the land is prone to different levels of erosion. In Turkey some 109.124 km^2 are desert, and some 374.441 km^2 are in danger of desertification and deshabitation, the latter due to illnesses of the respiratory system as well as unemployment because the agricultural land has become useless. There are moderate effects as a result of improper farming and natural causes such as wind erosion and flooding. About 73 % of cultivated land and 68 % of prime agricultural land is prone to erosion in Turkey. Taking into account the huge areas affected by soil erosion, more preventive measures and additional financial resources are needed [17]. One of the most important issues concerning the management of natural resources in Turkey is the lack of up to date and scientific data on natural resources.

The different ecosystems in Lebanon are threatened mainly by deforestation, over-grazing, urban development, road development, bad agricultural techniques, excessive use of chemical products, over-hunting, and industrial development [24]. Those countries affected by drought and desertification were asked to sign and ratify the Convention to Combat Desertification (CCD).

Desertification is a major constraint on development in Syria, particularly over most of the Syrian Badia Rangelands (Steppe zone) and the marginal zone which is located between the Badia zone and the agricultural lands. Both zones, which basically have fragile ecosystems, are suffering from mismanagement and misuse of their natural resources. In addition, there has been a severe change in the ecological system since the end of the 1950s. Besides this other factors are still playing a major role in the deterioration of the environment overall in both zones such as drought which has been affecting the area for many years [16].

Table 1. The arid regions and their assessments for the study region

P/PET	Region	Assessment
0.05-0.20	Arid	Open to desertification (not in Turkey, but in Iraq and Syria)
0.20-0.50	Semi-arid	Open to desertification (Turkey- Konya Plain and Iğdır sub-region, Iraq, Syria, Lebanon)
0.50-0.65	Dry sub-humid	Open to desertification (Turkey-South-eastern and Central Anatolia regions, Lebanon, Syria)
0.65-0.80	Semi-humid	Open to desertification (Turkey- in west and around dry sub-humid areas)
0.80-1.00	Semi-humid	May be vulnerable to desertification
1.00-2.00	Humid	Non risk for desertification (Turkey- Black Sea Region)
> 2.00	Very humid (wet)	Non risk for desertification (Turkey- at Rize and Hopa district)

P/PET: Aridity Index; P, precipitation (mm); PET, evapotranspiration (mm).
Source: UNEP 1993 [25]

In Iraq, areas subject to desertification are estimated to exceed 92% of the total surface area. Since 1981 this percentage has increased, especially since military operations have damaged both soil and plants and had other negative consequences detrimental to the environment [6]. Desertification in Iraq can be classified in various categories according to the intensity and the areas affected by it. The problem of desertification is the most important problem besetting Iraq. It has had several consequences: loss of productive lands and their conversion into barren lands; an increase in the surface of sand dunes with their mounting negative effects; diminishing forms of biota; an increase in air pollution and sand movement; and increasing pressure on groundwater.

The vulnerability to desertification in the study area is given in Table 1.

2. Profile of the Study Area

The prospective study area is located between Asia, Europe and Africa. It has a total surface area of 1.413.232 sq km and is bound by the Black Sea in the north, by the Aegean Sea and the Mediterranean Sea in the west, by the Persian (Basra) Gulf in the south and by Iran in the east. Geographic coordinates are 30-42 latitude, 26-48 longitudes. It includes Turkey, Lebanon, Syria, and Iraq.

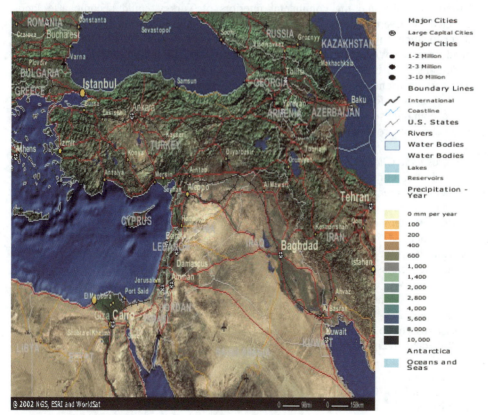

3. Main Desertification Factors

3.1. CLIMATE

Desertification is related to climate in many ways. Climate is a variable phenomenon in that climatic conditions may change in response to external forces and because of its own natural variability. It should be expected that, as in the climatic changes in the past, regional and temporal differences will occur in the future.

3.1. 1. Temperature Change

Climatic factors that may lead to desertification in Turkey were investigated by analysis of the spatial and temporal variations of the precipitation and aridity index series for the period 1930-1993 [21]. [21] Annual and winter precipitation totals have decreased at many stations. Severe and widespread dry conditions occurred, especially in 1973, 1977, 1984, 1989, and 1990. The decrease of winter precipitation may have resulted in a degradation of the soil moisture content and a depletion of the ground water level. There has also been a general tendency for a shift from the 1960s to the dry sub-humid climatic conditions of the early 1990s in the aridity index series of Turkey. South-eastern Anatolia and the continental interiors of Turkey could become arid lands, affected by desertification processes, owing to climatic factors that may lead to their desertification. Significant trends towards a drier than normal climate in term of annual conditions and winter precipitation, and towards dry sub-humid or semi-arid climatic conditions have been the increasing climatic factors leading to desertification in the Mediterranean and Aegean regions of Turkey. When other natural and anthropogenic factors, such as forest fires and the recent misuse of agricultural lands, are also taken into account, these regions could be considered as areas that may be more vulnerable to desertification processes in the future.

3.1.2. Climate Change Scenarios

Scenarios of climate change were developed in order to estimate their effects on temperature, precipitation, runoff, crop yield, and water stress in Turkey [22]. The climate change scenarios in Turkey for the 2080's are the following;
  Projected temperature change
  * About 3-4 °C increase in the annual average temperature in Turkey from the present day (average for the period 1961-1990) to the 2080s, resulting from the unmitigated emissions scenario.
  * About 2-3 °C increase in the annual average temperature by the 2080s, resulting from the emissions scenario that stabilizes CO_2 concentration at 750 ppm.
  * About 2-3 °C increase in the annual average temperature by the 2080s, with emission scenario leading to the stabilization of CO_2 concentration at 550 ppm.
  Projected precipitation change
  * About 0 to -1 mm/day change in the annual average precipitation by the 2080s, resulting from the unmitigated emissions scenario;
  * About 0 to -0.5 mm/day change in the annual average precipitation by the 2080s, with two stabilization scenarios of CO_2 concentration at 750 ppm and at 550 ppm.
  Projected vegetation biomass change

* No considerable change in the vegetation biomass (kg/m^2) by the 2080s in response to climate change, due to the unmitigated emissions scenario and the two-stabilization scenarios of CO_2 concentration at 550 ppm.

Projected % change in annual runoff by major river basins relative to 1961-1990

* About 20 to 50% decrease in annual runoff by the 2080s under the unmitigated emissions scenario;

* About 5 to 25% decrease in the annual runoff by the 2080s under an emissions scenario leading to the stabilization of CO_2 concentration at 750 ppm;

* About 5 to 25% decrease in the annual runoff by the 2080s under an emissions scenario leading to the stabilization of CO_2 concentration at 550 ppm;

Projected % change in crop yield

* About 0 to 2.5% change (a decrease) in the crop yield with the unmitigated emissions scenario;

* About 0 to 2.5% change (a decrease) in the crop yield, with the emissions scenarios that lead to stabilization of CO_2 at 750 ppm and at 550 ppm.

It is also estimated that the annual average temperature in Turkey increased by approximately 1-3 °C during the last century [11]

There is no available data about Lebanon, Syria or Iraq.

3.1.3. Greenhouse Gas Emissions

The degradation of vegetation cover decreases the carbon sequestration capacity of dry land, thus increasing the emission of carbon dioxide into the atmosphere. But the carbon storage capacity of dry lands is poorly documented. The predicted global climate warming resulting from the build-up of greenhouse gases in the atmosphere is expected to have profound impacts on global biodiversity at levels that may compromise the sustainability of human development on the planet. Climate warming will cause higher evaporation rates and lower rainfall, both of which are major determinants of dry land ecological processes. Simulation models of climate change predict shifts in species distribution and reduced productivity in dry lands.

Within the framework of national climate change studies and activities in Turkey, The State Institute of Statistics (SIS) calculates greenhouse gas emissions. Consumption figures for the 1990-2000 period and projections for the 2000-2020 period show that, as is the case today, there will be a large increase in the amounts of greenhouse gases caused by fuel consumption in the future. CO_2 emissions are by far the highest compared to other greenhouse gas emissions. When the sectoral distribution of greenhouse gas emissions caused by fuel consumption is compared with present and projected consumption levels, it is observed that, while the ratios for some sectors are on the rise, there is an evident decrease in others. In 2000, 34 % of CO_2 emissions were caused by electricity generation, 32 % by industry, 17 % by transportation, and 16 % by other sectors. However by the year 2020, it is estimated that 41 % of CO_2 emissions will be caused by electricity generation, 33 % by industry, 13 % by transportation, and 13 % by other sectors. When compared with other countries of the world in terms of basic CO_2 indicators, it is seen that Turkey is ranked 23rd in total CO_2 emissions, 75th in CO_2 emissions per capita, 60th in the ratio of CO_2 emissions to the Gross Domestic Product (GDP), and 55th in the ratio of CO_2 emissions to the GDP, measured on the basis of purchasing power parity [19].

Studies based on 1994 data indicate that most of the air pollution in Lebanon originates from the transport and energy sector. Lebanon's per capita CO_2 emissions are 4.55 tons, which is 3 times more than the average for India. The CO_2 emission of vehicles was calculated to be 1.030.275 t/year, and CO_2 emissions from power plants to be 1 225 750 t/year in 1993/4 [15].

The amounts of carbon dioxide emissions for Syria and Iraq are 3,3, and 4,1 metric tons per capita, respectively.

3.1.4. Sand and Dust Storm

Arid lands are significant contributors of dust. The phenomenon of sand dunes is considered to be one of the most dangerous consequences of desertification, due to its negative impact on every vital aspect of life.

Dust storms are an unpleasant feature of the central plains region in Iraq and Syria. The main manifestations are an increase in sand and dust storms, increased soil salinity and waterlogging, and widespread rangeland degradation. Sand dunes are located mostly in the central and southern regions and are shifted by wind force. Undoubtedly, sand and dust storms over the central and southern regions pollute the environment and affect human health and agricultural production. Sand and dust storms disrupt the physiological functions of plants, especially during pollination and inflorescence. Sand storms blow from the dune fields in central and southern regions. Their incidence has increased during recent years [20]. The problems have become worse since the imposition of economic sanctions in 1990. Poor soil/water management and severe climatological factors have changed extensive agricultural lands in Iraq's alluvial plain into the present bare, water logged soils covered with aeolian sand sheets and pseudo sand dunes. Iraq faces a severe desertification problem that jeopardizes its food security through the effects of soil salinity, water logging, loss of vegetative cover, shifting sand dunes and severe sand/dust storms. All of these problems need to be addressed to halt the threat. To combat these problems Iraq has launched programmes to rectify soil salinity, to develop natural vegetative cover and to halt the encroachment of sand dunes, as well as reduce the frequency and severity of sand and dust storms.

Dust, dust storms, sand accumulation on roads, railroads, formation of sand sheets, sand hummock, and sand dunes are the main environmental consequences resulting from the introduction and expansion of rain fed agriculture in the Syrian steppe. Dust frequency and intensity have increased remarkably during the last few years in the eastern part of the country. An equation based on visibility during dust storms has been used to calculate the amount of soil loss by a single dust storm in summer 1987. The result was about 570 million-tons of soil.

The frequency and amount of sand and dust storms in Turkey and Lebanon is less than in Iraq and Syria.

3.2. LAND DEGRADATION

Land degradation has been a dominant problem throughout the past decade. Most land is either desertified or vulnerable to desertification.

In Turkey, agricultural lands that are classified 1st and 2nd grade in terms of soil quality have been opened up as new areas for urban development and taken over in order to supply the non-agricultural sectors with the required raw materials. This has caused the conversion of close to 5 million ha of agricultural land to non-agricultural land-uses. Close to 172.000 ha of high quality land, classified as 1-4 grades in terms of soil quality, are used for settlements and industrial areas. Over the past 20 years, the occupation of agricultural lands by settlements and commercial and industrial facilities has considerably accelerated. This situation has reduced the productivity of the agricultural sector while, at the same time, increasing the likelihood of floods and flash floods. Furthermore, overuse of the natural pasture systems for grazing has had a great effect on the amount and species of the natural grass vegetation. These are the most powerful factors of land degradation related to land degradation in Turkey [19].

The percentage of desertified land ranges from 10 in Syria to nearly 100 in the United Arab Emirates. In Iraq and Syria, desertification has affected wide areas of rangelands. It is estimated that the cost of soil degradation in Syria is equivalent to about 12 % of the value of the country's agricultural output or about 2.5 % of the total GNP. In Lebanon degradation is serious on steppe mountainous land.

The main causes of land degradation in the study area are soil erosion, alkalinisation/alkalinisation, urbanisation, and soil sealing.

3.2.1. Soil Erosion
The topographical structure of Turkey is generally sloping, high land, and mountains. In Turkey nearly 63 % of the land has slopes steeper than 15 % on average, even in the coastal areas. The percentage of the land with steep slopes is very high and 48.0 % of the land has more than 20 % slopes. Step slopes are associated with soil erosion which is one of the major problems in Turkey. About 73 % of cultivated land and 68 % of prime agricultural land are prone to erosion in Turkey [18].

In areas with moderate or high rates of precipitation erosion, caused largely by water soil erosion, is the biggest problem for Turkey. Soil erosion distribution of the Turkish soils is given in Table 2. It can be seen from Table 2 that 58,7 % of land is exposed to severe and very severe soil erosion. In this land soil fertility is reduced by about 50 % and they cannot be used economically [10].

The main causes of soil erosion in Lebanon are deforestation, overgrazing, and the deterioration of the agricultural terraces. Soil erosion affects approximately 25% of Lebanon's total surface area. With the increasing move of the rural population to the urban centers, many of the terraces that once covered the Lebanese mountains are collapsing. Recently, grazing is being curtailed with an expected positive effect on the forest ecosystem.

Mountainous regions with a humid and sub-humid climate in Syria are, probably, the regions most susceptible to water erosion due to their natural conditions such as steep and long slopes, shallow soil cover, high rainfall average (800 – 1 500 mm), and frequent rain storms. Semi-dense forests cover most of the area. About 8,000 ha have been subject to fire during the period 1985-1993.

It has been estimated that more than 50% of the soils of Syria are extremely susceptible to erosion. Another human activity leading to water erosion is farming. According to the last official report about 2,440 ha of natural forest were converted to agricultural lands during the period 1985-1992 [16]. Iraqi lands are affected by water and wind erosion (Table 2). Wind speed averages 8 km/h. Wind erosion is evidenced by the presence of widespread sand sheets in the arid interior.

Table 2. Soil erosion distribution of the Turkish and Iraqi Soils

		Turkish soil		Iraqi soil		
Type	Intensity	Area affected (ha)	%	Intensity	Area affected (ha)	%
Wind erosion	total	506 309	0.65	Light-medium	1 431 000	3.37
				Very strong	635 000	1.49
Water erosion	none-slight	10 778 519	13.8	Light-medium	4 691 000	11.06
	Moderate-severe	61 294 146	78.7	Strong-very strong	5 007 000	11.86

Sources: General Management Planning of Turkey 1987[10]; El Faragi 1994[6]

Water erosion of a geological nature is obvious, particularly in the bare areas around the Euphrates Valley, and is entirely natural even in the arid interior which is crisscrossed by large Wadi systems and other seasonal drainage lines. Large areas affected by accelerated water erosion can be observed in the south and southwest of Jebel Bishri. Gully patterns that extend over the farm blocks indicate that after the farms were abandoned, accelerated erosion took place.

3.2.2. Alkalinisation/Salinisation
Salt affected soils in Turkey are especially located in the Central South, Central North and Mediterranean regions. The salt content of the Turkish soils is less than 0.15 % in 95.51 % of the soils. The rest of the soils (4.49 %) are to some extent exposed to the salt problem. In the Central Anatolia region, the climate and topographic properties of the soils, and parent material, aid formation of salt which has affected the soils. The low amount of precipitation and the dry and very high temperatures during the summer seasons affect the soils. Bad management practices are the major causes of salinisation in other regions [5].

Desertification processes due to salinisation in the GAP (Southeastern Anatolia Project) have become a very serious problem. Irrigation started in the Harran plain in 1996. The Harran Plain is the biggest plain of GAP in Turkey and has 152,000 ha of irrigable land. Due to excess irrigation with minor control, salinisation problems are becoming prevalent in the plain. Furthermore, soils have heavy texture in the tillage layer, and approximately 85% of irrigable land belongs to cotton. The study investigated an increasing trend of salinisation over 13 years. It was noticed that the increase was almost double after the commencement of surface irrigation on the plain. Results showed that the total salinized area was 5,550 ha in 1987, 7,498 ha in 1997, and 11,403 ha in 2000, respectively [2].

Salinity and waterlogging are extensively present in irrigated agriculture in Syria. It is estimated that 532,000 ha or about 40 % of the present total irrigated areas are salt-affected soil to varying degrees. At present 60.000 ha of previously fertile soils have been excluded from production and 100.000 ha have only 50 % of their potential production. The gypsic lands are estimated to cover 20 % of the total country area. The random pumping of groundwater for irrigation in marginal land has resulted in salinity increment that has led to land salinisation. The dominance of gypsiferous soils along the water resources of the Euphrates, Balikh, and Khabour rivers causes pollution and salinisation of fresh waters [8].

In Iraq, the soil has a high saline content. It is estimated that nearly 61 % of agricultural land in Iraq is threatened by salinity problems. The salinity rate is 8 % in Iraq. It means that the all-Iraqi soils will be saline after 12 years if a drainage system is not applied. As the water table rises through flooding or through irrigation, salt rises into the topsoil, rendering agricultural land sterile. Drainage thus becomes very important. However Iraq's terrain is very flat; Baghdad for example, although 550 kilometers from the Persian (Basra) Gulf, is only 34 meters above sea level. Salinisation of agricultural lands has become acute in Iraq due to an irrational use of water, as well as poor irrigation and drainage networks. This has led to a rise in ground water and salt accumulation in soils. Thus, the plains area, renowned for the fertility of its lands, has turned into salinized land. Productivity in large areas of the nation has fallen to nearly zero. It is estimated that about 100 thousand dunums (250,000 thousand square meters) per year suffer from salinisation [6].

[4]13%, 7%, 17%, and 71% of the irrigated lands are degraded in Turkey, Lebanon, Syria and Iraq, respectively. Salinisation is the main degradation factor in these areas, and the rate of desertification is very serious. The estimated degradation status of the irrigated land in Turkey, Lebanon, Syria and Iraq is presented in Table 3.[4] Many of the figures come from the answers to a questionnaire sent out by the UNEP in 1982. The answers to that questionnaire probably mean very little. Experts even from sophisticated governments say they had great difficulty answering the questions. They had little of the data that they were asked for [9].

Table 3. The estimated degradation status of the irrigated land in Turkey, Lebanon, Syria and Iraq

Country	Total Area 1000 ha	Desertification				1000 ha	% desertified
		Slight	Moderate	Severe	Very Severe	Total (Moderate +)	
Turkey	2 150	1 860	250	30	10	290	13
Lebanon	86	80	6	0	0	6	7
Syria	652	542	70	30	10	110	17
Iraq	1 750	500	750	300	200	1 250	71

Source: Dregne and Chou 1992 [4]

3.2.3. Urbanisation and Soil Sealing

Soil sealing has become a serious problem on fertile agricultural land. Rates of soil loss due to growth in urbanisation and transport infrastructure are high in these Eastern Mediterranean countries. In addition, consumption of soil resources due to urbanisation is becoming a problem in these countries, especially in Turkey due to the tourist industry development in the valuable coastal zone.

Urbanisation in the Mashriq countries has occurred as a result of the slow shift of population from agriculture to industry and services in such well-established urban centers as Damascus and Baghdad. In the Mashriq sub-region, the Lebanon urbanisation growth rate was 8.14 % during 1950-1955 but it decreased dramatically to 1.18 % during 1985-1990, reflecting the effects of the protracted Lebanese war and political instability. Ambitious projects such as the coastal highway, the airport, the reconstruction of the downtown area, and other construction projects needed huge amounts of sand and rocks in Lebanon. Over 700 quarries (only 45% of them licensed) all over the country dug big holes in the once picturesque mountains and diminished the tourist value of the countryside [1]

The same situation can be observed in Turkey. Urban development, road construction, and industrial development do not take into consideration ecological and agricultural values and soil sources. The degraded good quality agricultural soils are 573,239 ha due to uncontrolled urbanisation in Turkey with an increase rate of 333% in the last decades.

In Syria and Iraq, urbanisation kept pace there with a slow but steady economic growth. The urban population in Iraq increased from 35.1 % in 1950 to 74.5 % in 1995. In Syria, the comparable figures are 30.50 % in 1950 and 52.2 % in 1995. Cities consume natural resources from both near and distant sources. In doing so they generate large amounts of waste that are disposed of within and outside the urban areas, causing widespread environmental problems. Estimated municipal waste generation increased in the Mashriq sub-region. These figures are 185 kg per capita per year in Syria and 285 in Iraq.

3.3. ECOSYSTEMS

3.3.1. Forest Ecosystem

Forest degradation can be due to the unregulated, excessive and extensive tree felling in forests. Shifting agriculture, overgrazing of the forestlands, forest fires, diseased and insect ridden trees and poor environmental awareness of the importance of natural resources are leading to the forest degradation.

About 27 % (20,763,248 ha) of the land area of Turkey is officially recognized as forestland. In the forest ecosystem of Turkey, degraded forests and coppice-land make up close to 52 % (10,735,680 ha) of the total forestlands of the country. There are more than 8 million forest villagers living in 17,797 forest villages in Turkey. Studies show that between the years of 1937-1995, 20,000 ha of forestland (close to 1 % of total forest) have been cleared and converted into farmland and 27,000 ha of forestland have been converted into settlement areas [19]. However, according to the report of the FAO in 1998 [7], the value of the total forest area in Turkey has decreased by 13,3% of the total land area (Table 4).

The Ministry of Forestry of Turkey aims to protect an area of 1,865,000 ha against soil erosion by 2003. In addition, an area of 285,000 ha is to be reclaimed by pasture plantation. 8,5 million ha of the irrigable areas is to be improved by the General Directorate of State Hydraulic Work and the General Directorate of Rural Affairs of Turkey. In total, an area of 150,000 ha should be expropriated every year. The total amount of irrigated and non-irrigated areas which should be left for natural forest development [3] is approximately 5,115,000 ha.

At present the wooded areas in Lebanon cover some 60,000 hectares, approximately 5.7 % of the area of Lebanon. This percentage is dangerously low, for it is recommended that a country's forest area should be close to 20 %. The current reforestation rate in Lebanon (5 to 7%) is definitely insufficient for a country in which mountains cover 73% of the territory. Most of Lebanon has been deforested. The great cedar forests have largely disappeared except in the higher mountains [24].

Syria has large areas of semi-arid and desert environments, and natural forest covers only about 2% of the landscape. Syria lost around 10% of its forest cover between 1990 and 1995 due to desertification, overgrazing, poor agricultural practices, and uncontrolled tree felling.

Forest resources and management of Turkey, Lebanon, Syria, and Iraq are given in Table 4.

Table 4. Forest resources and management of Turkey, Lebanon, Syria and Iraq

Country/ Area	Land area	Forest area 2000					Area change		Volume and above-ground		Forest under management	
		Natural forest	Forest plantation	Total forest			1990-2000 (total forest)		Biomass (total forest)		Plan	
	000 ha	000 ha	000 ha	000 ha	%	Ha/ Capita	000 ha/year	%	m³/ha	T/ha	000 ha	%
Turkey	76963	8371	1854	10225	13,3	0,2	22	0,2	136	74	9954	97
Lebanon	1024	34	2	36	3,5	n.s.	n.s.	0,4	23	22	-	-
Syria	18377	232	229	461	2,5	n.s.	n.s.	n.s.	29	28	-	-
Iraq	43737	789	10	799	1,8	n.s.	n.s.	n.s.	29	28	-	-

Source: FAO 1988[7]

Forest covers about 1.8 % of Iraq's surface area. According to the FAO reports, the data indicated that national forest is about 789,000 ha and forest plantation is 10,000 ha. Forests in Iraq used to cover all mountain areas in the north and the northeast. In 1970, forests covered 1.851 million hectares but in 1978 they only covered 1.5 million hectares. What can be seen nowadays are only odd forests of oak trees in the most remote areas. Forest degradation, resulting from excessive woodcutting, overgrazing and fires, has led to deteriorating plant cover in the forests of the north of Iraq. This has in turn increased water erosion causing the fertile layer of the soil to disappear. Furthermore, it has had a negative impact on the stocking capacity of dams and the operational efficiency of irrigation works, and it increases maintenance costs [6].

3.3.2. Steppe Ecosystem
As a result of the accelerated destruction of forests, steppe flora has gradually become dominant in Anatolia. During the last 50 years, on the other hand, due to human-induced problems, such as clearing land for agricultural activities, erroneous irrigation methods, and inappropriate land-use, a large section of steppe areas has also been irreversibly destroyed, and the remainder has become degraded with overgrazing.

The meadows are an important component of the steppe ecosystem and constitute 28 % (21,745,000 ha) of the land area of Turkey. This figure was 44,300,000 ha in 1935 and 37,000,000 ha in 1950. Meadows have been destroyed by policies that allowed these lands to be converted into farmland in order to meet the food demand of a growing population. Today the total area covered by steppe ecosystems, which includes meadows and marginal lands not suitable for agriculture, is 28,000,000 ha. The reasons for the destruction of steppe lands and their ecosystems in Anatolia can be listed as follows; high population growth over the last 50 years, overgrazing in the absence of meadow management, conversion of meadows into agricultural lands, erroneous agricultural practices, unregulated hunting, stubble burning, pollution, increased soil erosion, highway and dump construction, excessive gathering of plants of high economic value using unsustainable methods, and improper mining activities [19].

There are no natural pastures in Lebanon except for some farms where a small area is used as pasture. From this we can conclude that, for Lebanon, the default value of 77% for cattle in pasture range and paddock system is high and that the default values 3 and zero presented in solid storage and dry lot system for dairy and non-dairy cattle respectively are very low.

On the overgrazed and denuded valley slopes, soil resources are quite limited in Syria. Most of the topsoil has been eroded and natural vegetation survives only in small niches between stones and rocks. Large pockets of soil, though, lie preserved under the stones. While the current land use results in low productivity on the slopes, potential exists for fruit-tree and shrub plantations with appropriate soil- and moisture-conservation measures and nutrient management.

Degradation of pastures, which constitute about 70-75 % of Iraq's total area, results from the following factors: early overgrazing without taking into account the capacity of pastures; cutting and uprooting of forage plants for fuel purposes; exploitation of pasture land for agricultural purposes by cultivating land where average rainfall is below 200 millimeters/year; unregulated distribution of water resources causing cattle concentration in areas where water is available, with a negative impact on plant cover [6].

3.4. WATER USE

Turkey has a national action program under the FAO's International Action Program for Water and Sustainable Agriculture Development. It mainly covers topics such as water resource policies, improving water use efficiency, environmental protection, and soil and water conservation while introducing bottom-up planning and farm development.

Turkey has an average annual renewable water potential of 205 billion cubic meters, or about 3,150 cubic meters/person per year, which is far below the 10,000 cubic meter parameter needed to classify a country as water rich. Taking into consideration the economically usable water potential of the country (110 billion cubic meters), the available annual per capita water is reduced to about 1,700 cubic meter, which would make Turkey a water-stressed country. Furthermore, rapid population growth, industrialization, and rising standards of living will decrease the annual per capita renewable water potential to 2,500 cubic meters by 2000 and to 2,000 cubic meters by 2010. If we estimate the economically usable per capita annual water potential, we can project a severe situation in which available water is reduced to 1,580 cubic meters/person per year, or even less, by 2000. As can be seen from this data, Turkey's water resources are far from abundant. Average annual precipitation is 643mm in Turkey. In the middle, eastern, and southeastern parts of Anatolia, a very large portion of Turkey, the weather is usually drier than in the other regions and annual precipitation is 250mm; because of aridity this area is vulnerable to desertification [18].

With its topography and physiography, along with its edaphic conditions, Lebanon constitutes a good water reservoir. However, the degradation of the soil and vegetation cover causes most of the rainfall to be lost through surface runoff. Despite this fact, Lebanon is one of the richest countries in the area in water. Most water sources used in Lebanon are springs and wells. In addition, catchments wells and reservoirs used for storing rain represent major supplementary sources of usable water, especially in the dry season with the increased water demands. The total amount of available water is 3,992 million cubic meters in Lebanon. The volume of potentially usable groundwater is 600 million cubic meters/year but only 160 million cubic meters of this is used. In 1966, the domestic and industrial sectors consumed 94 million cubic meters of water, and the agricultural sector consumed 400 million cubic meters. By the mid-1990s, Lebanon was estimated to consume at least 890 million cubic meters/year of water, close to 50% of which was drawn from aquifers. Daily domestic water consumption was estimated at 165 L per capita in the mid-1990s. This figure is expected to reach 215 L by 2000 and 260 L by 2015 [12]. Of all the arable land in Lebanon, 146,000 ha was rain fed in 1996. The irrigated area was 23,000 ha in 1956 (10% of the then cultivated area), 54,000 ha in 1966, 48,000 ha in the early 1970s, and 87,500 ha by 1993. According to studies conducted by the Food and Agriculture Organization of the nations and by the United Nations Development Programme, the irrigated area of Lebanon is expected to rise to 170,000 ha by 2015 [1].

Syria has 25.79 cubic km renewable fresh water potential per year and the available fresh water amount per capita will be decreased from 2,089 cubic meter in 1990 to 546 cubic meters in 2050 [16].

Iraq has more water than most Middle Eastern countries, which led to the establishment of one of the world's earliest irrigation systems. The level of ground water can rise while the soil surface settles, due to unregulated irrigation, to low-lying ground, or to a shortage of drainage networks. More than half of the irrigated agricultural lands in Iraq are estimated to be affected either by salinisation or waterlogging. About one-fifth of Iraq's territory consists of farmland. About half of this total cultivated area is in the northeastern plains and mountain valleys, where sufficient rain falls to sustain agriculture. The remainder of the cultivated land is in the valleys of the Euphrates and Tigris rivers, which receive scant rainfall and rely instead on water from the rivers. Both rivers are fed by snow pack and rainfall in eastern Turkey and in northwest Iran. The flow of the rivers varies considerably every year. Geographic factors have contributed to Iraq's water problems [6].

Table 5. Population versus per capita water availability

Country	Total Renewable fresh water per year km^3	1955 Population Thousands	1955 Availability m^3	1990 Population Thousands	1990 Availability m^3	2025[a] Population Thousands	2025[a] Availability m^3	2050[a] Population Thousands	2050[a] Availability m^3
Lebanon	4,98	1613	3084	2555	1949	4424	1126	5189	960
Syria	25,79	3967	6501	12348	2089	33505	770	47212	546
Turkey	203,00	23859	8508	56098	3619	90937	2232	106284	1910
Iraq	n.s.	8000	n.s.	15000	n.s.	34000	n.s.	77000	n.s.

[a] *UN medium projection*
Source: UN Population Division 1994 [23]

The water resource stress of Turkey is due to climate change. By the year 2080, Turkey and the Middle East region will be among the most stressed areas of the world with an increase in water stress, with unmitigated emissions scenario and the two emissions stabilization scenarios of CO_2 concentration at 750 ppm and at 550 ppm.

In 1955 fresh water availability cubic meter / inhabitant in Lebanon, Syria, and Turkey were 3,084 6,501 and 8,508 cubic meters per capita, respectively. These values were changed to 1,949, 2,089 and 3,619 cubic meters in 1990, and the estimated values for 2025 and 2050 are 1,126,770 and 2,232 cubic meters and 960, 546 and 1,910 cubic meters, respectively [23]. Population versus per capita water availability is given in Table 5.

3.5. BIODIVERSITY

Loss of biodiversity (genetic erosion) is occurring at a rapid rate in many areas of the world. The physical processes of land degradation, biodiversity evolution or extinction, and climate change are intimately inter-twined, especially in dryland [14]. Land degradation reduces natural vegetation cover, and affects productivity of crops, livestock and wildlife. The protection of biodiversity safeguards the potential economic benefits we may derive from species in the future. Besides the benefits for today's population, the protection of biodiversity for future generations is an ethical responsibility.

In terms of biodiversity, Turkey is one of the rich countries of Europe and the Middle East. Turkey contains 5% of the plant species found in Europe. Studies indicate that there are 163 plant families covering 1,225 types, which in turn cover about 9,000 species. These grow naturally and about one third are endemic. Research on seedless plant groups however is still extremely inadequate. Due to Turkey's geographical position as a natural bridge connecting Asia, Europe and Africa, Turkey's fauna exhibits an immense richness. Among the small mammals, however, whose territories are quite limited, endemic species of vertebrates may be encountered in the lakes and marshy areas of closed basins. The mammals found in Turkey consist of 120 species which are members of 31 families attached to eight orders and are presented by several subspecies regarded as endemic. Problems regarding fauna are pollution, degeneration of habitat, the introduction of exotic breeds and wild species, and hunting [17].

Lebanon has over 1,000 plants with medicinal properties and more uses for plants may be discovered. Generally, the areas richest in biodiversity are the middle elevations of the western slopes of the Lebanon Mountains. Many changes have been observed in the last 30 years in the marine ecosystem. New species, some of them originating in the Red Sea, have become prominent, while previously abundant species have become rare. As everywhere, loss of habitat due to human activities is the greatest threat to biodiversity in Lebanon. In addition, unsustainable practices such as fishing with dynamite or poison and over-fishing have greatly exhausted fish stocks. Excessive hunting has reduced the number of local and migratory bird species and has placed many of the remaining large mammals on the brink of extinction. 19 mammal species are already extinct. Of the terrestrial flora, 38 species are considered rare or endangered due to flower picking, urban development or overgrazing. The misuse of pesticides is another hazard to biodiversity, although there is no alternative to their use at the present time.

The extensive exploitation of sand for construction has aggravated the problem of seawater intrusion and destroyed the habitats of many coastal and marine biota, including marine turtles, along the Lebanese and Syrian coasts.

Reclamation and infilling of inertial areas and marshes in Iraq is destroying habitats and jeopardizing their biological diversity.

4. Conclusion

It is believed that all the agricultural structure, from government to public, has to be re-organized for a sustainable use of natural resources. A comprehensive new agenda is needed to protect the sustainability of the ecosystem and to prevent land degradation and desertification.

The relevant countries need rational land use planning at a national level with the purpose of the conservation of the high quality lands and landscape. However, the lack of information on land use planning in the four regional countries should be urgently solved. There should be further legislation for preventing desertification and to develop an increased awareness for the participation of the people in combating desertification. Education of farmers, particularly for the implementation of environmental friendly agricultural practices, should be undertaken as specified by the OECD. Thus, with appropriate legislative precautions and economic subsidies, the soil and water resources will most probably be sustainably used in spite of the pressure of an increasing population. However, the negative factors governing the development of desertification seem to be hard to cope with. This eventually leads to the increasing hazards of erosion, salinity, and depletion of nutrients. Thus, the degradation of the agro-ecosystem causes the loss of soil and land in the above-mentioned countries. The recent decrease in the traditionally high wheat production of Turkey is a good example of the drawbacks of increasing land degradation.

According to the data of the Ministry of Agriculture of Turkey, there will be a 2% decrease in the wheat production of Turkey in the next decade. Although corn is a fundamental component of the annual cereal production of Turkey, no trend has been noticed in the increase of its production. Turkey has long been known as a country in which agriculture production plays a significant role in its trade and which is self-sufficient in its demands for agricultural products, and has even been known as an exporting country. [13] Especially for wheat and maize, Turkey is becoming an importing country and not an exporting one, indicating that it will become more dependent on outside countries with respect to the main agricultural products over the next ten years. This means that Turkey will be a net wheat importing country from 2003.

Military activities in all forms, be they bombing, building trenches, or the use of heavy machinery, have damaged the surface layer of soil in all parts of Iraq and especially in its desert areas. Plant cover in the desert, which has grown over the years, has been damaged. This plant cover cannot be restored without launching scientific programmes on a wide scale [6]. The mass migration (ca. 500.000 people) from northern Iraq following the Gulf War in 1991 has been reported by the Government to have created severe degradation of the natural resources of Eastern Turkey.

Turkey, Lebanon, Syria and Iraq are countries which have considerable experience in the development of land reclamation processes (i.e. pasture plantation and reforestation). However, these activities have not been adequately performed due to the lack of socio-economic conditions. The analyses of global climate changes have indicated that in the next fifty years, the factors increasing desertification will be more effective than the efforts to mitigate desertification. Within the framework of the international desertification program, priority should be given to concepts such as cooperation between countries, and providing financial, technical and technological support. Moreover, research focusing on the development of water saving technologies against desertification should be encouraged.

In future, desalination of drainage water, development of irrigation technologies, and the usage of salt-tolerant genetically modified plant species will be extremely important in the attempts to preventing land degradation.

References

1. Amery, H.A. (2000) Assessing Lebanon's Water Balance, in Brooks, D. B., and Mehmet, O. (eds.), *Water Balances in the Eastern Mediterranean*, International Development Research Center, Ottawa, 4-13.
2. Çullu, M.A., Atmaca, A., Sahin, Y., and Aydemir, S. (2002) *Application of GIS monitoring soil salinisation in the Harran plain*, International Conference of Sustainable Land Use and Management, Çanakkale.
3. DPT (2001) *Su havzalari kullanimi ve yönetimi özel ihtisas komisyon raporu. Sekizinci 5 yillik kalkinma plani yayin no DPT-2555-ÖIK*, Ankara.
4. Dregne, H. E., and N.T. Chou (1992) Global desertification and costs, in *Degradation and restoration of arid lands*, Lubbock Texas Tech. University, Texas, 127-163
5. Eyüpoglu, F. (1999) Soil Fertility of Turkish Soil, General Directorate of Rural Services. *Technical Publication T-67*, Ankara.
6. El Faragi, A.F. (1994) *The Iraqi experience in combating desertification and its impacts*, Chief of the Division of Desertification Control, Ministry of Agriculture and Irrigation, Bagdad.
7. FAO. (1998) *Overview and opportunities for the implementation of national forest programmers in the Near East*, Damascus, 6-9 December, Secretariat note, Near East Forestry Commission 13th session.
8. FAO-AGL (2000) *Land and plant nutrition management service. extent and causes of salt-affected soils in participating countries*, http://www.fao.org/ag/agl/agll/spush/topic2.htm
9. Forse, B. (1989) The myth of the marching desert, *New Scientist 4* (1650), 31-32.
10. General Management Planning of Turkey, (1987) *Soil Conservation Main Plan* Ministry of Agriculture, Forestry and Villages, General Directorate of Rural Services, Ankara.
11. Houghton, J.T., Ding Y., Griggs, D.J., Noguer, M., Van der Linden, P.J., Dai, X., Maskell,K., and Johnson, C.A. (2001) *Climate Change 2001: The scientific basis*, Cambridge University Press, Cambridge.
12. Jaber, B. (1997) *Water in Lebanon: problems and solutions*, Public lecture given in the Department of Hydrology, Purdue University, Lafayette, April.
13. Koç, A., Uzunlu, V., and Bayaner, A., (2001) *Projections of Turkey agricultural crop production in 2000-2010*, Project report, General Directorate of Rural Affairs of Turkey, Ankara.
14. Lean, G. (1995) *A simplified guide to the convention to combat desertification, why it is necessary and what is important and different about it*. The Center for Our Common Future, Geneva.
15. Masri, R. (1995) *The human impact on the environment in Lebanon*, Seminar, Massachusetts Institute of Technology, Massachusetts.
16. National Report of Syria (2000) www.unccd.int/cop/reports/asia/national/2000/syria-summary-eng.pdf.
17. National Report of Turkey (1999) Ministry of Environment, Ankara.
18. National Report of Turkey (2000) Ministry of Environment, Ankara.
19. National Report on Sustainable Development (2002) Ministry of Environment, Ankara.
20. Think,G. (1998) *Climate Change. Brochure of the Climate Change Enabling Activity*, Ministry of Environment, Ankara.
21. Türkes, M. (1999) Vulnerability of Turkey to Desertification With Respect to Precipitation and Aridity Conditions, *J. Engineering and Environmental Science 23*, 363-380 (in Turkish).
22. Türkes, M. (2000) *Climate change studies and activities in Turkey, Advanced Seminar on Climatic Change: Effects on Agriculture in the Mediterranean Region*, Mediterranean Agronomic Institute of Zaragoza, 25-29 September.
23. UN Population Division (1994) *World population prospects (Sustaining water, an update)*, The United Nations, New York.
24. UNCCD (2000) *National Report on the Implementation of the UNCCD in Lebanon*, Ministry of Agriculture, Beirut.
25. UNEP (1993) *World Atlas of Desertification*, Edward Arnold, London

Chapter 10

Prospects for desertification impacts in Southern Europe

J. Puigdefabregas & T.Mendizabal[1]
Estación Experimental de Zonas Aridas (EEZA-CSIC)
General Segura 1, 04001 Almeria (Spain)

Consejo Superior de Investigaciones Científicas (CSIC-Central Org.)
Serrano 117, 28006-Madrid (Spain)

Prospective model-based approaches to desertification are hindered by a lack of unifying process specifications underlying the phenomenon, and by the difficulty of forecasting trends in its driving forces. Elementary concepts for a dynamic systems approach to desertification are discussed as a groundwork for the prospective task. Emphasis on the differentiation between 'relict' and 'current' desertification enables successful treatment programs to be planned. The prospective task focuses on the Southern European Union countries (Greece, Italy, France, Spain and Portugal), which share more than 95% of the southern European Mediterranean climate area. The effort to control relict desertification has been evaluated using the evolution of forest-woodland and agricultural areas over a period of time, together with the total reforested area. Current risk of desertification was approached in three steps: (i) analysis of the temporal evolution of a driver-level indicator such as the per capita Net Internal Agricultural Produce (ANIP); (ii) examination of the associated changes in land-use systems indicators, and (iii) exploring the impact of these changes on natural resources. Results show that in Southern EU countries, the older traditional agriculture will continue to be transformed into that of an open market. This process involves the development of site-selective and technically sophisticated irrigated agriculture. The main consequence is a build-up of large water deficits, particularly if climate change scenarios are considered, with severe impact on: (i) water and soil quality and (ii) interregional conflicts triggered by, *affecting large areas of irrigation* water suppliers. Prospects for EU agricultural policies indicate a shift from supporting production to helping integrated development and sustainability. Within this scenario, desertification factors, such as overstocking rangelands and tree crops encroaching in marginal areas, are unlikely to persist while restoration of threats of relict desertification is expected to increase in the near future.

[1] The present study was supported by the Spanish 'Comisión Interministerial de Ciencia y Tecnología' HISPASED Project (REN2000-1507-CO3-01/GLO) and by the EU Commission LADAMER Project (EVK2-CT-2002-00179). The authors are indebted to Mrs Laura Barrios from the 'Centro Técnico de Informática (CSIC) for her support with the statistical design and data computing.

A. Marquina (ed.), Environmental Challenges in the Mediterranean 2000-2050, 155–172.
© 2004 *Kluwer Academic Publishers. Printed in the Netherlands.*

1. Introduction

In a previous contribution of the authors to a NATO sponsored symposium on the Mediterranean [10] it was shown that, during the 20th century, purely climatic factors were rarely responsible for desertification in the Mediterranean region, because droughts were relatively short-lived. Natural and agricultural ecosystems may be affected, but in most cases they recover easily. Socio-economic disturbances, particularly when they occur combined with climatic fluctuations, become the main drivers of desertification in the area. They affect water balances and land degradation through changes in land-use patterns.

Based on this review, the present contribution focuses on a more prospective approach to desertification phenomena in the Southern European Union countries, Greece (62%), Italy (40%), France (16%), Spain (64%) and Portugal (62%). The percentage of their territory sharing Mediterranean climate is indicated in brackets, and altogether includes more than the 95% of the southern European Mediterranean climate area. Prospective study in desertification is a difficult task that faces a dual challenge: that of identifying a unified underlying desertification process and of forecasting trends in driving desertification forces.

Desertification is not a lay perception. It is a technical concept applied by scientists and policy makers after droughts threatened the Sahel in the last quarter of the 20th century. It was defined by the United Nations Convention to Combat Desertification (UNCCD) as 'the degradation of the land in arid, semi-arid and dry-sub-humid areas, as a result of several factors, including climatic change and human activities'. The dominant symptomatic character of this definition does not account for the underlying processes of the phenomenon. The consequence is that the popular meaning of desertification is often associated with a catalogue of environmental calamities rather than specific sadness in the human population-renewable resources system.   In such conditions both prospects and mitigation become extremely uncertain.

Among the forces driving desertification, those related to climate are relatively easy to predict in the frame of existing models (i.e., long term climatic fluctuations, anthropogenic climate change scenarios, etc.). However, the unpredictability of the socio-economic forces (i.e. markets, political changes) is far greater, and adds significant difficulty to determining prospective desertification.

In the following, elementary concepts for a dynamic-systems approach to desertification will be discussed as a way to assist in the forecasting task. Afterwards, the temporal variability of key variables that integrate climatic and socioeconomic effects will be analysed, as well as associated changes in land-use and their implications for land condition.

2. A process approach to desertification

In a particular area of interest, human population and natural renewable resources may be considered as two linked elements in a single system, which is affected by climatic or socio-economic disturbances (Fig 1). The former include droughts, rain spells, etc. The latter involves demographic, political, market and technological changes that enable or hinder access to those resources.

Under steady-state conditions, the intensity and duration of disturbances remain within the range of those that have appeared throughout the history of the system. They have been incorporated into its own evolution, in such a way that it recovers quickly after they have ceased. However, a new or very extreme disturbance or combination of disturbances may occur that takes the system beyond its threshold of sustainability. This may occur as an increased availability of resources (i.e. a humid period, the introduction of a new technology), an increased demand for products (i.e., higher prices, local increase in agricultural population), or on the contrary, as a reduction of available resources (i.e., extreme drought).

Figure 1. A perceptual model of desertification

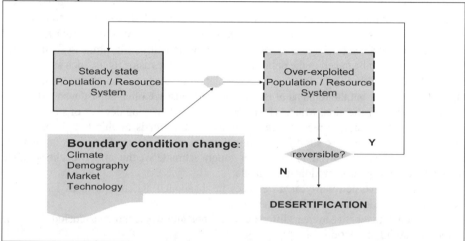

In both cases, resources become over-exploited. If the system is endowed with feedback mechanisms to reverse this condition, it can recover and return to the steady state. Otherwise it falls into an over-exploitation loop that leads to its own extinction. This process, when it happens in drylands, may be considered the core of desertification [24].

The sequences of increase-decrease resource availability (i.e., rain spell followed by drought) are very efficient triggers of desertification. The former attracts people and investment while the latter gives rise to over-exploitation. This process structure has been described in most desertification hot spots around the world [23]. The most outstanding example comes from the Sahel, where a rain spell in 1955-65, was followed by a drought in the seventies. People that migrated into the Saharan border settling agriculture and livestock were unable to escape from the subsequent drought by leaving towards the south [30]. Structurally similar cases have been described for the central US, northern Patagonia, north-west China, and also in the Mediterranean basin [22].

Such disturbances or desertification drivers may continue working to date or not. In the first case we are dealing with 'current' desertification. In the second, the forces that drove desertification in the past are no longer at work today. If resilience thresholds of natural resources have not been exceeded, natural recovery is possible, if they have (i.e., extreme soil erosion), we are dealing with 'relict' desertification. In the latter case, the imprints of past desertification are observable today, even after the disappearance of the underlying factors.

Distinguishing between current and relict desertification is crucial for designing treatment programs [22]. The former requires either relieving driving forces or providing the affected systems with capacity for adaptation. The latter needs only ecological and economically sound restoration.

3. Assessing the effort to control 'relict' desertification

The last widespread desertification event in the Mediterranean area was triggered by an increase of rural population that peaked at the beginning of the last century [4] and extended until the late fifties, aided by national food-security policies [27]. As a result, rain-fed cereal agriculture encroached on rangelands, often on shallow and hilly soils, prone to erosion. The final outcomes can be traced in the delta accretion rates of the major rivers, such as the Ebro, Rône, Piave [4]. When, mid-century, rural population drained into the cities, this marginal agriculture ceased, but the remains of its harm to the landscape often persist even today as a form of relict desertification, which is being controlled with varying degrees of success.

There is no reliable direct source for assessing the trends of this last process as a ground for prospecting the short-term national and regional future. Therefore, an indirect approach has been followed to provide a rough estimate of the effort to check relict desertification. Four assumptions were made:

The cumulative between-year differences in abandoned agricultural areas include the relict desertification generated during the period.

The cumulative between-year differences of forest and woodland areas include the area reforested during the period.

The degree that the latter approaches the former, or even surpasses it, may be considered an over-estimate of the restoration effort carried out during the period.

Most forest plantation is in the abandoned agricultural areas. Therefore, this amount may be used to refine the above-described estimate.

The bulked values for all the southern EU countries (Fig 2) show that until the eighties, the forest and woodland area followed the track of abandoned agricultural land and in some periods even extended beyond it. Since the eighties this trend has been reversed and abandoned agricultural land now increases faster than the forested area. The accumulated values in the year 2000 (Table 1) show that, for the entire southern EU, forest & woodland recovery amounts to 75% of that abandoned by agriculture. Only 40% of it, which amounts to one third of the area of abandoned agricultural land, can be allocated to plantation, and the rest is natural regrowth.

A look at the different countries leads to the conclusion that the restoration effort decreases from Spain through France to Italy, where planted forest amounts to 60%, 21% and 3% of the area of abandoned agriculture, respectively. Greece and Portugal deviate from the general pattern because, until recently, their agricultural area has been growing. Therefore, their forest plantation was mostly in either degraded rangelands or areas where agriculture was abandoned before 1960. In any case, forest plantation activity has been greater in Portugal than in Greece, where it amounts to 63% and 45%, respectively, of total forest and woodland recovery.

Figure 2. Cumulative changes in abandoned Agricultural Area and Forest & Woodland Area in Southern EU countries.

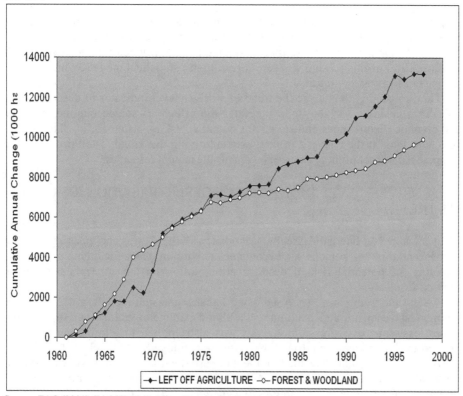

Source: FAO (2001) FAOSTAT Database Results. FAO, Rome

These results suggest that the forthcoming restoration effort is expected to increase in most of the southern EU countries, particularly in Italy and France, less in Spain and probably at a later date in Greece and Portugal. The relationship of this restoration effort with checking relict desertification is closer in Spain, Greece and Portugal, with Mediterranean climate in more than 60% of their territory, than for Italy and France, where this share amounts to 40% and 16% respectively.

Table 1. Accumulated changes up to year 2000 of key land use types relevant in the marginal areas (10^3 ha).

Country	a Planted forests	b Aband. Agric.	c Forest&Wood.	a/b	a/c	c/b
FR	961	4595	3644	0.21	0.26	0.79
GR	120	-181	266	-0.66	0.45	-1.47
IT	133	5306	1080	0.03	0.12	0.20
PO	834	303	1330	2.75	0.63	4.39
ES	1904	3150	3581	0.60	0.53	1.14
S- EU	3952	13173	9901	0.30	0.40	0.75

Source: FAO (2001) FAOSTAT Database Results. FAO, Rome

4. Assessing trends of 'current' desertification

If process-based models for desertification are available, they can be used to develop prospecting tools. Unfortunately this is not the case for the study area. Such models are rare and those that exist have been developed for specific types of land use [9]. For this reason, we were forced to use statistical models, which rely on the past to forecast the future.

The first step is to analyse the trend of a driver-level indicator of desertification risk. This means a variable that integrates the effects of socioeconomic, political and climatic disturbances and stands for potential risk to sustainability. The second step is to look at the changes in land use underlying this trend. The third step is to examine the impact of these changes on renewable natural resources.

4.1. A DRIVER-LEVEL INDICATOR OF DESERTIFICATION RISK AND ITS TREND.

The per capita Net Internal Agricultural Produce (ANIP) [3] was chosen as an indicator of concentration of persons and economic activity in the agricultural sector, and therefore, of potential risk of desertification, following the concepts outlined in Section 2.

ANIP time trends were analysed using linear regression models, with first order autoregressive errors and maximum likelihood parameter estimation. In addition, linear least-squares regression analysis was also applied to allow comparison of general trends between countries. The software employed was SPSS 11.0. [11].

A preliminary analysis using simple and partial autocorrelations showed that no seasonality was noticeable in the ANIP series for any of the countries studied. Therefore, predictions of autoregressive models could only rely on the previous year's value (AR 1) and with time as an independent variable. This strongly limits the confidence in forecasting to any number of years beyond the observed series. Hence, the estimates below should be taken only as rough approximations of the most probable trends based on the latest experimental data and in the absence of events unrecorded to date.

ANIP time series (Fig 3) show that France and Italy, with Mediterranean climate in 16% and 40% of their territory respectively [7], behave quite differently from the remaining countries, where this proportion is over 60%. The ANIP rate of increase, estimated as the slope of the linear regression over time, is lower for France (1.029) and Italy (0.793) than in Greece (1.556) or Spain (1.785). This may be interpreted as a consequence of the more developed, stable agricultural systems, which prevail in the first two countries, mostly with humid temperate climates, little affected by water shortage.

Greece, Spain and Portugal show larger ANIP variability along their main trends. Thus the residual variance, obtained from the autoregressive models, increases from France (12.2) and Italy (12.8) to Greece (19.8) Spain (24.0) and Portugal (55.7). This can be associated with specific climatic, economic or political events. Thus the major trough in the middle nineties recorded in Spain coincides with an exceptional drought;

Figure 3. Time trends of Net Internal Agricultural Produce (ANIP) in South-EU countries showing recorded values (thick line), estimated values (thin line) and upper - lower 95% confidence limits (dashed lines). AR(1): coefficient expressing the auto-regressive effect; YEAR: the slope over time; CONSTANT: the independent term. Bold triangle: Drought; Empty triangle: Joining the European Community; Empty star: 'Pink revolution in Portugal'.

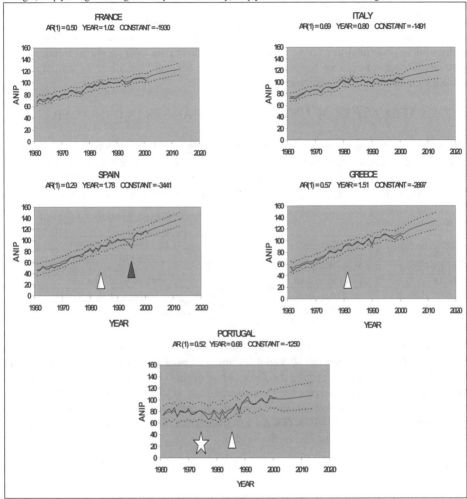

Source: FAO (2001) FAOSTAT Database Results. FAO, Rome.

the large depression of the seventies and early eighties in Portugal matches the political changes that followed the 'pink revolution' in that country, and the consequences of joining the European Community were not homogeneous: while in Spain it was followed by an increase in between-year variability, in Greece and Portugal it led to a spell of positive anomaly. These results suggest that, compared to temperate zones, agriculture in the Mediterranean climate shows greater sensitivity to climatic and socioeconomic disturbances, but recovers quickly after their disappearance. Its ANIP displays faster rates of increase, and it is potentially more aggressive in triggering desertification.

Mediterranean between-year climate variability has been associated with the so-called 'Mediterranean Oscillation', an east-west flip-flop over a period of around 22 years [16]. The lack of seasonality in the ANIP time series suggests that economic and political factors, such as those mentioned above, dominate the effects of this climatic variability and only the impact of heavy drought is perceived.

In summary, it may be concluded that the performance of Mediterranean agricultural systems is driven by a long-term, increasing socio-economical trend, and by the occurrence of shorter 'events', either socio-economic, political or climatic, that cause sudden troughs in this trend.

4.2. EXAMINATION OF UNDERLYING CHANGES IN LAND USE

In order to gain an insight into the agricultural changes underlying ANIP trends, a set of variables concerned with the state of agricultural systems were selected (Table 2). These variables cover various aspects of demography (AGRP, RURP, TOTP) technology (TRACTOR, FERCON, IRRIGA) and land use (CERP, OLIVP, ALMP).

Table 2. List of variables in CATPCA obtained from F.A.O.

Acronym	Variable description	Acronym	Variable description
AGRP	Agrar. pop./Tot. pop.	IRRIGA	Irrigated area/Agr. area
RURP	Rural pop./Tot. pop.	CERP	Cereal prod./Agr. area
TOTP	Total pop./Agr. Area	OLIVP	Olive prod./Agr. area
TRACTOR	Tractor n°/Agr. Area	ALMP	Almond prod./Agr. area
FERCON	Fertilizer cons./Agr.area	ANIP	Agric. Net Internal Prod./Agr. area

Source: FAO (2001) FAOSTAT Database Results. FAO, Rome

The above variables in each country were averaged for the periods 1961-65, 1978-82 and 1995-99. A matrix of 10 variables and 15 cases (5 countries x 3 periods) was thus obtained. In order to reduce the dimensions of this data set, factorial analysis with Principal Component Analysis for Categorical Data (CATPCA) was applied to avoid uncontrolled deviations from normality, given the small number of cases. The software employed was SPSS 11.0 (CATPCA: 1.0 from Data Theory Scaling System Group (DTSS), Leiden University, The Netherlands).

Table 3. Explained variance in CATPCA

Dimension	Explained Variance	
	Total (Eigenvalues)	Relative (%)
1	5.682	57
2	2.733	27
Total	8.475	84

The first two dimensions explain 84% of the variance (Table 3). Plotting the variables in the space of these two dimensions (Fig 4a) shows that DIM1 opposes field crop intensification -technological development (TRACTOR, FERCON), and yield increase (CERP) against 'ruralisation' or traditional agriculture (high fraction of agrarian AGRP and rural population RURP). DIM2 differentiates high-tech development of typical Mediterranean crops, such as olive and almond orchards (OLIVP, ALMP) and irrigation developments (IRRIGA).

Plotting the cases (country averages in each of the three periods) in the space of the first two dimensions (Fig 4b) shows that countries migrate toward high DIM1 values. This means the shifting of their agriculture from the more traditional patterns of closed markets fostered by national food-security policies in the first half of the past century, to open market agriculture encouraged by the consolidation of the European Union. This trend matches an increase in ANIP.

Figure 4. Plot of the variables and cases (countries x periods) in the space of the first dimensions two.

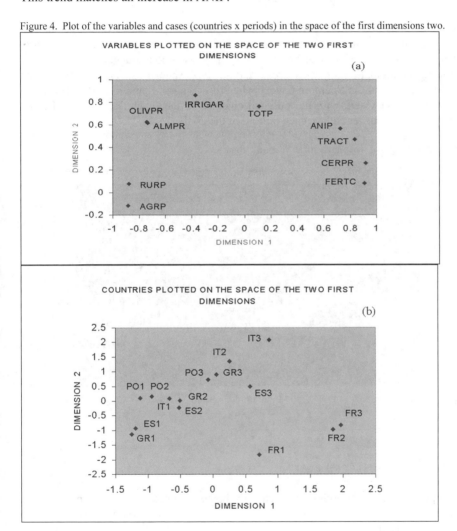

In France, where temperate agriculture is of greater importance, this migration involves a small increase in DIM2 (limited growth of irrigation IRRIGA and total population supported per unit of agricultural area TOTP). However, the remaining countries, with a Mediterranean climate in a larger share of their territory, in order to progress along DIM1, must also advance along DIM2. This means that the transformation and intensification of their agriculture necessarily goes through concentration in the best suited areas, as indicated by the increase in people supported per unit of agricultural area (TOTP), development of irrigation (IRRIGA), and expansion of high-tech-managed traditional tree crops (OLIVP, ALMP).

4.3. THE CASE OF SHEEP BREEDING

Sheep raising has been specifically considered because its available time series do not match those included in the set for the above factorial analysis. Since the sixties, there has been a general trend to decrease stock (Fig 5) as a consequence of the already progressing depopulation of rural areas and lack of shepherds. This decrease was particularly significant in Spain, in Greece where it was slower, and in Portugal where it occurred later, associated with the political changes of the seventies, while in France and Italy it started earlier and touched bottom in the sixties.

Figure 5. Time changes of sheep stock in the South-EU countries.

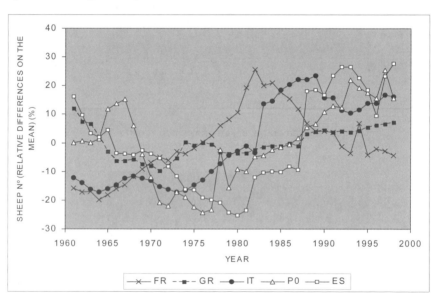

Source: FAO (2001) FAOSTAT Database Results. FAO, Rome

European agricultural policies significantly helped the recovery of sheep raising as an alternative to the market saturation of milk and meat from dairy farming. Sheep stocks started rising during the seventies and up to the eighties and nineties in France and Italy, respectively. In Spain, Portugal and Greece, the increase in sheep matched integration of these countries in the European Community. This happened in 1981 for

Greece and 1986 for Portugal and Spain. In the case of Greece, such a recovery was moderate because migration from rural to urban areas was slower, and stock losses in the previous period were smaller.

Mid-term prospects do not forecast maintenance of the current increase in sheep stock for much longer, even with supportive EU policies. In France and Italy, a decrease has already begun, while in most of the countries, free grazing grounds are often over-stocked, and the current increase is associated with the consumption of crop debris from intensive agriculture [21].

4.4. IMPLICATIONS FOR DESERTIFICATION OF TRENDS OBSERVED IN AGRICULTURAL SYSTEMS

It has been shown that in terms of increasing ANIP, the evolution of Mediterranean agricultural systems involves the development of irrigation and innovative management of traditional fruit tree crops, as well as maintaining free grazing systems with large numbers of sheep. The three issues are ways of adapting agriculture to the current climatic and socioeconomic conditions, and do not intrinsically mean risk of desertification. However, the three involve the attraction of people and investments that may easily lead to over-exploitation of land resources beyond their thresholds for resilience. In the following we try to review information on how this happens.

4.4.1. Further development of irrigation
Total water resources and abstractions in the different countries and scenarios are shown in Fig 6 (a, b). Resources and current abstractions (1997) were obtained from Eurostat [2] while 2025 projections without effects of climate change come from Margat [8]. Current abstractions in France and Italy were found to be greater than the 2025 projections given by Margat. Following this author, we have assumed that no further change is to be expected in these countries.

The impact of climate change on water resources has been approached considering both the increase in water consumption by irrigation and the decrease in natural runoff (Q). The first was evaluated as $1000 \text{ m}^3.\text{ha}^{-1}. \text{ y}^{-1}$, an intermediate rate between the measured values of $500 \text{ m}^3. \text{ ha}^{-1}. \text{ y}^{-1}$ and $1500 \text{ m}^3.\text{ha}^{-1}.\text{y}^{-1}$ [8]. This supplementary rate has been applied to the irrigated areas in 1998, acquired from Eurostat [2]. For the second, a very rough estimate of -9% has been used.

This value was derived from the figures proposed by Palutikof et al [17] for the Mediterranean basin, with GIS-based scenarios using the Hadley Centre GCM simulation HadCM2 [5]. The estimated changes in average annual values between 1970-79 and 2030-39 are $+1.1°C$ for temperature (t) and -1.2 mm for rainfall (P). Given the large spatial and seasonal variability included in this prospective change of annual rainfall, and given its low value, it has been assumed that no overall change throughout the basin is to be expected. Obviously, this does not preclude the occurrence of significant positive or negative changes in some areas and in some seasons.

Annual temperature change was converted into annual potential evapotranspiration change (PET) using $PET=68.64.t$ as proposed by Le Houerou [7]. The corresponding reduction in runoff (Q) was estimated using the regional relation between Q/PET and P/PET proposed by Budyko [1]. The figure obtained (-9%) is between the -5% and -

14% estimates considered by the 'Ministry of the Environment' [12] for the whole of Spain, and seems rather conservative for reduction in runoff.

Most of the projected increase in water demand, without considering climate change, originates from the expansion of irrigation. Bulked country values show abstraction intensity rates between 30% and 40% of the total resources for Spain and Italy (Fig 6 b). If the impact of climate change is further taken into account, Spain would reach 50%. Mediterranean agriculture is still underperforming in terms of the water consumption rate (WCR) . Some of the presented WCR values may be too high, particularly for Italy, due to statistical inaccuracies in the irrigated area. However, it is well known that, for example, drip irrigation allows water consumption to be reduced to 4000 m^3 .ha^{-1}. y^{-1} in southern Spain. This means that upgrading water-saving technologies might cancel out the effects of climatic change at the country scale, although at the sub-national scale, some regions will be threatened by more severe water shortages.

In any case, abstraction intensities reported in Fig 6b approach and even exceed safe thresholds for exploitability [8], particularly if between-year rainfall and within-country spatial variability are taken into account. The outcomes are serious water stress and risk of supply failure to be expected, not only in the southern regions of Spain and Italy, but also in Portugal and Greece.

Figure 6. Water resources, total abstractions and abstraction intensities in several scenarios in the South-EU countries.

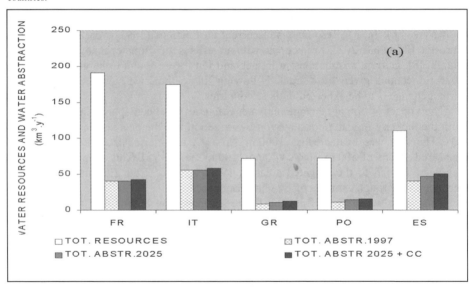

Sources: Water resources and abstractions, EUROSTAT (2000) *Environment & Energy. Water Resources – Long term annual average* (LTAA).Office for Official Publication of the European Communities, Luxemburg; projections without climate change effect, Margat, J. (2002) *Are water shortages a long-range outlook in Mediterranean Europe?. First Synthesis Report for the Mediterranean Commission for Sustainable Development (MCSD).* Plan Bleu Regional   Activity Centre, Sophia Antipolis [8].

Table 4. Water abstractions by agriculture and water use efficiency of irrigated land.

| Country | Water abstracted (1997) | | Water abstracted (2025)* | | WCR** (1997) |
	$km^3.y^{-1}$	% of total abstraction	$km^3.y^{-1}$	% of total abstraction	$(m^3.ha^{-1}.y^{-1})$
FR	5	12	7	16	2 486
IT	26	46	29	48	9 582
GR	8	87	11	91	5 345
PO	9	79	13	85	13 872
ES	28	68	38	74	7 655

(*) Socioeconomic and climate change effects are included
(**) Water Consumption Rate of irrigation
Data sources for estimations: 1997 abstractions, EUROSTAT (2000) *Environment & Energy. Water Resources – Long term annual average* (LTAA).Office for Official Publication of the European Communities, Luxemburg; 2025 abstractions, Margat, J. (2002) *Are water shortages a long-range outlook in Mediterranean Europe?. First Synthesis Report for the Mediterranean Commission for Sustainable Development (MCSD)*. Plan Bleu Regional   Activity Centre, Sophia Antipolis [8].

Unfortunately, most mitigation policies rely more on providing external water input than on increasing the capacity of the affected agricultural systems to limit their own growth. External water input can be supplied by making other areas as water providers and by water desalinization. The first option means affecting a 15-20-unit area per each unit increase of irrigated area, and it is therefore a source of within-country political conflict that will be exacerbated in a climate change scenario. The second option is limited to particular conditions in coastal areas. It puts agricultural systems at the risk of energy market uncertainties, and requires being careful of associated environmental threats, such as brine deposition and carbon emissions into the atmosphere.

The threat of expanding irrigation does not only concern the amount of total water resources, but also affects land degradation in several ways. The most widespread impact is soil and water salinization, as a result of marine and continental salt intrusion in coastal and inland areas respectively. The former is driven by an unbounded water demand that depletes aquifers and lowers water tables. The latter is caused by the mobilisation of salts from evaporitic rocks through irrigation, and their spread over soils and rivers. This anthropogenic salinization adds to a purely climatic one which, driven by climate change, is expected to grow [29].

A second much-reported impact of irrigated agriculture increasing beyond the capacity of water resources is the change of hydrological patterns due to groundwater overdraft. In threatened areas, rivers and wetlands are often fed by shallow water tables, whose outflow varies little with time. The drop in the water table leaves the drainage systems to be fed only by storm runoff. The flow becomes episodic, which causes peat oxidation and disappearance of the former river and wetland ecosystems.

Some general features can be drawn regarding the impact of unsustainable irrigation from the perspective of the desertification process outlined in Section 2. Irrigation developments are difficult to reconvert because of the large investment they require. This means that they can persist long after sustainability thresholds have been surpassed. Irrigated coastal agriculture is mainly market-driven (fruits and vegetables) and is therefore prone to unbound growth. Irrigated inland agriculture is restricted to maize and alfalfa because of the winter cold. Therefore it is mostly policy-driven, and may persist even when such policies are wrong, due to the inertia of large investment and reconversion that were made in their day.

4.4.2. Development of traditional tree crops

Rain-fed fruit-tree crops have traditionally been a common adaptation to the dry summer Mediterranean climate that takes advantage of their ability to use water from deep soil layers. The most widespread are vineyards, olive and almond orchards. The former underwent significant expansion in hilly areas of marginal agriculture during the 18th-19th centuries. They often took part in very long-term rotation with pinyon pine plantations (i.e., north-east Spain) and sustained farmers providing substantial income. However, during the second half of the 19th century, Mediterranean vineyards were hit by the American plant louse (*Philoxera*), which spread across Europe destroying all the vines and ruining many farmers. Most of them were forced to leave their farms, but those who remained replanted their vines using resistant American rootstocks and, in the long term, they even succeeded in improving their vineyards. Since then, in general, vine crops have been restricted to suitable areas regulated by a policy that looks more to quality than quantity.

Figure 7. Time changes and trends of olive production in the South-EU countries.

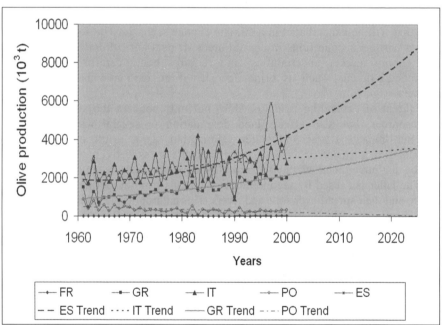

Source: FAO (2001) FAOSTAT Database Results. FAO, Rome

Olive and almond tree crops underwent dramatic changes during the second half of the 20th century. During 1960-70, the demand for olive oil was almost internal to southern European countries, and was unable to overcome the competition of other vegetable-oil sources. The consequence was a progressive reduction in olive orchards and tree dieback. Since the early eighties, olive production has started to increase dramatically in Spain and, at a lesser rate, in Greece (Fig 7). In Italy, production increase was recorded earlier at a moderate rate, while in Portugal and France, it is quantitatively less relevant.

Two kinds of events underlie these changes. The first is the European Agricultural Policy supporting olive crops. The second is a substantial increase in demand due to the growth in popularity of the so called 'Mediterranean diet' of which olive oil is its most genuine representative. In all cases, the production time series is markedly seasonal because of the effect of between-year climatic variability and fruit-bearing alternance. However, the 2025 projections must be taken with caution because demand is expected to become saturated in the relatively short term [26].

The case of almond trees is slightly different. In Spain, production increased step by step, by around 65%, in the early seventies, and has remained stable thereafter. In Greece the rate of increase is slower, while in Italy it has decreased. As for olive production, Portugal and France have a far less significant production. Subsidising almond tree crops has been a national policy in Spain since the seventies, and an EU policy since the eighties, as an aid to stabilising the rural population.

Most of the increase in olive and almond production comes from the expansion of both crops over hilly areas, and is associated with heavy soil erosion. Soil-loss rates of between $60\,t\,ha^{-1}\,y^{-1}$ and $100\,t\,ha^{-1}\,y^{-1}$, have been reported in [6] for olive trees, and equivalents of 50%-60% of bare ground rates were published [15] for almond trees. These rates widely exceed the thresholds for sustainable soil conservation [10].

Unfortunately, as both crops are driven by agricultural policies and market conditions, they are managed with a view to short-term benefit rather than long-term sustainability.

4.4.3. Overstocking rangelands

In contrast to either continental or tropical rangelands, Mediterranean ecosystems have a long history of grazing pressure concentrated in the more favourable seasons. Wild herbivores of tropical origin have been present since the Middle Pleistocene, around 500,000 years ago [25].

The consequence has been the evolution of a wide range of plant adaptations to heavy grazing, such as deterrent (e.g. spines, chemical repulsion) and high-regeneration capacity strategies. The latter are particularly important because they have contributed to the evolution of the plant communities adapted to continuous grazing which are specific to the Mediterranean region [13], [14].

In this way, livestock breeding, including not only animals but also the whole technical and economic system that ensures their management, helped to shape the Mediterranean landscape during centuries. Its impact has been strong at periods with high population density in rural areas or adverse political conditions, while it was milder when the human population was smaller or during periods of political and economic stability [20].

Under strong impact conditions there is overgrazing, and serious land degradation may be triggered, not only by the livestock itself, but also by its management (e.g., using fire to control inedible plants). Landscape damage associated with overgrazing concerns vegetation, soil and hydrological patterns. Grasses may be substituted by unpalatable shrubs and tree renewal can be impaired by the elimination of seedlings. Trampling and reduction of plant cover promote overland flow and soil erosion. During drought, livestock looking for fresh forage concentrate in bottomlands and triggers gully initiation [28].

Such damage by overgrazing fostered a hostile attitude toward livestock by conservationists, particularly foresters. As a consequence, policy regulations borrowed from central Europe and the US were introduced (e.g., livestock exclusion from forests, severe limitations on goat stock) without taking into account the specific conditions of the Mediterranean rangelands summarised at the beginning of this section.

Since the seventies, research findings have contributed to modifying this attitude as it is realised that moderate grazing pressure does not reduce, but even increases range productivity, helps to maintain higher levels of ecological diversity and keeps down risk of forest fires [20] [31]. With proper management, even goats can bring specific benefits to forests [19]. Instead of excluding livestock from forests and other Mediterranean ecosystems, a new viewpoint is growing that encourages the harmonisation of livestock with forestry and agriculture [20].

Unfortunately, recent EU agricultural policy is more attuned to maintaining rural life than to the subtleties of the relationship of grazing with landscape, and as a consequence, sheep stocks are growing in many areas of the southern EU (see Section 4.3). A per-head subsidy rate stimulates sheep breeders to overstock their rangelands. Animals are supplemented with concentrated food, but suffer from a fibre deficit and overgraze the rangelands in search of it. The outcome is the land degradation patterns described above.

5. What conclusions can be arrived at concerning prospects?

In order to maintain national and EU social and economic integration, Southern EU agriculture must steadily increase its net internal produce (ANIP). This means accelerating the transformation from older traditional closed-market to open-market agriculture. To achieve this goal, a site-selective and technically sophisticated irrigated agriculture is being developed. Innovative management of traditional horticulture and tree crops make up the core of this development, which will continue in the near future.

The main consequence of this trend is the build-up of large water deficits in some areas that call for large-scale water transfers and associated interregional conflicts. Other foreseeable outcomes are the loss of water resource quality and spreading soil salinization.

A second outcome concerns rangelands and old marginal areas surrounding the most active sites of agricultural development. These areas will be particularly threatened by new encroaching agricultural settlements, and by agricultural policies aiming to mitigate the socio-economic setback of the hinterland (e.g. subsides for sheep or almond trees). Unfortunately, none of these external forces are endowed with the necessary environmental concern to avoid short-term land degradation. Moreover, the ecological rangeland restoration effort is insufficient to reverse the relict desertification accumulated in them.

In order to achieve the EU integration trend Southern EU agriculture is faced with large scale climatic and socioeconomic changes, with expected impacts which may be summarized as follows.

a) Water demand for irrigation will be triggered by socio-economic demand and increased by climate change. Acute water stress is expected to arise at the sub-national scale, and it will not be overcome by water conservation policies alone, but by developing internal feedback for self-limiting agricultural growth.

b) External water input is likely to be more and more called for, increasing the associated risks of political conflict or energy dependence. Water availability constraints are expected to limit the long-term growth of agriculture in the southern EU (ANIP). It is true that there will be increasing competition with southern Mediterranean countries for horticultural crops. However, southern EU countries will probably have a better near-future response in innovation and product quality. Furthermore, if southern EU agricultural systems do not succeed in adapting to water limitation, a widespread impact on desertification, in terms of soil and aquifer salinization, is likely.

c) Prospects for EU agricultural policies indicate a shift from supporting the production of some target crops to helping integrated rural development together with ecological and socio-economic sustainability. Hopefully this process will also be helped by the work of Regional Implementation Annex 4 (Northern Mediterranean) of the United Nations Convention to Combat Desertification (UNCCD). Under such a scenario, overstocking of rangelands and encroaching tree crops in marginal areas are unlikely to persist, particularly if the olive market becomes saturated.

d) The integration of the ten Acceding States into the EU will probably force Southern EU countries to increase their specialization in off-season horticultural crops, and to abandon rainfed crops in extensive inland areas. The expected outcomes are rising water deficit and its associated desertification risk, and growing up marginal lands with management requirements to cope with relict desertification.

References

1.   Budyko, M.I. (1974) *Climate and Life*. Academic Press, London.
2.   EUROSTAT (2000) *Environment & Energy. Water Resources – Long term annual average* (LTAA). Office for Official Publication of the European Communities, Luxemburg.
3.   Food and Agriculture Organization (FAO) (2001) FAOSTAT Database Results. FAO, Rome.
4.   Grove, A.T., Rackham, O. (2001) *The Nature of Mediterranean Europe. An Ecological History*. Yale University Press, New Haven and London.
5.   Johns, T.C., Carnell, R.E., Crossley, J.F., Gregory, J.M., Mitchell, J.F.B., Senior, C.A., Tett, S.F.B. and Wood, R.A. (1997) The Second Hadley Centre coupled ocean-atmosphere GCM: model description, spinup and validation. *Climate Dynamics 13*, 103-134.
6.   Laguna, A., and Giráldez, J.V. (1990) *Soil erosion under conventional management systems of olive tree culture. Proceedings of the Seminar on Interaction between Agricultural Systems and Soil Conservation in the Mediterranean Belt*, European Society Soil Conservation, Oeiras. 82-89.
7.   Le Houérou, H.N. (1992) Vegetation and land use in the Mediterranean Basin by the year 2050: a prospective study, in Jefftic, J., Milliman, J.D., and Sestini, G. (eds.), *Climatic Change and the Mediterranean*, Edward Arnold Publishers, London, 175-232.
8.   Margat, J. (2002) *Are water shortages a long-range outlook in Mediterranean Europe?. First Synthesis Report for the Mediterranean Commission for Sustainable Development (MCSD)*. Plan Bleu Regional Activity Centre, Sophia Antipolis.

9. Martínez, Vicente, S., Martínez, Valderrama, J., and Ibáñez, Puerta, J. (2000) A simulation model of desertification by salinisation in the Coastal Irrigated Agricultura, in Puigdefabregas, J., and del Barrio, G. (eds.) *A Surveillance System for Assessing and Monitoring of Desertification (SURMODES)*. Project nº 902, Expo 2000 Hannover –Projects around the World. CDROM, EEZA (CSIC), Almeria (www.eeza.csic.es/surmodes).
10. Mendizábal, T., and Puigdefabregas, J. (2003) Population and land use changes: impacts on desertification in Southern Europe and in the Maghreb, in: Brauch, H.G., Liotta P.H., Marquina, A., Rogers, P., Selim, M., (eds.) *Security and Environment in the Mediterranean – Conceptualising Security and Environmental Conflict*, Springer, Heidelberg, pp. 687-711.
11. Meulman, J.J., and Heiser, W.J. (2001) *SPSS Categorías 11.0*. SPSS Inc., Chicago.
12. Ministerio de Medio Ambiente (1998) *Libro Blanco del Agua en España*. Dirección General de Obras Hidráulicas y Calidad de las Aguas & CEDES, Madrid.
13. Montserrat, P. (1990) *Pastoralism and desertification, in Strategies to combat desertification in Mediterranean Europe*, Commission of the European Communities, Luxembourg, pp. 85-103.
14. Montserrat, P., and Fillat, F. (1990) *The systems of grassland management, in Spain Managed Grasslands* (Regional Studies). Elsevier, Amsterdam, pp. 37-70
15. Moreira, Madueño, J.M. (1991) *Capacidad de uso y erosión de suelos; una aproximación a la evaluación de tierras en Andalucía*, Junta de Andalucía, Agencia de Medio Ambiente, Sevilla
16. Palutikof, J.P., Conte, M., Casimiro, Mendes, J., Goodess, C.M., and Espirito, Santo, F. (1996) Climate and climate change, in Brandt, J., and Thornes, J. (eds.) *Mediterranean Desertification and Land Use*, Wiley, J.C. pp. 43-86.
17. Palutikof, J.P., Agnew, M.D., Watkins, S.J. (1999) Statistical downscaling of model data. Medalus 3, Project 2: Regional Indicators. Final Report, Contract ENV4-CT95-021, King's College, London, pp. 365-412.
18. Papanastasis, V.P., Peter, D. (eds.) (1997) *Ecological basis of livestock grazing in Mediterranean Ecosystems*, Commission of European Communities, Bruxelles.
19. Papanastasis, V.P. (1986) Integrating goats into Mediterranean forests, *Unasylva, 154, 38*, 44-52.
20. Papanastasis, V.P. (1998) Livestock grazing in Mediterranean ecosystems: an historical and policy perspective, in Papanastasis V.P., Peter, D. (eds.) (1997) *Ecological basis of livestock grazing in Mediterranean Ecosystems*, Commission of European Communities, Bruxelles, pp.5-9.
21. Puigdefabregas, J., del Barrio, G. (2000) (eds.) A Surveillance System for Assessing and Monitoring of Desertification (SURMODES). Project nº 902, Expo 2000 Hannover –Projects around the World. CDROM, EEZA (CSIC), Almería, Spain (www.eeza.csic.es/surmodes).
22. Puigdefabregas, J., Mendizábal, T. (1998) Perspectives on desertification: Western Mediterranean, *Journal of Arid Environment 39*, 209-224.
23. Puigdefabregas, J. (1995) Desertification: Stress beyond resilience, exploring a unifying process structure, *Ambio 24, 5*, 311-313.
24. Puigdefabregas, J. (1998) Ecological impacts of global change on drylands and their implications on desertification, *Land Degradation and Rehabilitation ,9,5*, 393.
25. Rackham, O., Moody, J. (1996) *The Making of the Cretan Landscape*. Manchester University Press, Manchester.
26. Rallo, L. (1998) Sistemas frutícolas de secano: El olivar, in: Jiménez Díaz, R., Lamo de Espinosa, J. (eds.) *Agricultura Sostenible*, Ediciones Mundi-Prensa, Madrid, pp. 471-487.
27. Roxo, Mª.J., Casimiro, P.C., Mourao, J. (2000) Politicas agrícolas, mudanzas de uso do solo e degradaçáo dos recursos naturais-Baxo Alentejo Interior. *Revista Mediterrâneo 12/13*, 167-190. Instituto Mediterrâneo, F.C.S.H., U.N.L.
28. Schnabel, S., (1997) Soil erosion and runoff production in a small watershed under silvo-pastoral land use (dehesas) in *Extremadura*, Geoforma Ediciones, Logroño.
29. Szabolcs, I. (1990) Effects of predicted climatic changes on European soils, with particular regard to salinization, in Landscape Ecological Impact of Climatic Change, Boer, M.M. and De Groot, R.S. (eds) IOS, Amsterdam, pp. 177-193.
30. Thebaud, B. (1993) Causes et consequences de la desertification au Sahel de l'Ouest: Essai d'interpretation. International Panel of Experts. Secretariat of the Intergovernmental Negotiating Committee for a Convention to Combat Desertification. INCCD-UN, Geneva, 7.
31. Trabaud, L. (1991) Le feu est-il un facteur de changement pour les systèmes écologiques du bassin méditerranéen?, *Secheresse 2, 3*, 163-174.

Section IV

Water Availability

Chapter 11

From Water Scarcity to Water Security in the Maghreb Region: The Moroccan Case

M. Ait Kadi
President of the General Council of Agricultural Development, Rabat , Morocco

The Maghreb region covers an area of 5 countries (Mauritania, Morocco, Algeria, Tunisia and Libya) with different water situations. Each country has individual water resources characteristics and water management history. Nevertheless, three of them, namely Morocco, Algeria and Tunisia, share a number of similarities that constitute a good basis for a generalisation using Morocco as a case study as in this paper. Indeed, water scarcity is becoming a serious challenge affecting development in these countries. Population growth and urbanisation, industrialisation and tourism, decreasing precipitation and higher frequency of droughts increase the pressure on the resources. Present water use patterns and withdrawals are not sustainable. Thus, the three countries have no choice. They must make a decisive break from past policies and management practices to embrace a holistic water sector approach that is economically, socially and environmentally sustainable.

Driven by these challenges and consistent with a worldwide movement towards integrated water resources management, the three countries have embarked on reforming their water sector. A change in thinking and action in water management is slowly taking place as is illustrated by Morocco's case study presented below. So far a strategic approach to water planning and management is present in these countries.

1. Water Scarcity: A Serious Development Challenge Facing Morocco

The emphasis in Moroccan development planning for the last three decades has been on maximizing the capture of the country's surface water resources and providing for their optimal use in irrigated agriculture, potable water supplies, industry and energy generation. Enormous capital resources have been invested in the essential infrastructure to control surface water flows. Infrastructure to capture and utilize about two-thirds of the surface water potential is operating and a number of other major infrastructure projects are in advanced stages of planning and/or construction to capture most of the remaining potential.

As Morocco nears the end of the infrastructure phase of its national development plan, emphasis is beginning to shift to the more sophisticated and difficult task of ensuring socially and technically efficient allocation of the existing water resources among competing consumer groups on a sustainable basis. This task is ever more complex given Morocco's relatively high population growth and the higher rate of

A. Marquina (ed.), Environmental Challenges in the Mediterranean 2000-2050, 175–185.
© 2004 *Kluwer Academic Publishers. Printed in the Netherlands.*

immigration from rural to urban areas. As depicted in figure 1, in the 20th century population has greatly increased. It has multiplied by six from 5 million inhabitants at the beginning of the century to 30 million inhabitants now. But the country has begun the second phase of its demographic transition due to several socio-economic factors related to an evolutionary process of modernisation and urbanisation. The population growth rate was about 2.8 % during the period 1952 – 1960; then it stabilized around 2.6% between 1982 and 1994 with a decreasing trend since then to reach the present figure of 1.7%. It will further decrease to 0.8% by the year 2050 when the total population is expected to approach 46.2 million inhabitants. Moreover, urbanisation has been increasing fast. The urban population which was only 29.2 % of the total population in 1960, now represents 53% and will reach approximately 72% by the year 2050. This growth will be sustained by rural migration estimated at 220,000 rural migrants per year during the decade 2000 – 2010.

Figure 1. Population and Urbanisation Growth in Morocco

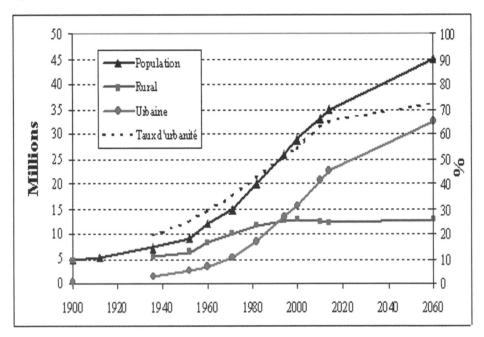

The consequence of a rapidly growing population, accentuated by a progressive shift from rural to urban living, is the growth in the requirements for both the quantity and quality of water resources, and their more intensive and comprehensive use.

The demand for potable water is expected to increase by 2% to 3% per annum. The result of this is growing competition between agricultural and domestic water supply, raising a number of fundamental questions that are at the heart of the water debate in the country. These questions are:

- How much more irrigation does Morocco need to meet the future needs of a growing population?

- How do we restructure consumption patterns from the present wasteful low value water intensive uses?
- How can farmers achieve a better livelihood from every drop of water?
- What will the side effects on the rural community be of transferring water to cities? And what are the implications for food security?
- Does the import of "virtual water" support food security and a more equitable and efficient allocation of water?

Obviously the answers to these questions lie in both a supply management strategy involving highly selective development and exploitation of new water supplies (conventional or non-conventional such as desalination and reuse of sewerage water), and a more vigorous demand management involving comprehensive reforms and actions to make better use of existing supplies. In this context, a particularly difficult challenge is to improve the efficiency of agricultural water use to maintain crop productivity growth while at the same time allowing for the reallocation of water for other uses. This is a difficult task because agriculture is still important as a source of income and employment. Morocco is becoming a large grain importer but has comparative advantages in fruit and vegetables where it can make the best use of irrigation water. Thus, in the Euro-Mediterranean context this situation provides an opportunity to develop a regional partnership that transcends "assistance" premises and mercantile considerations in order to embrace opportunities for a mutually beneficial socio-economic development that draws on the existing complementarities between the North and the South of the Mediterranean sea.

2. The Moroccan Water Sector: Issues and Constraints

Despite remarkable achievements, Morocco faces a growing challenge in the water sector. The main issues and constraints can be summarized as follows:

2.1. DECLINE IN AVAILABLE WATER RESOURCES

The mean annual rainfall throughout Morocco under average seasonal conditions is estimated to total 150 billion cubic meters (figure 2). However, the renewable water resources do not exceed 29 billion cubic meters (BCM). Taking into account potential storage sites and groundwater development possibilities only 20 BCM are divertible annually, 16 BCM from surface water and 4 BCM from groundwater.

Some 103 large dams have been built increasing the storage capacity from 2.3 billion cubic meters in 1967 to 16 BCM in 2003. It has required a major investment effort estimated at 2.5% to 3% of the country's GDP or 18% of the public investments.

Morocco is also endowed with groundwater resources. Some 32 deep aquifers and more than 46 shallow ones, scattered all over the country, have been inventoried. Groundwater withdrawals have increased from 1.5 BCM in 1960 to 3.6 BCM in 2003.

Some 11 BCM are now committed to agriculture, domestic and industrial uses. As population increases, coupled with demands for high per capita domestic and industrial consumption resulting from improved standards of living, the sustainable upper limit or "carrying capacity" of water resources utilisation will be approached by the year 2020.

Figure 2. Morocco's water resources

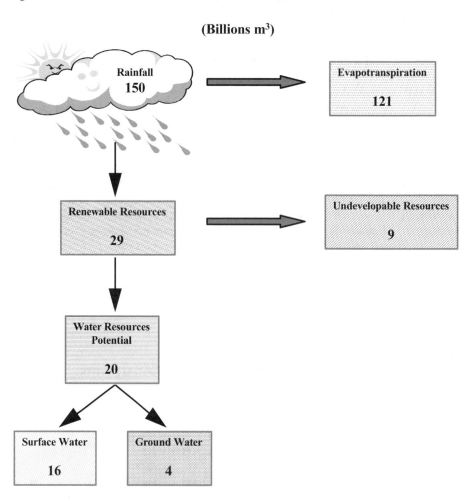

Thus, a growing scarcity is anticipated when, as a result of rising demands resulting from the expansion of irrigated areas and urban development, and a slowing of the growth of available supplies, together with the depletion of aquifers and the pollution of available resources, per capita renewable water resources are expected to fall from 850 cubic meters to 410 cubic meters in 2020 when all renewable resources are projected to be mobilized. So in 2020 Morocco will move from being defined as a « water stressed » to being a « chronically water stressed » country. A number of river basins are already experiencing water shortages which will impose costly inter-basin transfers (figure 3).

Some of the more intensively used aquifers are now considered to be under stress. The unsustainable abstraction of groundwater and the depletion of groundwater aquifers is a major problem. Uncontrolled groundwater extraction is leading to seawater intrusion along the coastal line.

Figure3. River basins water balance

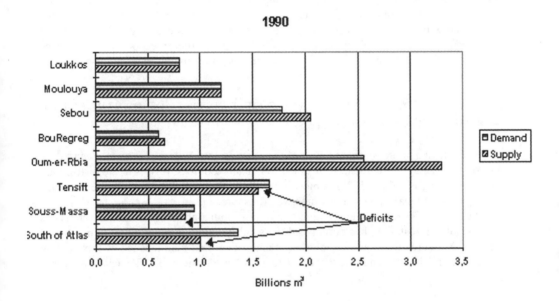

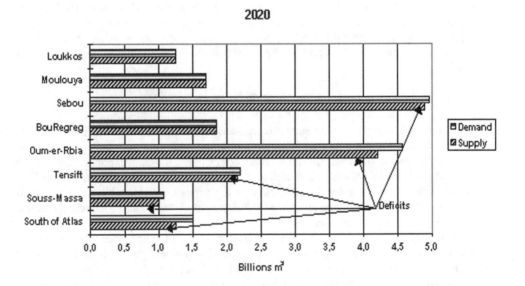

2.2. RAPID DEGRADATION OF WATER QUALITY

This is due mainly to the under-developed sewerage subsector. Sanitation infrastructure has not kept pace with the drinking water supply, and urban effluent is currently a major contributor to the pollution of surface, coastal and groundwater. Management of the rapidly growing volume of wastewater is becoming a serious challenge. The volume of urban effluents has multiplied by 10 since 1960. If the Moroccan economy is to proceed along its path of growth, water quality requirements will grow faster than those of water resources.

2.3. INADEQUATE MAINTENANCE OF EXISTING INFRASTRUCTURE AND SILTING OF RESERVOIRS

A key challenge is the ineffective use of the existing hydraulic infrastructure caused by inadequate maintenance due to insufficient financial resources. Losses in the distribution systems are high (30% to 50%) in most cities. Reducing unaccounted for water in urban networks is a crucial step to save scarce water resources.

The silting of dams, due to the high rate of erosion mainly in the northern watersheds because of deforestation, has diminished the available storage capacity by as much as 8 percent. Measures to improve watershed management and maintenance of the existing infrastructural investments are expected to result in considerable financial and water savings.

2.4. LOW LEVEL OF PROVISION OF POTABLE WATER TO RURAL POPULATION

Despite Morocco's success in providing water to virtually all its urban inhabitants, the provision of rural potable supplies remains an immediate challenge. Half of the population live in rural areas and barely 32 per cent had access to safe and reliable supplies of water in the early 1990s. An important program has been launched with the aim of providing potable water to 80 per cent of the rural population before the year 2010.

2.5. LOW WATER USE EFFICIENCY IN IRRIGATION

Irrigation is the largest water consumer. It accounts for 88 % of the water used compared to 8 % for domestic use and 4 % for industry. Increasing water resources development costs, along with severe financial constraints and competition for scarce public funds, have fostered a substantial change in attitudes to water conservation, and serious questions have been directed towards water use efficiency in the irrigation sector while recognizing its strategic role in the economic and social development of the country. Indeed, the whole irrigation sub-sector currently represents only 10 per cent of the arable land but contributes 45 % of the agricultural added value and produces 75 % of the agricultural exports. In addition to boosting food production, irrigation development has also increased rural employment, promoted agroindustry and helped stabilize domestic production. It has also raised productivity and incomes significantly by bringing modern agriculture to small farm families. It has, in several areas, reversed the flow of people from rural to urban areas. Furthermore, it has also contributed to the conservation of natural resources by reducing the pressure on areas with a fragile ecology.

2.6. THE EFFECT OF CLIMATE UNCERTAINTIES

Decreasing precipitation, higher frequency of extreme rainfalls and droughts are a reality with the consequent social and economic effects. Severe droughts were experienced in the periods 1980-85, 1991-95 and 1999-2001. They have aggravated the water deficit as the run off was substantially reduced. A recent study shows the average run off countrywide has been reduced by as much as 35% from a comparison between the run off series of 1945-1970 and 1970-2000. Climate change is considered a long-term risk yet it is not being confronted with adequate risk management measures.

2.7. GLOBALISATION AND TRADE LIBERALISATION

Globalisation and trade liberalisation carry with them the risk of economic and political instability, worsening inequality and the vulnerability of the poor. In this context Morocco's agricultural strategy has moved from the food self sufficiency objective to the food security objective. This means that domestic food needs will be met through strategic levels of national agricultural production and the gap will be covered by relying on the international market in order to make the best use of the country's competitive advantages given the water resources constraint. Of course, the liberalisation of agricultural trade widens the entire spectrum of economic possibilities offering Morocco the potential to make efficient allocation of its water resources and to make the most of its comparative advantages. The challenge is to identify agricultural opportunities and start moving forward. Morocco is already on this track. But appropriate trade policies require astute understanding of the underlying economic competitiveness in any agricultural sector in liberalized world markets and a strong government policy that encourages its development. The other essential ingredients are immense investments in infrastructure and technology to increase productivity and improve quality. But it is important to realize that there are both economic and social transitional difficulties and adjustment costs. Indeed in Morocco agriculture still has an important role in the national economy as it contributes to 17 % of the GDP and still provides employment for a large portion of the population (50%).

3. From Scarcity to Security: Holistic Water Sector Reform

To meet the challenges posed by the growing water scarcity, Morocco has adopted an integrated approach to water resources management through mutually reinforcing policy and institutional reforms as well as the development of a long term investment programme:

3.1. WATER SECTOR DEVELOPMENT POLICY

The major policy reforms adopted are the following:

- The adoption of a long term strategy for integrated water resources management. The National Water Plan will be the vehicle for strategy implementation and will serve as the framework for investment programmes until the year 2020;
- The development of a new legal and institutional framework to promote decentralized management and increase stakeholder participation;
- The introduction of economic incentives in water allocation decisions through rational tariff and cost recovery;
- The taking of capacity-enhancing measures to meet institutional challenges for the management of water resources; and
- The establishment of effective monitoring and control of water quality to reduce environmental degradation.

All these policy features are embedded in the new legal and institutional framework.

3.2. THE NEW LEGAL AND INSTITUTIONAL FRAMEWORK

A new water law was promulgated in 1995. It provides a comprehensive framework for integrated water management. Some of the salient features of this law are:
- Water resources are public property;
- The law provides for the establishment of River Basin Agencies in individual or groups of river catchments. It clarifies the mandates, functions and responsibilities of the institutions involved in water management. In particular, the status and the role of the High Water and Climate Council has been enhanced as a higher advisory body and a forum on national water policies and programmes. All the stakeholders from public and private sectors including water users associations sit on this council;
- The law provides for the elaboration of national and river basin master plans;
- It has established a mechanism for the recovery of costs through charges for water abstraction and the introduction of a water pollution tax based on the principle « user - pays » and « polluter-pays ».
- The law reinforces water quality protection by defining environmental mandates and enforcing sanctions and penalties.

Concerning the institutional set up, the major change is the establishment of River Basin Agencies empowered to manage individual or groups of river basins. The three principal responsibilities of these agencies consist in the development of water resources, the allocation of water as defined by the master plan and the control of water quality. The agencies reinforce the network of existing institutions in charge of the different water management functions. But the capacity building requirements are high to make this network of institutions capable of an integrated water management.

3.3. THE INVESTMENT PROGRAMME

Enormous capital resources have been invested in essential infrastructure in order to capture and utilize two-thirds of the surface water potential, to develop hydropower, and to make water available for urban and rural potable use and for irrigation. As shown in figure 4, during the 1990-95 period, total expenditure in the water sector (investments plus operation and maintenance) reached 2.6 to 3.2 % of the country's GDP. This portion will increase in the future as efforts will be continued to:

- capture most of the remaining potential and develop the accompanying hydropower infrastructure in order to reduce energy imports;
- meet the government objective for the potable water supply sub-sector which is to supply virtually the entire rural and urban population by the year 2020; and
- continue the on-going expansion of modern irrigation which is expected to bring the irrigated area to 1.35 million hectares by 2020. However, in view of the current concerns about impeding critical water shortages, conserving water and improving efficiency , productivity, cost effectiveness and the sustainability of irrigated agriculture are increasingly necessary if Morocco's economic growth is to continue. In this context, the National Irrigation Program (1993-2000) adopted an integrated approach to carry out, along with the expansion investments, improvements in three major, interrelated areas which are: (i) improving the hydraulic efficiency of the irrigation systems, (ii) strengthening the irrigation agencies' managerial capacities and (iii) increasing productivity.

Figure 4. Total expenditure in the water sector (Investments + Operations & Maintenance)
In millions of US Dollars

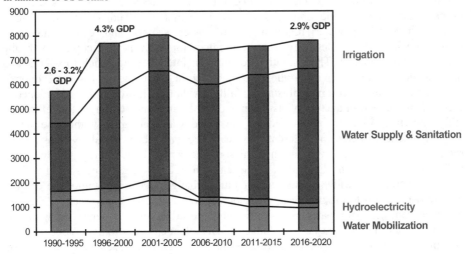

There are various sources of funding for water in Morocco. The national budget is currently the major funder of investment. Cash flow from water revenues covers recurrent costs (Operation and Maintenance) but only rarely contributes to funding investment, though Morocco has a well established cost recovery system both in the irrigation sector and the water supply and sanitation sector. In irrigation, the Agricultural Investment Code provides a legal and institutional framework for the significant recovery of both investment and operating costs. It calls for the full recovery of Operations and Maintenance costs and up to 40 % of the initial investment costs. Similar principles apply to potable water pricing. In this sector the tariffs are differentiated between production and distribution, between cities and between different categories of users through a progressive pricing structure. A solidarity tax paid by urban users has been established to support investment efforts for improving access to potable water in the rural areas. Furthermore, Morocco has embarked on a series of concessions to private international companies for the development and management of the water supply and sanitation systems in a number of major cities (Casablanca, Rabat, Tangier-Tetouan). A public – private partnership in the form of a BOT (Build, Operate, Transfer) is in the process of being agreed for the development and management of an irrigation project. Designing this agreement has been a complex task given the limited precedents. The major issues relate to the appropriate risk allocation.

4. Conclusion

The Moroccan example shows that new strategies for water development and management are urgently needed in the Maghreb region to avert severe water scarcities that will depress agricultural production, affect the household, tourism and the industrial sectors and damage the environment. In spite of the seriousness of the situation, the understanding of the water cycle-related linkages between different societal sectors is still weak. Goal conflicts remain unattended. Fundamental trade-offs need to be clarified in particular between the consumptive use of water involved in expanded food production and the water requirements of other sectors and ecosystems. The conventional, compartmentalized supply-oriented approach is not coping with the present water problems. Their solution requires an integrated approach to water, land use and ecosystems addressing the role of water in the context of socio-economic development and environmental sustainability. Although Morocco has developed a comprehensive reform agenda of its water sector, the implementation of the reforms is still slow. It is indeed complicated by the place water occupies within the set of interdependencies between different physical, economic and social systems. Thus, reforms cannot be seen in isolation from socio-economic development policy, urban policy and urban design, land use and agricultural policies etc... They should be also considered from the wider political and cultural changes occurring within the country mainly with regard to the progress of democracy and distributive governance. Transfer of water from the main water user – agriculture- to urban areas is perceived as having high social and political costs. Non conventional waters can contribute to cope with the declining water resources as wastewater is in most cases not adequately treated and reused and desalination is becoming a competitive technology for urban water supply in coastal areas where supply from conventional sources is limited.

References

1. Ait Kadi, M. (1998) *A comprehensive Water Conservation Programme*, Paper presented at the Regional Conference on Water Resources Management in the Mediterranean Countries, Valence, June
2. Ait Kadi, M. (1998) *Irrigation Water Pricing Policy in Morocco's Large Scale Irrigation Projects* Paper presented at the World Bank Sponsored Workshop on Political Economy of Water Pricing Implementation, Washington D.C., November 3-5

3. Ait Kadi, M. (2000) *Les Politiques de l'Eau et la Sécurité Alimentaire au Maroc à l'aube du XXIème Siècle* (Exposé Introductif) Proceedings of the Academy of the Kingdom of Morocco, Autumn Session 2000, 20-22 November

4. Ait Kadi, M. (1998) *Water Challenges for Low Income Countries with High Water Stress – The need for a Holistic Response : Morocco's Example* Proceedings of the Seminar on the 20th Anniversary of Mar del Plata, SIWI, Stockholm, August 10-16

5. Direction Générale de l'Hydraulique, (2002*) Plan National de l'Eau – Rapport de concertation*, juin

6. Association Marocaine de Prospective, *L'économie de l'eau au Maroc*, actes du colloque organisé les 18, 19 et 20 mai 2001, Université AL Akhawayne- Ifrane

Chapter 12

Water Scarcity Prospects in Egypt 2000 – 2050

S. T. Abdel-Gawadh, M. Kandil and T. M. Sadek
Ministry of Water Resources and Irrigation
El-Warak – Giza, Egypt

The challenges faced by many countries in their efforts for economic and social development are usually related to water. Population increase on the one hand and water shortages and quality deterioration on the other are among the problems which require greater attention and action. Therefore, immediate measures should be endorsed and actions should be taken to mitigate the serious destructive impact of water shortages and deterioration of water quality. Integrated Water Resources Management (IWRM) is recognized as a process that can assist countries in their endeavor to deal with water issues in a cost-effective and sustainable way.

In this regard, all endeavours and activities pertaining to water issues and their eventual foreseen crisis in Egypt have revealed a strong emphasis on the urgent need to formulate a comprehensive national water resources plan to be endorsed into the twenty first century.

Egypt has been well acquainted with pressing water related problems and the various challenges of the 21st century. During the past two decades, the Ministry of Water Resources and Irrigation (MWRI) has developed water policies such as Water Master Plan (1981) and Water Security projects (1993) to define guidelines of management of water resources (both water supply and demands) for short and long terms in Egypt. The current National Water Resources Plan (NWRP) which is being developed now, utilizes an integrated management approach to develop the future national water policy up to the year 2017. Its main objective is: "To develop the National Water Resources Plan that describes how Egypt will safeguard its water resources in the future, both with respect to quantity and quality, and how it will use these resources in the best way from a socio-economic and environmental point of view".

1. Water Resources Availability in Egypt

1.1. CONVENTIONAL WATER RESOURCES

The conventional water resources in Egypt are limited mainly to the Nile River and groundwater in the Nile Delta, the deserts and Sinai. Limited rainfall and flash floods are also viable sources of water.

The Nile, which is the main and almost exclusive source of fresh water in Egypt, originates outside the country's international borders. The country relies on the available water stored in Lake Nasser to meet needs within the limit of Egypt's annual share of water, which is fixed at 55.5 Billion Cubic Meters (BCM) by an agreement with Sudan in 1959. Being the most downstream country in the Nile basin, this makes cooperation with other Nile basin countries essential.

A. Marquina (ed.), Environmental Challenges in the Mediterranean 2000-2050, 187–203.
© 2004 *Kluwer Academic Publishers. Printed in the Netherlands.*

The renewable groundwater aquifer of the Nile basin covers the large spatial areas of the Nile Valley and Delta. This aquifer is recharged from excess irrigation water as well as leakages from the Nile and the distribution network. The approach of the Ministry of Water Resources and Irrigation (MWRI) in utilizing this aquifer is based on sustainability rather than mining, i.e. that annual withdrawal rates do not exceed those of the recharge, implying the dependence of this resource on the Nile water. Nevertheless, the fact remains that this aquifer stores vast amounts of water that may be mined if deemed necessary [3]. Currently, abstraction from the Nile aquifer is about 4.8 BMC/yr. By the year 2017, an amount of 7.5 BCM/yr is expected to be abstracted.

Groundwater in the Western Desert is in the Nubian sandstone aquifer that extends below the vast area of the Governorate of the New Valley and its sub-region, east of Awaynaat. This aquifer stores large amounts of water, estimated in hundreds of billions of cubic meters. However, the groundwater is stored at great depths and the aquifer is generally non-renewable. The total extraction potential of fresh water in the Western Desert is estimated at about 3.5 BCM/year. Considering the planned horizontal expansion in the Western Desert this potential will be fully utilised by 2017[4]. Most of the development (more than 85%) will take place in East Oweinat and the Farafra and Dakhla oases (Figure 1).

There are vast amounts of unpolluted brackish groundwater of varying salinity in the Nubian sandstone and Moghra aquifers. Potential use of this largely untapped resource includes use in agriculture (salt tolerant crops), aquaculture, and industry. Depending on the distance to settlements, brackish water could also be used as a raw water source for public water supply, since desalination of brackish water is cheaper than desalination of seawater. This source can be utilised for further developments in desert areas from the year 2017 to 2050.

Apart from the Western Desert, there is also potential for further groundwater development in the Sinai (some 200 to 300 MCM/YR) and the Eastern Desert (roughly 50 to 100 MCM/yr). The water is found in a mixture of aquifers that receive recharge (wadi aquifers and basement rocks) and aquifers that contain fossil water (Nubian sandstone and carbonate rocks).

Rainfall on the Mediterranean coastal strip decreases eastward from 200 mm/year at Alexandria to 75 mm/year at Port Said, and declines inland to about 25 mm/year near Cairo. The total amount of rainfall may reach 1.5 BCM/YR. Rainfall occurs only during the winter season in the form of scattered showers, and therefore cannot be considered a dependable source of water. Nevertheless, people living on the northern coast to the west of Alexandria and in Sinai utilize these small amounts of water in some seasonal rain-fed agriculture.

Floods due, in part, to short periods of heavy storms and terrain attributes are a source of environmental damage, especially in the Red Sea area and southern Sinai. Harvesting of flash floods in, for example, the Sinai or the Eastern Desert, through the development of small reservoirs and subsequent infiltration to recharge groundwater, is only considered viable if these reservoirs also have an important flood protection function. This is the case in some areas of the Sinai. A major problem is that rainfall is highly irregular in time and variable in magnitude. Small reservoirs have a higher yield/storage ratio but they rapidly fill with sediment. Larger reservoirs have a lower yield/storage ratio but their benefit/cost ratio is considered too low.

Figure 1. Map of Egypt

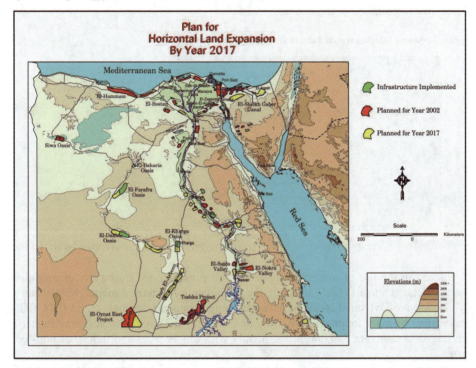

More feasibility studies to define potential locations for such harvesting, in combination with flood protection are being carried out.

1.2. NON-CONVENTIONAL WATER RESOURCES

One way of using scarce water resources more efficiently is to reuse low quality water for agriculture, industry or non-potable use. This includes agricultural drainage water, treated wastewater, and desalination of brackish and seawater. Each resource has its use limitations. These limits relate to quantity, quality, space, time, and/or cost of use. These types of sources should be used and managed with care, and their environmental and health impacts must be evaluated carefully.

Agricultural drainage water has emerged as the most attractive type of unconventional resource in Egypt in supplementing available water resources. In fact, the reuse of agricultural drainage water has been adopted as a national policy since the late seventies. The policy calls for pumping drainage water and mixing it with fresh water. Mixing criteria are based on the sustainability of the blend for irrigation of all crops. Currently, an amount of 5 BCM/yr of drainage water in the Nile Delta is now reused directly or after mixing with fresh water (Figure 2). The plan for horizontal land expansion is to reuse 8.4 BCM/yr by the year 2017. In view of cost-benefit considerations and possible adverse quality effects of drainage water reuse, priorities have been set for areas where drainage water would otherwise flow outside the system to sea and deserts, also where no harm is done to other downstream users and groundwater is least vulnerable to pollution in these areas.

Figure 2. Annual drainage reused water in the Delta

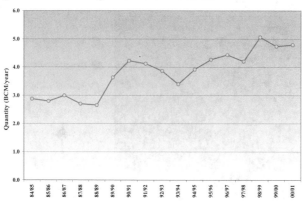

Interest in the use of treated wastewater, as a substitute for freshwater in irrigation, has accelerated since 1980. Currently, the total quantity of reused treated wastewater is 0.7 BCM/yr of which 0.26 BCM is undergoing secondary treatment and 0.44 BCM undergoing primary treatment. By the year 2017, 2.5 BCM/yr of treated wastewater is envisaged to cultivate 280,000 acres, mainly with timber trees and industrial non-food crops.

Desalination of seawater in Egypt, as a source of water, has been given low priority due, in part, to its high cost (Egyptian Pounds £E 3-7/m^3). Nevertheless, it may be feasible to use this method to produce and supply drinking water, particularly in remote areas where the cost of constructing pipelines to deliver Nile water is relatively high. It is expected that desalination plants for drinking water and industrial use in areas where no other cheaper resources are available, will be developed as the demands grow up to 2050. However if brackish (ground)water is available nearby in sufficient quantities, this may be the preferred source for desalination, depending on the distance to source.

2. Water Uses

2.1. WATER FOR AGRICULTURE

The agricultural sector is the largest user and consumer of water in Egypt, with its share exceeding 82% of the total gross demand for water. On a consumptive use basis, the share of agricultural demand is even higher at more than 95%. Consumptive use has been steadily increasing from an estimated 29.4 BCM/yr in 1980 [8] to 41.4 BCM/yr in 1999/2000. This increase has been evident due to an increase in drainage water reuse and groundwater abstraction, and a decrease in the fresh water outflow to the sea. The total water diverted to agriculture from all sources that includes conveyance, distribution and application losses is estimated to be about 55.2 BCM/year (Table 1).

During the period (1980-1996/97) cultivated land increased from 5.8 million acres to 7.8 million acres; the amount of water consumed per acre increased from 5,070 to 5,230 m³. The rise was due to increased crop intensity (about 180% in 1996) and/or increase in the cultivation of high water consuming crops (such as rice and sugarcane). Including the new horizontal expansion areas, the total cultivated area was about 8.5 million acres in the year 2000 (about 5.5% of Egypt's area).

The future increase in overall irrigation supply will depend on changes in the (priority) demands for the municipal and industrial sectors, the development of new groundwater resources, and measures to reduce the outflow (terminal drainage) from the Nile system. Any water becoming additionally available will be used primarily to irrigate new development areas and not to increase the supply to existing land.

2.2. POTABLE WATER SUPPLY

Despite the rapid population growth in Egypt, the percentage of the population with access to municipal water supply has increased over the past two decades due to large investments in the water sector. An estimated 97% of households in urban areas and almost 70% of households in rural areas had access to piped water in 1995. In populous cities, such as Cairo, Alexandria, Port Said and Suez, 99% of households have access to piped water; whereas this is the case for 90% of urban households in Upper Egypt. Rural areas, and especially those in Upper Egypt, are the most inadequately served. Only 55% of households in rural Upper Egypt have access to piped water. The parts of the population that have no access to piped water obtain their water from public standpipes (often connected to groundwater wells), street vendors or directly from canals and the Nile river. It is expected that all Egypt will be covered by potable water supply by the year 2017.

Coverage rates for sanitation systems, which is the percentage of population connected to sanitation services compared to the total population, are much less than those for water supply, but have improved for urban areas. According to the Institute of National Planning (1996), 97% of urban households and 70% of rural households have access to sanitation in the form of some kind of toilet facilities. Urban Governorates (Cairo, Alexandria, Port Said and Suez) have the highest sanitation coverage (98%), whereas the sanitation coverage in the rural areas of Upper Egypt is the lowest (57%). Only a fraction of households with access to sanitation are actually connected to sewer systems. Regional variations in sewerage connections are profound. At present, the nationwide coverage for household sewerage connections in urban areas ranges between 30 and 70% with Cairo having the highest level of coverage.

These rates have been sustained over the past decades despite a large population growth, particularly in Cairo. Although rural population densities are often high, there is almost no sewerage coverage in rural areas, which have an estimated coverage of less than 5% with sewerage network. In areas without sewerage network, wastewater is often collected in septic tanks or other forms of on-site disposal systems. These installations are frequently leaking and overflow due to poor construction and lack of maintenance. They are a major source of water pollution and unhygienic living conditions.

The municipal water demand is currently in the order of 4.5 BCM/yr. This figure is expected to increase significantly due to the relatively high annual population growth rate (2.1%) (Table 1). The resulting total municipal demand for the year 2017 is projected to be 6.6 BCM/yr. It should be mentioned that the drinking water sector receives higher priority in water allocation than the agricultural sector in Egypt.

2.3. INDUSTRIAL WATER CONSUMPTION

Industrial water demand constitutes a considerable part of the total water demand in Egypt. The total present demand is estimated at 7.6 BCM/yr. A small portion of this quantity (only 0.79 BCM) is consumed through evaporation during industrial processes, while the rest returns back to the system in a polluted form. Future industrial water requirements are estimated by taking into account planned growth rates of different industrial sub-sectors. The projected industrial water demand for the year 2017 is estimated at 10.6 BCM/yr (Table 1).

Table 1. Present and future water requirements

Requirements	Quantity (BCM/year)	
	2000	2017
Agriculture	55.20	70.70
Municipal	4.50	6.60
Industry	7.60	10.60
Navigation	0.26	0.00
Evaporation	3.00	3.00
Total	70.56	90.90

2.4. FRESH WATER TO THE SEA

This is the amount of water that is released during the low-flow period (January – February) to maintain sufficient water depths for navigation in the main Nile. Most of this fresh water spills directly into the sea through the Edfina barrage on the Rosetta branch (Figure 3). During the year 1999/2000, the fresh water discharged to the sea was 0.26 BCM.

2.5. EVAPORATION

Annual evaporation from open water surfaces is estimated to be about 3.0 BCM/yr using the total water surface area of the Nile River inside Egypt and the irrigation network and an average annual rate of evaporation. This amount varies slightly from one year to another according to climatic conditions as well as the rate of infection of canals and drains with aquatic weeds.

Figure 3. Schematic diagram of the Nile and its branches with main control structures

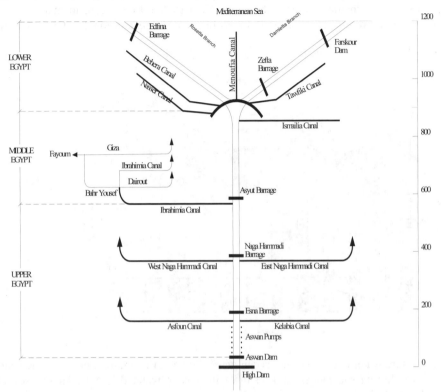

3. Key Issues and Challenges in the Water Sector

Egypt is confronting numerous constraints and challenges impeding the achievement of sustainable water development and management. The key issues are synthesized below [2].

3.1. POPULATION GROWTH AND WATER SCARCITY

The most important challenge facing Egypt is the expected population growth from 69 million in 2000 to 83 million in 2017 (Figure 4) and the related water demand for public water supply and economic activities, in particular agriculture [5]. To relieve the population pressure in the Nile Delta and Nile Valley, the government has embarked on an ambitious program to increase the inhabited area in Egypt from the present 5.5% to about 25%. It is obvious that the demand for water resources will increase. Industrial growth, the need to feed the growing population and hence a growing demand for water by agriculture, horizontal expansion in the desert areas, etc. will cause a growing demand for water resources. At the same time the available fresh water resources are expected to remain more or less the same. This is therefore urgent to make a more efficient use of present resources and, if possible, to develop additional water resources.

Figure 4. Population growth and water availability per capita in Egypt

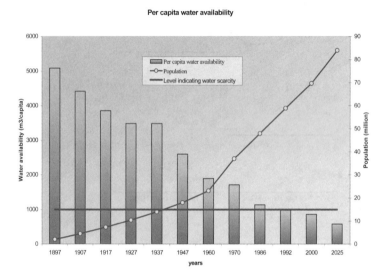

The future water quality problems can generally be attributed to high (and increasing) loads of polluting substances as a result of increasing water demands. The causes are described as follows:

- Institutional causes
    - Co-ordination between ministries and institutions involved in distribution and treatment of water and wastewater is insufficient. As a consequence investments in treatment or reuse pump stations do not always lead to the optimal benefits.
    - Implementation and enforcement of laws and regulations are ineffective, and are generally more aimed at the letter of the law rather than the purpose of the law. As a consequence, proposals for limited treatment have not been given permits. Thus the situation of no treatment in some places continues.
    - Information dissemination is insufficient. Data is rarely distributed outside the collecting agencies. As a consequence many ministries and institutions have to work without sufficient insight into the actual status of the system.
- Water quantity
    - Reduced flows in the Nile (due to increased demand in new horizontal expansion areas) and drains (due to increased efficiencies and reuse) will lead to less dilution of the loads and will thus lead to increased concentrations.
    - Increased reuse will bring more polluted water into the Irrigation Canal system.

- Pollution loads
  - Increased population and industrial production will increase the loads of pollution in the system. As the treatment plants to be built in the coming years will not be able to cope with the increase, the remaining load that will reach the water system is still expected to grow significantly in the coming years.
  - In addition to increases in industrial and domestic loads, agricultural loads also are expected to grow as a result of the increase in the agricultural area and an intensification of agricultural production.
- Economic / financial causes
  - The lack of sufficient financial resources is limiting the implementation of the required measures of the future strategy. This is partly caused by the fact that a large portion of the finances are presently expected to come from central government. For that part of the finances that is presently recovered from beneficiaries of the services, there is a problem of being able to obtain social acceptance of the actual costs, as well as problems related to billing and metering.
  - Problems with cost recovery limit possibilities of implementing BOT (Build Operate Transfer) and further involvement of the private sector in financing measures.

3.2. CHALLENGES RELATED TO INSTITUTIONAL FRAMEWORK

The Government of Egypt aims to promote sustainable growth through enhanced private sector involvement in production and in-service provision. The assumption underlying this aim is that less Government involvement in the day-to-day management of water allows for a greater economic efficiency.

Likewise, the water sector's economic efficiency is expected to increase through attributing a greater role to water users in the management of Egypt's waters. Or in other words: the same (or higher) levels of service and production would be achieved at a lower cost.

Moreover, with its present large role in water management, the government assumes also a substantial financial burden. To reduce the financial burden for investments, operation and maintenance of the water-related infrastructure, the Government aims to share the costs with the primary stakeholders.

The MWRI has taken the above concerns to heart and has embarked on pilot activities on water management transfer. The MWRI has developed a Vision Statement – with a deadline to 2025 – which defines the future institutional set-up for irrigation water management and which draws up a program of measures to reach that situation. Meanwhile, an Institutional Reform Unit (IRU) has been set-up to be the motor behind the management transfer and related measures in the MWRI.

The Vision stands on three pillars:

- Decentralisation of water management responsibility from MWRI to Water Boards. These are user organisations that are mandated to govern local water management within their jurisdiction.
- Privatisation of services in water management, in which specific MWRI activities are targeted for transfer to the private sector.

- Concentration of the MWRI on the management of creating the environment for decentralised and privatised water management by defining and enforcing the legal framework for sustainable water management, by ensuring adequate management of the main irrigation system and by concentrating its remaining services at Governorate level.

These institutional reforms are to be effected over a period of 20 years. The MWRI has formed an Institutional Reform Unit, which is charged with facilitating the implementation of the envisaged institutional reform. It would do so by:

- Co-ordinating institutional reform in the irrigated agricultural sector at strategic and operational levels.
- Increasing the internal and external awareness and acceptance of institutional reform in water management.
- Assessing the organisational effects of institutional reform in the irrigated agricultural water sector on MWRI and the wider economy.
- Stimulating, facilitating, supporting, monitoring, evaluating and reporting on decentralisation of Government responsibilities, private sector participation and privatisation of various assets and services of MWRI.

4. The National Water Policy Up To 2017

One of the major challenges facing the water sector in Egypt is closing the rapidly increasing gap between the limited water resources and the escalating demand for water in the municipal, industrial and agricultural sectors.

To meet the increasing demand, and at the same time ensure the sustainable use of water resources, Egypt has to optimize the use of conventional resources as well as investigate the means for making non-conventional sources of water feasible.

The MWRI is mandated to control and manage all fresh water resources in Egypt including surface and ground water. In addition to construction, supervision, operation, and maintenance for water structures, wells, irrigation and drainage networks and pump stations, the Ministry is also responsible for meeting the demands of all other sectors for good quality fresh water on time.

During the last decades, noteworthy developments with respect to water management have been initiated in Egypt [2]. A detailed water policy was formulated in 1975 that discussed available resources and development potentials. A water master plan was prepared in 1981 [8] that included a framework for water policies, investment programs and operational procedures. A water security plan was formulated in 1993 that defined guidelines of water resources management (both water supply and demands) for the short and long terms [9].

However, recent developments regarding water scarcity, infrastructural needs and environmental and socio-economic considerations necessitate a revision of planning and management practices. Currently, the MWRI is finalizing a National Water Resources Plan that describes how Egypt will safeguard its water resources (quantity and quality) and how it will use the resources in the best way from the socio-economic and environmental points of view up to 2017. The new water policy, which will be completed by the end of 2003, considered the following [7]:

Enhancing the approach of MWRI

The new policy advocated that water management is not an aim in itself but that it should support the achievement of other governmental goals like social, economic and environmental goals. This means that an integrated water management should consider all interests, such as the interests of agriculture, ecology, industrial development, transport, recreation, fisheries and public water supplies. And, these interests are not more or less conditions such as in the previous policy document but now they are considered as policy objectives that deserve to be achieved. This is a major step in the direction of a more developed integrated water management. This step can be taken because there are now simulation models available by which the quantitative effects of actions can be demonstrated, enabling a trade-off between different objectives.

Stakeholder involvement

The integration with related policies requires co-operation between representatives of different groups, in other words stakeholders' involvement. These stakeholders are not to be restricted to organisations of public administration, like other ministries and governorates. The private sector has its own responsibility as a water user. This sector should also have a task and role in an efficient use of water resources, the development of new water resources and the protection of water quality. The new strategy contains proposals to enhance the involvement of representatives of all stakeholders' organisations to continue the monitoring and evaluation of the implementation process of the national water policy.

Institutional change

Another major step forward towards achieving water policy objectives is the overall government's policy of improving the performance of the public sector by enhancing the private sector. The new policy will elaborate further the institutional reform policy in water management that aims at an improvement in the performance of the irrigation and drainage system by transferring public responsibilities to the private sector. The present vision is that the government should remain fully responsible for the main parts of the irrigation and drainage system, while the private sector, and especially agriculture, should be more involved in the operation and maintenance of the lower parts of the system, such as the branch canals and the district canals. This strategy advocates some further steps towards such an involvement.

Implementability and Implementation Plan

Implementability is a key characteristic of this strategy. During the process of choosing the required measures to formulate the future water strategy, special attention was given to the question as to whether the measures could be implemented in terms of costs, necessary institutional capacity and public support.

Next, as part of the National Water Resources Plan an Implementation Plan will be made that will describe who will be responsible for the implementation of the various measures, how the measures will be financed and what kind of co-ordination and stakeholder involvement will take place for this implementation.

Update of resource assessment and use of new planning tools
The water sector has experienced rapid developments, and demands for water have changed significantly. Moreover, new tools have become available to estimate the demands and assess the availability of water. The new policy will take the new insight in these aspects into account, where possible based on quantitative assessments provided by computational planning tools.

The new policy included measures that have been classified according to the following three major categories:

- Develop additional resources. Examples are fresh groundwater development, rainfall harvesting, desalination.
- Making better use of the existing water resources. This means an improvement of the efficiency of water use such as improvement of irrigation, influencing the consumption of drinking water, etc.
- Protect environment and health, e.g. pollution control.

In addition, measures have been defined that focus on institutional changes. These measures will be needed to provide the conditions for a successful implementation of the more technical measures (Section 6). The three categories of the new policy are explained as follows [6]:

1) Development of new water resources and cooperation with the Nile Basin Riparian countries

The possibilities of developing additional resources are limited. Deep groundwater withdrawal in the Western Desert can be developed up to an amount of 3.5 BCM/year, but, being fossil water, this is not a sustainable solution and should be carefully monitored. Small amounts of additional resources can be developed by rainfall and flash flood harvesting and the use of brackish groundwater.

Co-operation with the riparian countries of the Nile Basin will eventually lead to an additional inflow into Lake Nasser that can be utilised for more developments up to 2050. On the regional scale and recognizing that cooperative development holds the greatest prospect of bringing mutual benefits to the region, the Nile riparian countries, including Egypt, have established the Nile Basin Initiative (NBI).

Launched in February 1999, the Initiative includes all Nile countries and provides an agreed basin-wide framework to fight poverty and promote socio-economic development in the region. To raise broader donor support for the Nile Basin Initiative and its portfolio of cooperative projects, a first meeting of the International Consortium for Cooperation on the Nile (ICCON) took place June 26-28, 2001, in Geneva through which the donor community pledged around 140 million USD to support the NBI programs.

It is expected that Egypt's Nile Water share can increase through stimulation of such co-operative programs and the completion of the upper Nile Basin water conservation projects by the year 2050.

2) Making better use of the available water resources

Measures aiming at a better use of existing resources focus on improving the efficiency of the water resources system. The water use efficiency in agriculture can be improved by many measures, in particular by continuing the Irrigation Improvement

Project (IIP) activities and by reviewing the present drainage water reuse policy, e.g. by applying intermediate reuse (i.e. mixing the drainage water on the secondary less-polluted drains with fresh water in secondary canals rather than mixing on the main canals and drains level) and allowing a higher permissible salinity.

Other measures considered in the new policy are the introduction of new crop varieties such as early mature (e.g. short duration rice), salt tolerant seeds and the shifting of cropping patterns to less water-consuming crops which still achieve the food self-sufficiency and food security policies of the country.

Also, to ensure water allocation and a distribution system that will be based on equity and that will decrease the losses in the system, improvement of the physical infrastructure is being considered (e.g. installation of discharge control structures at key points and system rehabilitation where needed to lead to regional water allocation based on equal opportunities for farmers). MWRI also has programs to replace and rehabilitate existing grand barrages and control structures on the Nile and main canals to increase storage capacity of the system.

With respect to the municipal and industrial sectors, water use efficiency can be improved to reduce the losses in networks (e.g. demand management through installation/rehabilitation of the metering system and promotion of the application of water saving technologies in industry; and the reduction of leakage losses through leak detection and repair based on priorities for the most urgent rehabilitation work).

3) Protection of public health and environment

The strategy to protect public health and environment includes several infrastructural, financial and institutional measures. Priority is given to measures that prevent pollution. This includes the reduction of pollution by stimulating cleaner production technology, clean products and reallocation of certain industries. Agriculture will be encouraged to use more environmentally friendly methods and agro-chemicals. If pollution can not be prevented, treatment is the next option. This includes treatment of municipal wastewater.

Domestic sanitation in rural areas requires a specific approach. In both cases cost recovery is needed to maintain the services. The last resort will be to control the pollution by diverting the pollution away from urban or important ecological areas, and to reduce negative impacts on public health by raising awareness. Additional attention is required to protect sensitive areas, e.g. around groundwater wells and intakes of public water supply.

MWRI has established and operates a National Program of Water Quality Monitoring on the Nile, two branches, main canals and drains and Lake Nasser. On the other hand, the long term policies to control pollution include: coverage of the open conveyance system passing through the urban system to closed conduits; coordination committee with other concerned ministries to set and follow up on priorities for wastewater treatment plants; and the introduction of environmentally safe weed control methods (mechanical, biological and manual) and banning the use of chemical herbicides. Subsidies for fertilizers and pesticides were removed and some agricultural chemicals with long lasting effects were also banned. Public awareness programs are taking place about the importance of conserving Egypt's water resources in terms of quality and quantity.

5. Supporting Institutional and Legislative Elements

The National Water Policy also supports changes in the institutional and legal frameworks. The formulation of the relevant vision for these frameworks is given serious consideration and careful assessment in order to provide the proper environment for the application of the integrated water resources management.

5.1. STAKEHOLDERS PARTICIPATION

In order to achieve improved water management, some of the operational and maintenance responsibilities are now being gradually transferred to the water users and beneficiaries. Farmers are encouraged to form local water user associations and water boards through which planned programs can be implemented. This would result in a greater role for the users and the private sector in water management and services. Additionally, stakeholder participation in the decision making process would reduce water-related conflicts and provide more commitment and support to the water resources management and planning at all levels.

Egypt has slowly started studying and implementing the policy of cost recovery from beneficiaries as a mechanism for improving water use efficiency and introducing new methods and techniques that control the irrigation networks better and achieve fair water distribution [1]. For example, farmers already participate in the operation and maintenance cost of the field-level small canals (i.e. mesqas) as they pay £E 18 per acre per year and also through land tax as a contribution of labor cost recovery for operation and maintenance of the system.

5.2. PUBLIC/PRIVATE PARTNERSHIP IN MEGA PROJECTS

The Government of Egypt (GOE) has embarked on ambitious programs to increase habitable land from the current 5.5% to about 25%. The projected increase is planned through a series of mega land-reclamation projects of which the North Sinai and South Valley Development Projects rank top in terms of size and goals.

The South Valley Development Project aims at reclaiming an area of around 1 million acres in three regions: around the Toshka depression, in East Oweinat and in the New Valley Governorate Oases. Toshka and East Oweinat are considered the most important phases of the project since, combined, they count for over 750,000 acres of the total targeted area. Total investment required for developing the region is estimated at £E 300 billion over the coming 20 years. The government will finance up to 20-25% of the total expenses of constructing the main irrigation canal and branches, installing the pumping station and setting up a convenient infrastructure network. The rest will be financed by local and foreign private sectors in five major sectors: land development, new community development, industry, agriculture, and tourism.

The North Sinai Development is a multi-sectoral initiative. Total investment needed for the whole development program is projected to reach £E 64 billion by 2017. The private sector is expected to provide 62% of this amount. The development project includes land reclamation of 400,000 acres in Sinai. This amount includes financing the construction of new communities within the project site. Land holdings in the project area are a mixture of both small and large farms. All water resources needed for mega-projects are allocated from the current share from the Nile water which is 55.5 BCM/year (e.g. about 5 BMC/year will be allocated from the Toshka project when it is fully operational).

With greater participation of the private sector in these two mega projects, the GOE has established two holding companies to manage, operate and maintain the irrigation and drainage networks in both projects. The current vision for these companies is that they will offer mechanisms to provide appropriate services to both investors and small farmers. However, the main infrastructure including irrigation and drainage networks and pump stations will still be the property of the government, maintained and operated by MWRI. These companies will raise enough funds through the collection of service charges from beneficiaries in the projects and by selling new lands to investors. Each of the holding companies will comprise several smaller companies giving a greater role to the private sector from the beginning.

5.3. INSTITUTIONAL REFORMS

Institutional reforms in the water sector are an important premise in implementing the action plans and policies more effectively and efficiently. Existing water institutions need to be re-structured to undertake multi-disciplinary functions. The Government of Egypt is viewing the need for institutional reform in the water sector to promote economic efficiency and support sustainable water use in addition to environmental and ecological sustainability. This can be achieved by the decentralisation of the water management authority from MWRI to water boards and water user associations (i.e. farmer associations).

The MWRI vision on water management is also seeking to privatise selected Government responsibilities in water management. Building private sector capacity to assume these responsibilities requires good preparation. The MWRI is setting programs to assess private sector capabilities in water delivery on New Lands and in the Mega Projects; to identify capacity building needs (financial policy, management capacity, legal aspects, etc.); and to identify training needs. The Institutional Reform Unit (IRU) which was established in 2002 is playing an important role in assisting private investors to take over the operational and maintenance activities.

5.4. REVISION OF NATIONAL LAWS AND REGULATIONS

Regarding the management (operation and maintenance) of the Government-owned water management infrastructure, a modified version of the Governmental Law 12 for Irrigation and Drainage was drafted to provide the basis for government intervention and action and establish the context and framework for operation by non-governmental entities including water boards.

The modified Law provides the possibility of delegating part of the management or of transfering complete management to water boards/associations or to specialized companies. The level at which the water boards or specialized companies could operate is at the secondary (Branch) canal or above. This provision in the draft law is sufficient to accommodate the various conditions encountered in the irrigated areas and also allows adequate room for broadening the implementation of the concept of decentralized management and for a greater role of the private sector in the future. Through this law, water boards will be allowed to raise funds to cover operation and management (O&M) and other service expenses, including the cost of contracting service providing companies.

5.5. PROMOTION OF RESEARCH

The policy formulation process is based on a solid foundation of applied research findings and comprehensive planning studies. State-of-the-art technology, in various aspects of water resources planning and management, is employed by the MWRI. Several tools have been adopted and utilized such as remote sensing, modeling, GIS, monitoring networks and telemetry systems, which have produced a significant impact on the way the water resources are planned and managed. The MWRI is investing large resources in the National Water Research Center (NWRC) which was established in 1975 to carry out applied research in the various fields of the water sector and related subjects. This Center combines training with research on which the MWRI also places a growing emphasis.

However, research and development on wastewater treatment, desalinisation and water harvesting and conservation technologies are needed to develop new forms of appropriate, cheap and affordable technology that can be adopted and used widely. Cost, energy consumption and environmental quality are the main concerns for these water technologies.

6. The Road Ahead (from 2017-2050)

The complex dimensions of freshwater in Egypt, its fragility, its scarcity and its augmentation have been receiving considerable technical, scientific and political attention. Several programs and projects have been realized and others are on the way to alleviate water problems and to satisfy water needs. Some progress is now being felt. The most important challenge Egypt faces now is to continue on the road it has successfully embarked on so far with a sustainable water resource management scheme and a safer environment for the population of Egypt in mind.

Most of the efforts in water resources planning in Egypt have been focused on planning up to 2017 and 2025 as explained in the previous text. For water resources planning up to 2050, this will be based on strengthening cooperation at the regional level with the Nile Basin countries to increase the Nile water flow for the benefit of all countries sharing the resources. It is planned to stimulate water conservation projects in the Blue Nile basin to secure 12 BCM for the three countries Ethiopia, Sudan and Egypt. Also, efforts and studies will be conducted to develop and manage better the Nubian ground water aquifer, which is considered one of the largest aquifers in the World, in coordination with other countries sharing the same aquifer with Egypt. Furthermore, joint agricultural projects are being considered with the Nile Basin Countries or other African countries to optimize water use for food security and food self sufficiency in these countries.

At the national level, to secure more water resources up to 2050, more attention will be given to a greater development of deep and shallow groundwater in the desert and Nile valley aquifers and also to make the best use possible of brackish groundwater by desalination of the required quality needed for drinking, industrial developments and agricultural activities. More developments are planned with respect to desalination of seawater especially in remote and coastal areas as several studies are being carried out now to define the quantities required for developments in these areas. Additionally, efforts will be made to reduce evaporation losses from Lake Nasser by the reduction of surface areas at shallow reaches and to change the policy of operation of the reservoir and its impact on the water levels from upstream the High Aswan Dam. Other measures include increasing the storage capacity of the system by storing surplus flood water in the Eastern depressions of the Nile River, coastal areas of the Red Sea and by the recharging of groundwater aquifers in these areas. By applying these measures successfully it is expected that any deficit between supply and demand due to the rapid increase in population can be covered in 2050 by Egypt. Also a lower growth rate in the agricultural sector, which is the higher consumer of water, and a higher growth in industrial and services sectors, can be foreseen to meet any expected gap between the supply and demand of water resources in the future.

References

1. Abu Zeid M. and Hamdy A. (2002) *Water Vision and Food Security Perspectives for the Twenty First Century in the Arab World,* First Regional Conference on Perspectives of Arab Water Cooperation, Cairo, October 2002.
2. Attia, B. (1997) *Water Resources Policies in Egypt: Options and Evaluation*, IX World Water Congress of IWRA, A special session on Water Management under Scarcity Conditions: The Egyptian experience, Montreal, September 1997.
3. NWRP (2001a) Groundwater in the Nile Valley and Delta, *NWRP Technical Report 16*, June.
4. NWRP (2001b) Groundwater in the Western Desert, *NWRP Technical Report 15*, June.
5. NWRP (2001c) Population and Tourism Projections, *NWRP Technical Report 20*, June.
6. NWRP (2002) Water Management Measures, *NWRP Technical Report 24*, Draft version 2.0, May.
7. NWRP (2003) Facing the Challenge - Outline of a Draft Strategy, *Discussion paper 3*, version 2.3, January.
8. UNDP/IBRD (1981) *Water Master Plan*. Ministry of Water Resources and Irrigation. Cairo.
9. World Bank (1993) *The Demand for Water in Rural Areas: Determinants and Policy Implications,* The World Bank, Washington DC.

Chapter 13

Water worries in Jordan and Israel: What may the future hold?

Shlomi Dinar
The Paul H. Nitze School of Advanced International Studies (SAIS), The Johns Hopkins
University, 1740 Massachusetts Avenue, N.W., Washington, DC 20036

Rapid population growth, combined with inefficient and uncoordinated water utilisation and extraction practices are likely to exasperate the already keen water worries of the arid nations of the Middle East. This paper will discuss the water demands and practices of Israel and Jordan, two countries heavily dependent on the Jordan River Basin, which suffer from acute water shortages. The paper will also review the Treaty of Peace signed between the parties in 1994 that included a major settlement on the parties' historical water dispute and facilitated a cooperative framework for future water use and utilisation. Finally the paper will draw on future strategies and reflect on plans already in place by both states to attempt to alleviate their respective water worries.

1. Introduction

Among the many river basins that suffer from high degrees of water scarcity, the Jordan River Basin ranks high. The water deficit in the basin has had a great effect on both Israel and Jordan. While both countries signed a peace treaty in 1994 that included an agreement regarding their shared waters, the treaty did little to ameliorate their ever-increasing water needs. Furthermore, among all the other parties making claims to the still disputed waters of the Jordan River Basin, the Palestinians are perhaps most affected by the water shortage in the area. And while they are not a major focus of this paper, their water demands must be considered in any future agreement. Population growth will have the greatest effect on future water resources. By 2050 the population of both Israel and Jordan will double in size and that of the Palestinians will essentially quadruple (Table 1).

Table 1. Population trends of Israel, Jordan and the Palestinians (millions)

Year	Israel	Jordan	Palestinians
2002	6.6	5.3	3.5
2025	9.3	8.7	7.4
2050	11.0	11.8	11.2

Source: Population Reference Bureau

A. Marquina (ed.), Environmental Challenges in the Mediterranean 2000-2050, 205–231.
© *2004 Kluwer Academic Publishers. Printed in the Netherlands.*

To that extent, the gap between available water resources and the projected demand in the region is likely to increase to approximately 1230-1513 MCM/y[1] in 2020 and 2200 MCM/y in the year 2040 (Figure 1).

Figure 1: Regional Water Demand and Deficit (1994-2040) (MCM/y)

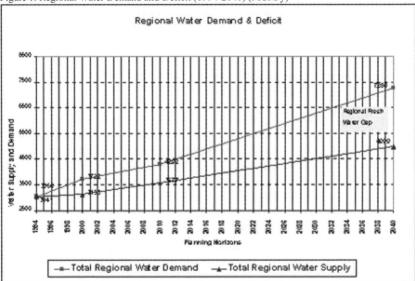

Source: Ministry of Regional Cooperation, The State of Israel, Israel and Jordan Launch Global Campaign to Save the Dead Sea, August 2002. [56]

Currently, water use for agricultural, domestic and industrial purposes varies throughout the region. Israel's use (between 1800-2000 MCM/y) is double that of Jordan's. Water usage on the West Bank and Gaza is the lowest, about 240-286 MCM/y [32, 67]. Within the next 20 to 30 years all of Israel's and Jordan's freshwater will be needed for drinking water demands alone if population growth, agricultural and industrial development continue at the present rate. This is likely to have an adverse affect on the agricultural sector—the sector that currently uses the majority of available freshwater.

In 1994, both Israel and Jordan signed a peace agreement. A large part of the Treaty of Peace focused on the parties' water dispute. Though, the water agreement concluded both institutionalized future and present cooperation, and for the most part settled particular grievances, water scarcity remains a serious issue for the region. While demand-side approaches are needed to alleviate the situation, supply-side mechanisms are also necessary. There are simply not enough natural water sources available in this arid region to satisfy the growing demand. Additional supplies will, therefore, need to be "created" to

---

[1] In addition to the figure, the totals are calculated using 360 MCM/y as a deficit for Jordan [88], 425 MCM/y for the Palestinians (with an estimated deficit of 425 MCM/y in 2023) [74], and 445-450 for Israel [79]. Other deficit estimates include: For the Palestinians—Between 30-220 MCM/y (if an autonomous Palestinian entity absorbs several hundreds of thousands of refugees the deficit is likely to double to between 60-440 MCM/y) [86]; 642-688 MCM/y (with Palestinian needs at 928 MCM/y and availability at 240-286 MCM/y) [69]; For Israel—800 MCM/y [86].

augment existing availability. It is likely that the latter approach will require subsidies from the international community and participation by the private sector.

2. Jordan

Hydrologically, Jordan can be divided into three major water resource sections: 1) the Jordan Valley—which includes the Yarmouk River, local wadis and groundwater wells in the same area (providing about 300 MCM/year); 2) the wadis draining into the Dead Sea, as well as associated groundwater sources in Amman, Zarka, Dhuliel Qastal and North of Zarka River (providing about 380 MCM/year); 3) other basins, oases and aquifers such as those in Wadi Araba, Jafir Basin, Disi, Azraq and the northeast region (providing a little more than 100 MCM/year). Approximately 80 percent of Jordan's known groundwater reserves are contained in three main aquifer systems: Amman-Wadi As Seer, Basalt, and Rum Aquifers (Figure 2). Estimates of Jordanian groundwater totals vary between 418 MCM/y and 530 MCM/y [30, 57]. The safe yield for nonrenewable aquifers is 275 MCM/y. Surface flows average about 692 MCM/y, with the collective long term average base flow for all basins at 359 MCM and flood flow at 334 MCM/y—only between 505-595 MCM/y can be economically developed [18, 57]. Jordan's overall total sustainable water supply has been estimated to be between 700-900 MCM/y [18, 30, 57].

Since Jordan is located in a semiarid and desert climate, it cannot solely rely on rainfall. The Kingdom rather depends on surface water, international and national aquifers. Current projections suggest that Jordan's annual water requirements are likely to reach 1230 MCM/y by 2005 and 1647 MCM/y in 2020 with supplies at 1042MCM/y and 1287 MCM/y, respectively [88]. The rise in water demand for the urban sector has been especially rapid— about 10 percent per year. In 2005 demand is estimated to be 400 MCM/y, while in 2020 it is estimated to be 757 MCM/y. According to World Bank projections the agricultural sector will continue to demand less varied amounts as opposed to the ever-increasing demands from the municipal and industrial (M&I) sectors (Table 2).

Table 2. Projected supply and requirements (1998-2020) (MCM/y)

Year	M&I		Agriculture		Total	
	Supply	Requirement	Supply	Requirement	Supply	Requirement
1998	275	342	623	863	898	1205
2005	363	463	679	858	1042[2]	1321
2010	486	533	764	904	1250	1436
2015	589	639	693	897	1283	1536
2020	660	757[3]	627	890[4]	1287[5]	1647[6]

Source: World Bank, 2001, 5 and 6. [88]

[2] Total supply is estimated to be 1030 MCM/y [58].
[3] Total demand is estimated to be 760 MCM/y [58].
[4] Total demand is estimated to be 900 MCM/y [58].
[5] Total supply is estimated to be 1260 MCM/y [58]; 1305 MCM/y (1165 MCM/y of renewable water resources plus 140 MCM/y of nonrenewable resources) [18].
[6] Total demand is estimated to be 1650 MCM/y [58]; 1742 MCM/y [18].

Figure 2. Jordan's underground water sources and flow

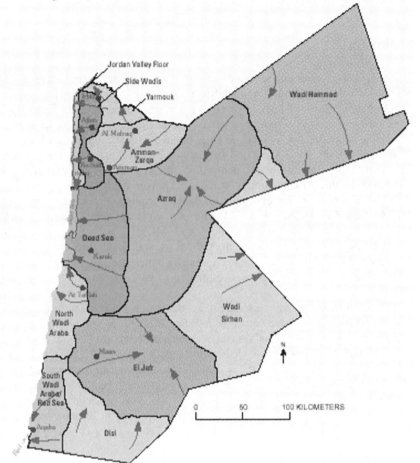

Source: Landers and Clarke, 1998. [42]
Note: Image is modified and international borders are not specified.

Interestingly, the supply figures are calculated on the assumption that particular projects are carried out and water deliveries are made with little delay. As current use already exceeds current supply and as delays in both the construction of projects and securing their financing is likely, the future is likely to bear the same results (Table 3).

Water demand will therefore grow due to increases in population, rise in living standards and an expanding industrial sector. Currently demand in urban circles can't be met because of the biased allocation of water to agriculture, inadequate infrastructure and inefficiency of water distribution.

Unaccounted for water in urban and industrial supplies exceeds 50 percent—due to inaccurate metering, pipe bursts, illegal connections and leakage [89]. In Amman, some residents often have to seek additional water during certain days of the summer. Certain residents are forced to make do without water, construct roof storage tanks for use during days of no available water, or purchase water from water trucks.

Table 3. Sector deficits and total deficits (1998-2020) (MCM/y)

Year	M&I	Agriculture	Total
1998	-67	-240	-307
2005	-100	-179	-279
2010	-47	-140	-187
2015	-50	-204	-254
2020	-97	-263	-360[7]

Most of the water in Jordan is used primarily for agriculture—the agricultural sector accounts for about 67- 77.5 percent of all water consumed [72, 88]. Smaller and medium sized farms are still using inefficient water practices, such as salt flushing and high use of water in winter times under the false pretense that additional water will increase yields.

Expansion in urbanisation and industry, however, is forcing the Kingdom to divert water from agriculture to municipal and industrial use and tap its non-renewable water sources [90]. The World Bank estimates that by 2020 the share of domestic and industrial use will constitute 52 percent of total water use [88].   With freshwater from renewable and nonrenewable aquifers going gradually to domestic uses, the agricultural sector will increasingly have to look for alternatives.

Jordan also suffers from groundwater over extraction, exceeding the safe yield by more than 100 percent in the majority of the identified renewable underground water sources [12, 89].[8] A similar grim picture can be painted for Jordan's nonrenewable aquifers. Already 50 MCM/y is being used from the Disi Aquifer by privately owned farms. Another 14 MCM/y is pumped out for domestic use in Aqaba [21]. Jordan also shares the Aquifer with Saudi Arabia. The safe yield of the aquifer is 125 MCM/y for 50 years, yet by 2020 Jordan plans to exploit about 140 MCM/y in fossil ground water from the Disi Aquifer [57, 58, 60].[9] Nonetheless, given Jordan's current over extraction from renewable underground aquifers, use of the Disi Aquifer, as a major future source of water, is considered the next best alternative by the Kingdom.

Due to the environmental and water quality ramifications of over drilling, contributing to the salinity of the water and the dire physical conditions of the aquifers, Jordan has already instituted several initiatives to reduce groundwater extraction. By 2020 the Jordanian government hopes to make use of 40 MCM/y of brackish water—saline water with a total dissolved solids of 1,000 to 10,000 parts per million [58]. Similarly, Jordan's reuse of wastewater in the Valley will also help to eliminate over-pumping. Treated

---

[7] Total deficit is estimated to be 400 MCM/y [58]; 437 MCM/y [18].

[8] Overdraft figure is at 300 MCM/y [12]; Over abstraction, including illegal wells, is about 225 MCM/y (with abstraction at 500 MCM/y and the safe yield at 275 MCM/y) [58]; Recent figures for 2000 indicate a deficit of 137 MCM (with abstraction from renewable sources at 412 MCM and safe yield at 275 MCM) [18].

[9] The plans call for a 195-mile long pipeline that will stretch from Disi, which lies near the Saudi border, to the capital, Amman. The project is to be completed over a five-year period. The capital cost of the Disi project is estimated at $625 million and recurring yearly costs at $18.2 million.

wastewater is expected to increase from 67 MCM in 1998[10] to 232 MCM in 2020—from 18 percent in 1998 to about 34 percent in 2020 (Table 4).

Most recently an agreement was signed between Jordan and Syria in April 2003 for the construction of the Wihdeh Dam on the Yarmouk River.

The Dam's capacity will be 80 MCM, and will supply greater Irbid as well as augment the Jordan Valley's irrigation reserves [4]. The water will eventually improve Amman's drinking water levels and will also control flooding from the Yarmouk River. Similarly, the planned Mujib Dam will have a capacity of 35 MCM, and will help meet possible shortages in the Southern Ghor Region if the winters end with small amounts of rainfall [60].[11] The dam will join three other recently constructed major dams: Karameh (55 MCM), Waleh (9.3 MCM) and Tannur (16.8 MCM) [57, 58].[12]

Table 4. Projected water reuse in the Jordan Rift Valley and the Highlands (MCM/y) (1998-2020)

Year	Jordan Rift Valley	Highlands	Total
1998	56	11	67
2005	65	43	108
2010	110	66	176[13]
2015	123	84	207
2020	137	95	232[14]

Source: World Bank, 2001, 10. [88]

Similarly, the government has already called for massive reductions in abstractions by Highland *pumpers* of renewable underground aquifers. About 90 percent of the wells are metered and new tariff methods are being instituted. Reduced abstraction by Highland agriculture is paramount (predicted 86 MCM/y by 2010 and 122 MCM/y by 2020) to preventing the loss of some aquifers due to the irreversible salinisation of groundwater stocks [88].[15] Stakeholder participation in the management and ownership of underground aquifers is being pursued to enhance the efficiency of and responsibility for aquifer use and upkeep [88].[16] Stakeholder participation is also enhanced through the facilitation of water markets. Already in the Jordan Valley, farmers own and manage surface water and sell it to one another. Private ownership of groundwater is also evident, though on a very small scale.

[10] In 2000 this figure was 75 MCM/y [18].

[11] Dam capacity to be 17 MCM [37, 58, 85].

[12] Other exclusively potable water projects (to augment Jordanian water reserves by 2011) include the above-mentioned Disi-Amman conveyor (100 MCM/y), the Karak-Tafila Water Supply project (5 MCM/y), the Tapqet Fahal-Irbid Phase I project (7 MCM/y) and other small projects (15 MCM/y); Some of the wastewater projects include Irbid Stage I (8 MCM/y), Amman-Zarqa Basin Phase I (40 MCM/y), Aqaba Water and Wastewater (5 MCM/y), Madaba Upgrading (3 MCM/y), Ramtha Upgrading (1 MCM/y) and Western Musa Water and Wastewater (4 MCM/y); Other small dams include Ibn Hammad (5 MCM), Meddein (1.46 MCM), Al-Karak (2.1 MCM) and Feedan (10.16 MCM) [60].

[13] Total wastewater reuse is estimated to be 140 MCM/y [58].

[14] Total wastewater reuse is estimated to be 245 MCM/y [18].

[15] During the period of 2004-2020, over abstraction of ground waters from renewable aquifers will be reduced to within the safe yield of the aquifers [18].

[16] Stakeholder participation has already been successfully instituted in Mexico, Australia and India, among other countries, and calls on users to directly be involved in the management, use and distribution of the water. The user-participation programs have lead to increased accountability, reduced costs and reduced pumping from aquifers (bringing inflow and outflow into balance).

Practical tariff policies have also been remarkably slow in their institution. Current estimates suggest that bills only reflect 80 percent of operation and management costs and user payments reflect only 66 percent of the costs of delivering irrigation services to farmers [88].

Stringent tariff policies must not only be pursued and intensified in the agricultural sector but must also take place in the municipal sector. In 1996 only 20 percent of operation and management costs of the Jordan Valley Authority were recovered—meaning that costs of delivery far exceed the tariffs levied on customers. Similarly subsidies were given to almost every customer of the Water Authority [89]. Nonetheless, large volume consumers of water have indeed been presented with higher water tariffs [88]. The same strategy should be applied to medium and small sized consumers of water as well.

The above mentioned multi-billion dollar investment plan will also entail upgrading Jordan's water distribution structures and reducing unaccounted for water instances. This will be financed largely through private sector participation (PSP). The main region targeted by this initiative will be Amman and the surrounding areas, which total about 1.6 million inhabitants—about 1/3 of Jordan's population [84].[17] Reduction of unaccounted for water will generate not only much needed water but also billable water. Similar projects being planned in Northern Jordan (areas such as Irbid, Jerash, Ajloun and Mafraq with a population estimated at 1.85 million by 2010), will include operation and maintenance of the water and sewage systems, rehabilitation of certain facilities, metering, and billing. By 2015, Jordan hopes to reduce municipal water network losses to 15 percent [18].

3. Israel

Israel's main sources of water include the Jordan River and its tributaries, the Sea of Galilee, the Coastal Aquifer and the Mountain Aquifer (Figure 3). Israel also extracts water from smaller sources such as in the Western Galilee, the Arava-Negev region and uses recycled and treated brackish water (Table 5).

Table 5. Total annual potential of renewable water (MCM/y)

Resource	Replenishable Quantities
Coastal Aquifer	320
Mountain Aquifer	370
Sea of Galilee	700
Additional Regional Resources	410
Total Average	1800 [18]

Source: Ministry of National Infrastructures, The State of Israel, The Water Sector: Water Resources and Water Availability, 2001. [48]

---

[17] The water network repair contracts are worth $210 million. For the purpose of the contract the Amman area has been divided into 43 zones.

[18] Figure probably does not take into account spills, leakage and unusable water due to salinity upon delivery. Also, in drought years the total is lower.

The majority of Israel's sources are vulnerable to climatic variability. The Sea of Galilee, for example, has had annual inflows ranging from as low as 100 MCM in drought years to a high of 1500 MCM.

Evaporation from the Sea of Galilee also affects Israel's water totals, as the Sea amounts to between a 1/4-1/3 of the country's water supplies. Leakage from pipes also takes its toll on the total amount of water delivered [41]. The Sea of Galilee's "red line," constituted as the mark of minimal levels, consistent with both water shortage and preventing environmental damages, has been consistently lowered to both adjust to the relative drought in the region and allow for needed extraction.

Figure 3. Israel's main underground aquifers

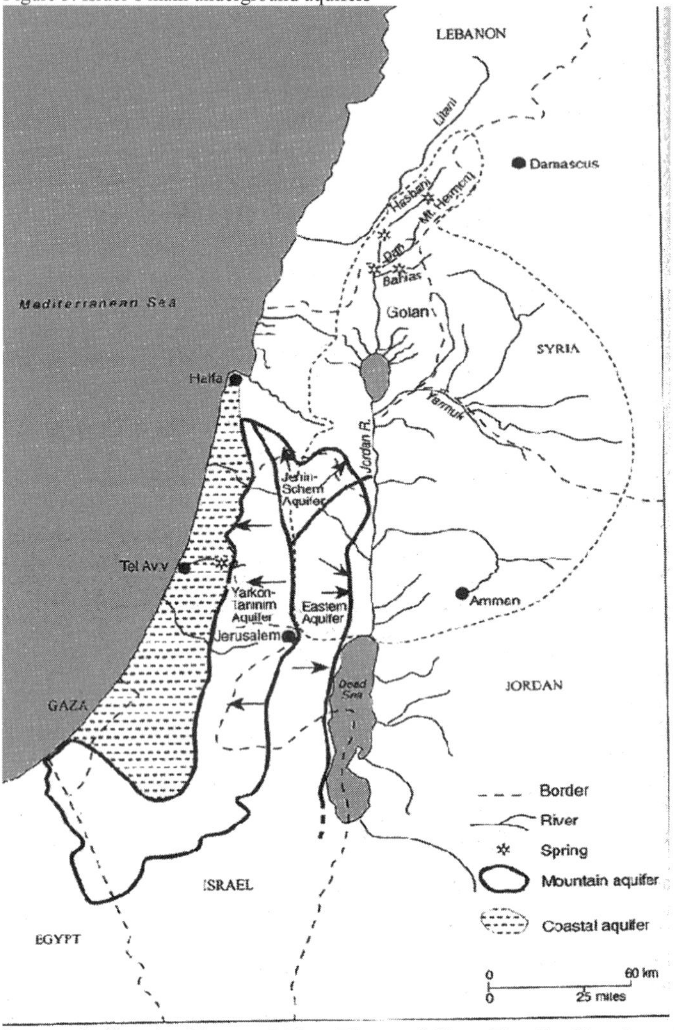

Source: Soffer, A., Rivers of Fire, Rivers of Fire: The Conflict Over Water in the Middle East, (Lanham: Rowman & Littlefield Publishers, Inc, 1999), pg. 102. [76]
Note: Certain aquifers under dispute with Palestinians

Drought, additional pumping and the lack of additional water supplies to augment current availability will only continue to lower the water levels in the Sea of Galilee and increase salinity (Figure 4).

Figure 4. Sea of Galilee water level and salinity (1969-2002)

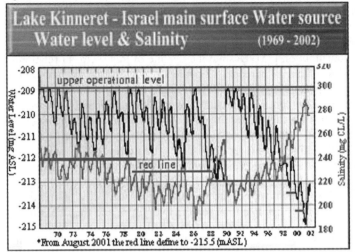

Source: Ministry of National Infrastructures, Water Commission, The State of Israel, Israel's Water Economy, August, 2002. [54]

Over-pumping from underground sources—especially the Coastal Aquifer and the Yarqon-Taneenim Aquifer relative to natural replenishment rates—has caused grave concern regarding dangerous chloride levels. Salinity both harms the quality of the water and the physical well being of the aquifer. Serious over-pumping has been occurring since the mid-1960s and portions of the above mentioned aquifers have been storing water that is above the internationally accepted standard of 250 mg/l. Salinity levels in other parts of the aquifers have reached 1000 mg/l [28]. Several other sources have affected the Aquifers, such as nitrogen seepage from agricultural fertilizers and sewage from urban areas. Average nitrate concentrations in Coastal Aquifer wells have greatly exceeded the international standards of 45 mg/l, reaching in some cases 70 mg/l [87]. Contamination from heavy metals and trace elements has been considered to be more of a localized issue but the presence of such contaminants is increasing [28, 45]. Yet, aside from the alarming salinity levels in particular parts of the Aquifers the situation is further complicated by the fact that some of these underground sources of water, specifically the Mountain Aquifer (comprised of the Yarqon-Taneenim Aquifer, the Northeastern Aquifer and the Eastern Aquifer), are not only shared but under dispute. Palestinians argue that they are entitled to an equitable share of the Mountain Aquifer as years of Israeli control of the West Bank have denied them this right [17]. Though Israel has recognized Palestinian water rights in the Mountain Aquifer, and interim allotments have been agreed upon, final allotments, division and redistribution of the water will be devised in a final settlement. Regardless of how the two parties settle their dispute over the Mountain Aquifer, a final settlement will not address the reality that the Aquifer can't sustain future demand. Today, Israel and the Palestinians face a situation where the water demanded and extracted dangerously exceeds the exploitable potential of the Aquifer. Recent findings from research projects in the region have found that yet unexploited water from the Eastern Aquifer, set in the Oslo II agreement as water to be used largely for Palestinian use, is highly saline and is therefore of poor quality [82].[19] Whether or not these recent findings will be vindicated by further research, both parties will have no choice but to search for other non-traditional alternatives to augment their water needs.

Currently agriculture is the dominant user of water in Israel—using between 60 to 72 percent of total water consumption [75]. Despite the decline of this sector in relation to the national economy (down from 11 percent to 2 percent of GDP since the founding of the state) and despite the virtual elimination of Israeli agricultural products for export, (down from 60 percent to 4 percent), agriculture has grown significantly in absolute terms, with important implications for water use [44]. Much of the consumption trends of the agricultural sector are due to the sector's strong political organisation and lobby power and to the subsidies given to water intensive agriculture. While government subsidies for farmer water costs have decreased (from 40 percent in 1993 to 23 percent in 1999) so have the associated capital costs, fixed costs and variable costs (from 36.4 cents/CM to 26.3 cents/CM, respectively). The ratios for consumer cost after subsidy divided by average costs are quite comparable, for the two years, and the difference is negligible—1:1.16 for 1993 versus 1:1.29 for 1999 [53].[20]

> 1993: 36.4 (average cost) * 1-0.40 (consumer pays after subsidy)= *21.840*
>
> 1999: 26.2 (average cost) * 1-0.23 (consumer pays after subsidy)= *20.174*[21]

The share of Israel's domestic and urban water use stands at about 30 percent and industrial use at 6 percent [44]. The Israel Water Commission predicts that domestic demand will increase to an estimated 1120 MCM/y in 2020 and that the agricultural sector's demand will stay at about the same level—1150 MCM/y. The shift within the agricultural sector in the year 2020 as compared to 2000, however, will take place in the increased use of recycled wastewater rather than freshwater [49, 79]—a 1.6:1 rather than a 1:3.25 ratio.

Though much is made of the agricultural sector's use of water (and for good reasons), positive assessments can also be made. Most recently, in 1999, the Israel Water Commission decided on a reduction in the water quotas for agriculture (1998 was set as the base year for the cut) by an average of 40 percent and by an average of 50 percent in the years 2000-2002. Extraction levies were also introduced in 1999, reflecting the country wide water shortage. Similarly, a relatively drastic reduction in freshwater consumption for agriculture was evident in 2001—a reduction of about 250 MCM/y compared to the average (844 MCM/y) of the prior eight years [54]. Similarly, Israel is intensifying its efforts to invest in agrotechnology, converting agriculture to using mainly low quality water. Floodwater is also being trapped for use in agriculture [54]. The agricultural sector has also incorporated several innovations to reduce reliance on

---

[19] The water amounts referred to by the Oslo II Agreement are the yet unexploited 78 MCM/y from the Eastern Aquifer.

[20] I am referring here to cited Mekorot Financial Statements.

[21] Average price the farmer pays is 25 cents per CM [40].

freshwater, including the famed drip irrigation technique and the use of untreated saline water for crop irrigation (eg. cotton, tomato and melon). Per person water available for agriculture during the 1960s (slightly above 500 CM/capita/y) is today slightly below 200 CM/capita/year. Despite this reduction agricultural productivity today is more than 150 percent of the quantity produced 40 years ago [40]. And finally, while the pricing scheme for farmer water costs requires some scrutiny, especially in comparison to earlier years, it should be noted that water prices for irrigation in Israel are one of the highest in the world [8].

Israel also has the most ambitious wastewater reuse effort under way in the world today. In 1999 about 65 percent (250-265 MCM/y) of the nation's sewage was treated and reused to irrigate large areas of agricultural land [50, 55]. In 2000, about 10-15 percent of the national water budget was derived from treated urban wastewater comprising about 15-20 percent of the water used for irrigation [2, 52].[22] In fact, the Ministry of National Infrastructures is working to advance the treatment of sewage water, its collection and purification, turning it into the main source of water for agriculture—replacing potable water [54]. The amount of reclaimed water is estimated to increase to between 570-600 MCM/y by the year 2020—about 81% percent of the nation's sewage [50, 55].[23] With little or no new sources of freshwater to tap, more intense and efficient reuse of wastewater will need to greatly expand in the years to come. The reclamation of brackish water is also utilized for both the agricultural and domestic sectors. Currently, about 166 MCM/y of brackish water are pumped from various boreholes and are desalinated. The Israel Water Commission has identified a total of 200 MCM/y of brackish water that can generate about 50 MCM/y of potable water [54].[24]

Thus, Israel considers the growing demand in domestic needs to be largely offset by water diverted from the agricultural sector and by increased wastewater recycling. According to the Israel Water Commission the difference between supply and demand will also be offset, in the years to come, by desalination (Table 6). In light of the above efforts, the Israel Water Commission has put forth an investment plan for the years 2002-2010 that includes the rehabilitation of saline polluted and depleted wells, treatment and reuse of sewage effluents and desalination of brackish water. Most importantly, however, the supply augmentation plan calls for the building of seawater desalination plants with a capacity of 400 MCM/y by 2006. The required investments for this campaign will total $4 billion—with $1.6 billion going to investment in seawater desalination [54].[25]

Projections of future deficits have been numerous and varied—some more pessimistic than others. The above figures by the Ministry of National Infrastructures, optimistic as they

---

[22] If all the urban wastewater of Israel could be treated the volume of new water available for agriculture would be about 500 MCM/y. Indeed with population increases expected for Israel and the Palestinians, the potential volumes of treated wastewater could reach 700 MCM/y with the potential of doubling with an additional population increase; At the end of 1999 an estimated 300 MCM (25%) of the total amount of water supplied for irrigation was in the form of reclaimed sewage effluents [50].

[23] The treated effluent flow by 2020 is estimated to be between 700-1000 MCM/y [8].

[24] About 35 MCM/y are used by industry and the rest by the agriculture sector. In Eilat and the Arava another 10 MCM/y of brackish water are desalinated and produce 7 MCM/y of potable water.

[25] The total amount also includes: $1 billion for treatment and reuse of sewage water, $600 million for water supply projects and $800 million for renovation and improvements.

are, assume that future deficits could be covered by treated wastewater and desalinated water—therefore assuming a balance. They also assume the availability of needed capital to fund such projects. Pessimistic figures predict a deficit of 2000 MCM/y by 2020, much of it due to the increased salinity of Israel's aquifers,

Table 6. Israel water supply and demand (1998-2020) (MCM/y)

Water Supply						
Year	Surface	Underground	Brackish	Treated	Desalinated	Total
1998	640	1050	140	260	10	2100
2010	645	1050	165	470	100	2430
2020	660	1075	180	565	200	2680
Water Demand						
Year	Urban Sector	Natural (Agricultural and Industrial)	Brackish (Agricultural and Industrial)	Wastewater Effluents (Agricultural and Industrial)		Total
1998	800	920	120	260		2100
2005	980	750	95	380		2430
2010	1060	680	75	490		2680
2020	1330	600	60	640		2680

Source: Ministry of National Infrastructures, The State of Israel, The Water Sector: Water Production and Consumption, 2001. [49]

which de facto reduces the amount of existing water stocks [20].[26] These estimates, in contrast to more optimistic figures, seem to be both exaggerated and take little account of increased use of reclaimed wastewater and desalination. Other estimates, which encompass not only Israel but also the West Bank and Gaza Strip, (include reclaimed wastewater but not necessarily desalination) present numbers that hint of relatively significant deficits for Israel and the Palestinians in the years to come (Table 7).

In contrast to the above projections, a study presented to the World Bank before the 1993 Oslo Peace Accords, estimated a deficit of 2000 MCM/y for Israel and the Palestinians, combined, by 2040 [19].[27]

Table 7. Israel, West Bank and Gaza Strip water balance projection (MCM/y)

Year	Urban Demand	Agricultural Demand	Total Supply (including treated wastewater)	Balance[28]
2010	1151	1317	2357	-111
2020	1440	1542	2560	-422
2040	2041	2017	2933	-1126

Source: Bar-El, 1995. [9]

[26] I am referring here to cited comments made by Dan Zaslavsky [20].

[27] I am referring here to a cited study spearheaded by Avishai Braverman in conjunction with Tahal, which was submitted to the World Bank.

[28] I am referring here to cited 1997 Government of Israel Documents (Partnership in Development 1998, Chapter 2, presented at the Middle East North Africa Economic Conference, Doha, Qatar) which predicted slightly higher figures. The break down for the years 2000, 2010,2020 and 2040 are as follows: -48 MCM/y, -187 MCM/y, -293 MCM, -570 MCM/y and –1266 MCM/y, respectively [44].

More recent projections from the Israel Water Commission, which take into account Israeli commitments of water deliveries to both Jordan and the Palestinians, predict a deficit of 445 MCM/y by 2020 (Table 8). Nonetheless, the Commission echoes the use of desalination and treated wastewater to provide the necessary water to restore the natural balance. The same policies will probably hold beyond 2020 [14, 70, 79].

Table 8. Freshwater demand, supply and deficit (MCM/y) (2000-2020)

									Supply	Deficit
			Demand							
Year	Domestic	Industrial		Agriculture		Total		Other[29]	Fresh Water[30]	
		Fresh	Other	Fresh	Effluent	Fresh	Other			
2000	720	100	35	880	270	1700	305	85	1555	230
2020	1120	150	60	530	620	1800	680	200	1555	445

Source: Tal, 2000. [79]

Indeed Table 8 makes reference to a phenomenon, which can't be sidelined—that is Israel's freshwater deficit as also being a function of present and future peace agreements with its immediate neighbors. While certain government authorities have argued that projected future deficits will largely be covered by non-traditional methods such as wastewater reclamation and desalination, the possibility arises that financial burdens, late project completion or the possible burdens of future negotiations gone unplanned could hamper these efforts or require their intensification. For example, the figures provided by the Israel Water Commission do not take into account a possible future agreement with Syria, which may place additional strain on Israel's water reserves.[31] Similarly, the figures do not account for additional Palestinian claims agreed upon in interim agreements or possible demands during final status negotiations. Israel will most likely, consider, with high regard, its water security as it negotiates. With the Palestinians, for example, Israel has been urging that non-traditional methods,[32] specifically a planned desalination plant in the Gaza Strip, which can produce 50 MCM/y of freshwater, be an integral part of future negotiations. In short, Israel wants to make desalination a strategy both parties will use to augment current supplies and to likewise reduce its own vulnerable position. The Israeli-Jordanian water agreement is discussed below.

---

[29] These figures include what Israel is to deliver to Jordan every year under its 1994 Treaty and what Israel estimates as the share to be given to the Palestinians from its own water system (Western Aquifer only).

[30] Actually there is a potential for 1820 MCM/y but because of spills/leakage and saline water diversions of an estimated 265 MCM/y, a total of 1555 MCM/y results [79].

[31] It is still unclear where Israel and Syria stand as some sources have claimed that during the 2000 failed negotiations between the two countries, Syria agreed to guarantee Israel's water rights if the United States would guarantee Syria's water rights from the Euphrates and Tigris Rivers vis-à-vis Turkey. On a similar note it is unclear, given the 2000 negotiations, where Syria will demand its presence (border with Israel) on Lake Tiberias to be, or if it recognizes full Israeli sovereignty over the Lake [15].

[32] See most recently an account of the 2000 Camp David Negotiations between Israel and the Palestinians [73].

4. Treaty of Peace Between Israel and Jordan [33]

The Israeli-Jordanian water agreement was part of a comprehensive peace treaty signed in 1994. Several key water allocations are reviewed below (Table 9).

Jordan was the main benefactor of these allocations. That said, there are conflicting views about Jordan's final allocations.

Table 9. 1994 Peace Treaty Water Allocations

Source	Israel	Jordan
Yarmouk River	12 MCM/y in winter and 13 MCM/y in summer (+)	Gets the remainder of the flow (50 MCM/y)[34]
	An additional 20 MCM/y from the winter flow and return it to Jordan in the summer on the Jordan River (+)	25 MCM/y[35]
Jordan River	Entitled to maintain current uses of the Jordan River waters between its confluence with the Yarmouk and its confluence with Tirat Zvi/Wadi Yabis	Entitled to an annual quantity equivalent to the share of Israel, provided that this will not hurt the quality and quantity of the above Israeli uses
		10 MCM/y in the form of desalinated water or from other sources (+) 20 MCM/y in summer period directly upstream from Deganya gates—part of concession made to Israel on the Yarmouk (+) 20 MCM/y entitlement of storage (+) 40 MCM/y brackish water[36]
Arava/Araba Valley	10 MCM/y (+)	
Other		50 MCM/y to be cooperatively created (+)

Source: Treaty of Peace Between The State of Israel and The Hashemite Kingdom of Jordan, October 26, 1994 (Article 6 and Annex II) [80]
Note: Some allocations taken from sources mentioned in footnotes.
(+) Specific amount mentioned in the agreement (In some cases the responsibility of one party to the other)

Figures range from 150 MCM/y [76][37] to 215 MCM/y [16, 18, 43, 26, 27]. Regardless, the agreement not only set water divisions but also established coordination

---

[33][80]; For an intricate account of the negotiations see [26].
[34] Cited calculated estimates [27].
[35] [27].
[36] [27].
[37] The agreement requires Israel to give Jordan 100 MCM/y as the additional 50 MCM/y is to be found cooperatively between the two countries (150 MCM/y in total for Jordan from the Treaty). But only 30 MCM/y to be transferred from Israel to Jordan are clearly specified. Both the storage entitlement of Jordan for 20 MCM/y and the Jordanian entitlement to an equivalent amount to that used by Israel (80 MCM/y) on the Jordan River between its confluence with the Yarmouk and Tirat Zvi/Wadi Yabis are less clear. The former does not state which party will build the storage facility to collect the flood water and what would happen in the case that little water was available for this venture. The latter means that new sources have to be found to supply Jordan so that Israel may

among the parties—particularly the joint need for storage facilities. Given that the Yarmouk River suffers from flow fluctuations from year to year, efficient use requires regulated flows. That is the flow needs to be stored during the flood season of winter for use in the summer [30]. Before the 1994 Treaty much of the flow escaped to the Dead Sea. Israel and Jordan, therefore, negotiated a diversion and storage clause allowing for a dam on the Yarmouk at Adassiya with additional storage on the Jordan River course and off the Jordan River course (on the side wadis) [25]. The flows guaranteed to Jordan from the Yarmouk would be diverted from the Adassiya Dam to the King Abdullah Canal, effectively utilizing the Yarmouk River's overflow.

The Treaty also called on the parties to form a Joint Water Committee (JWC) made up of Israeli and Jordanian experts charged with implementing the agreement and with resolving additional water related matters. The JWC was made up of two sub-committees. One was responsible for matters, relating to the Jordan/Yarmouk area (northern) while the other was responsible for matters dealing with the Arab/Arava region (southern). In fact, in the spirit of cooperation, the allocations of the agreement also provided for the inter-linking of the two countries' water systems. The 20 MCM/y owed to Jordan from the Jordan River would be transferred via a pipeline located partially in Israeli territory but built, financed and operated by Jordan. Similarly, it was also agreed that Israel would continue and expand (increasing its total yield from 8 to 18 MCM/y [16]) its use of wells with their associated systems that fall on the Jordanian side of the border. The two examples demonstrate a unique case of two states permitting entry of one country into the territory of the other for the purpose of water sharing [71].

Finally, the agreement addressed future supply and demand. Not only did this include an additional 50 MCM/y for Jordan but also mentioned the need for projects entailing regional and international cooperation.

5. Future Needs and Plans

The 1994 water agreement between Israel and Jordan was paramount to the Kingdom. The agreement guaranteed much needed water to augment the already meager Jordanian water supply. Yet the agreement also constituted a landmark event in the hydropolitical history of the Arab-Israeli conflict. As the main rivers and aquifers under dispute in the region tend to be governed by the status quo of historical use rather than negotiated agreements, cooperation and coordination over water is an important step.

There is no doubt that final agreements with the Palestinians, Syrians and Lebanese are necessary components of a comprehensive regional approach to water. Such agreements are necessary, not only to settle historical disputes and allow states to advance beyond sharing existing waters but also to cooperate over the exploration of additional sources.

---

not be harmed. Writing in 1997 when Israel assured Jordan it would receive 50 MCM/y from this latter source, Sofer concluded that Jordan is entitled to a total of 100 MCM/y from Israel.

5.1. EFFICIENCY IN WATER USE AND INCREASING SUPPLIES: A DUAL STRATEGY

Perhaps one of the most encouraging signs of the 1994 water agreement was its reference to cooperative plans and projects that augment existing water supplies for the two countries. While the agreement guarantees Jordan specific allocations of needed freshwater, these supplies are not even sufficient to cover the overdrawn water from the Highland Jordanian aquifers [89]. Similarly, due to the lingering regional drought both parties have not been able to make complete water deliveries to the other party [11, 35, 36, 37, 38, 43, 27]. Israel currently delivers 55 MCM/y to Jordan and Jordan delivers 26 MCM/y to Israel [18, 27].[38] Regardless, no matter what the final allocation Jordan is entitled to under the Treaty may be, the Kingdom would still be at a water deficit. Non-traditional means of augmenting supply will be needed.

While several strategies may be relevant, wastewater reclamation is the most likely alternative for improving water use efficiency and augmenting supplies. Though wastewater recycling undoubtedly must be accompanied by both political and technical reforms in the water sector, the water reclamation process can provide a large portion of the needed water for agriculture. Freshwater can then be used for much needed domestic purposes. While Jordan and Israel are already pursuing wastewater reclamation, the great potential of this strategy is not yet fully tapped, especially in Jordan. Similarly, the quality of treated wastewater will have to be the main concern for governments as farmers will be reluctant to use heavily saline water. In general there will be a potential for future utilisation of about 1794 MCM/y of wastewater for Israel, Jordan and the Palestinians [12].[39] Between 845-888 MCM/y of treated wastewater will be available by 2020 [12].[40]

Despite the available technology for wastewater reclamation, aggressively increasing regional freshwater supplies, has been the main fascination of the region—the means to do so has almost always been through seawater desalination.

It is clear that both wastewater reclamation and desalination can contribute to increasing freshwater supplies. In the report presented to the World Bank before the 1993 Oslo Peace Accords, it was argued that the gap between present and future water supply and demand in Israel, the West Bank and Gaza, at least in the short term, could be closed by increased use of reclaimed wastewater for agricultural use. That would require investment of about $550 million up to the year 2010 and a total of $1.8 billion up to the year 2040 in order to reach the highest possible level of wastewater reclamation for irrigation [19].

---

[38] The breakdown is as follows: 10 MCM/y brackish water or from other sources, 20 MCM/y which is stored for summer delivery and about 25-30 MCM/y until a desalination plant for brackish water (50 MCM/y) is set up. Out of the 80 MCM/y owed to Jordan from Israel only 55 MCM/y are being delivered and out of the 30 MCM/y owed to Israel from Jordan only 26 MCM/y are being delivered. It should be noted that while the 50 MCM/y of additional water guaranteed to Jordan under the agreement is to be developed cooperatively between the two countries, Israel is currently delivering a portion of that water from its own supplies [33, 36, 38].

[39] I am referring to the cited CES Consulting Engineers and GTZ, Middle East Regional Study on Water Supply and Demand Development, Phase 1, Regional Overview Committee on Sustainable Water Supplies for the Middle East.

[40] 43 MCM/y of wastewater are to be treated in the Palestinian territories in the future as compared to the current 20 MCM/y; For Israel—between 570-600 MCM/y [50, 55]; For Jordan—between 232-245 MCM/y [18, 88].

But according to the same report, the above strategy would supply only 58 percent of the total projected irrigation needs—leaving the need for freshwater to be used for both agricultural and domestic purposes [19]. Therefore, after 2010, desalination of both brackish water and seawater will have to be pursued in order to meet the water demands of Israel, the Palestinians and Jordan. Unlike wastewater reclamation, desalination can provide freshwater by turning unlimited seawater into potable water, which can then be used for domestic purposes. Similarly, on-going technological innovations have now pushed the price for desalinated seawater down, $6 a decade ago, with experts forecasting additional reductions [61]. Some estimates place total costs (including energy, labor, chemicals, filters and replacement parts) of seawater desalination at approximately $1.24 per cubic meter [29]. Lower estimates, mostly due to technological development and transition to larger units that enjoy the advantage of size, pinpoint the price of one cubic meter at 70-80 cents a cubic meter [24, 66]. The Israeli Foreign Ministry estimates the cost to be between 70 cents and $1—although the desalination plant in Ashkelon, Israel, described below, will supply water for about 50 cents a cubic meter [46].

5.2. SELECTED SEAWATER DESALINATION AND WATER AUGMENTING PROJECTS (FUTURE AND ON THE GROUND)

5.2.1.Israel

(a) Israel has several pilot and on the ground desalination plants both for water supply and improved water quality (Figure 5). Most of these plants desalinate brackish water using reverse osmosis technology. A small-scale seawater desalination plant already exists in Eilat and will produce 20,000 CM/day-27,000 CM/y (7.3-10 MCM/y) by 2003. Indeed plants installed in Israel over the last 30 years have a capacity of about 100,000 CM/day— production actually amounts to 60,000 CM/day (21.9 MCM/y), since some of the plants are not in operation [24, 75]. As for large-scale seawater desalination projects, Israel has considered the construction of several desalination plants that will desalinate 400 MCM/y, to be completed in the next 4 to 5 years. A plant in Ashkelon, governed by a BOT agreement, is already under construction and will produce between 50-100 MCM/y starting in 2005. The $150 million tender was provided by the Government Tenders Committee to the VID Group (Vivendi Environment, Dankner-Ellern Investments and Israel Desalination Enterprises).[41] The consortia set a record low price of $.527/CM [62, 83].[42] A 45 MCM plant in Ashdod as well as a series of smaller facilities along the Mediterranean coast are also planned [54, 62].[43] In summary, by mid-2002 four tenders with a total capacity of 305 MCM/y of potable

---

[41] The first 50 MCM/y will be delivered in 2004, while the second 50 MCM/y will be delivered in 2005 according to IDE Technologies Ltd. Constructed on a build-operate-transfer (BOT) basis, the plant will be built, owned and operated by VID Group for a period of about 25 years and then transferred to the State of Israel.

[42] The price per CM is relatively low because the government has guaranteed to buy the desalinated water for 25 years and a cheap supply of energy has also been promised. Given the BOT alternative Israel indirectly pays for the construction and operation of the plant through its water purchases for 25 years yet benefits from the completion of the plant in a relatively short time.

[43] Two tenders have been awarded for projects in Palmachim (15 miles south of Tel Aviv) and Haifa. Both will produce 30 MCM/y each and will cost about $140 million jointly. The Ministry of National Infrastructures, The State of Israel, hopes to have the two plants running by the end of 2005. Another 100 MCM/y plant near Caeserea is also planned.

water were published.[44] The desalinated water eventually produced could be shared with the Palestinians and Jordanians. A plant in Caeserea, for example, could be enlarged to produce more water and a pipeline could be extended to Jordan. While the political ramifications of such a project may presently be complex, Israel has the potential of playing the role of a regional water supplier or at least house the plants that provide desalinated water to its Arab neighbors [7].

(b) To augment its current supplies, Israel recently signed a treaty with Turkey that would export water from Turkey's Manavgat River to Israel. The deal would provide Israel with 50 MCM/y for a period of 20 years. While the Turkish water will only satiate 3% of Israel's needs during the 20-year period, Israel has already considered sharing this water with the Palestinians and Jordanians.

Figure 5. Israel's active and reserve sea and brackish water desalination installations (1999)

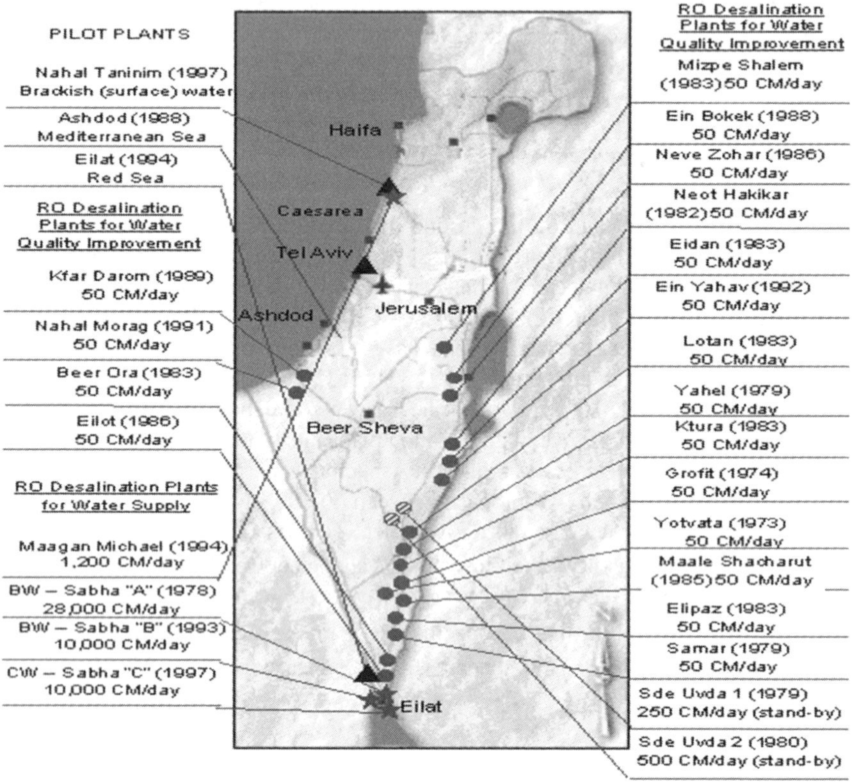

PILOT PLANTS

Nahal Taninim (1997)
Brackish (surface) water

Ashdod (1988)
Mediterranean Sea

Eilat (1994)
Red Sea

RO Desalination
Plants for Water
Quality Improvement

Kfar Darom (1989)
50 CM/day

Nahal Morag (1991)
50 CM/day

Beer Ora (1983)
50 CM/day

Eilot (1986)
50 CM/day

RO Desalination Plants
for Water Supply

Maagan Michael (1994)
1,200 CM/day

BW — Sabha "A" (1978)
28,000 CM/day

BW — Sabha "B" (1993)
10,000 CM/day

CW — Sabha "C" (1997)
10,000 CM/day

RO Desalination
Plants for Water
Quality Improvement

Mizpe Shalem
(1983) 50 CM/day

Ein Bokek (1988)
50 CM/day

Neve Zohar (1986)
50 CM/day

Neot Hakikar
(1982) 50 CM/day

Eidan (1983)
50 CM/day

Ein Yahav (1992)
50 CM/day

Lotan (1983)
50 CM/day

Yahel (1979)
50 CM/day

Ktura (1983)
50 CM/day

Grofit (1974)
50 CM/day

Yotvata (1973)
50 CM/day

Maale Shacharut
(1985) 50 CM/day

Elipaz (1983)
50 CM/day

Samar (1979)
50 CM/day

Sde Uvda 1 (1979)
250 CM/day (stand-by)

Sde Uvda 2 (1980)
500 CM/day (stand-by)

Haifa, Caesarea, Tel Aviv, Ashdod, Jerusalem, Beer Sheva, Eilat

Source: Glueckstern, 2001 [24]

5.2.2. Jordan

(a) Jordan has contemplated desalinating fossil brackish water in the Hisban area and the Jordan Valley—about 70 MCM/y [10]. The plan, which will cost about $70 million, and

---

[44] The Ashkelon plant, Palmachim, Haifa, Ashdod and Caeserea

entails digging about 17 wells with a delivery system to Amman, will make between 30-50 MCM/y of freshwater available [1, 18, 64].

(b) A desalination plant is also being considered for the port city of Aqaba. The current population of 74,000 is expected to increase to around 250,000 by 2025 with an increase in water demand from 15 MCM/y to 50 MCM/y. The region will probably need a desalination plant with a capacity of 5-10 MCM/y as water from the nearby Disi aquifer will be used to supplement supply in Amman [3, 18, 22, 59, 60].

5.2.3.Joint Jordan-Israel project
(a) A joint desalination project being considered between Jordan and Israel is the Red Sea-Dead Sea Peace Conduit and Desalination Project. Indeed the Project has been discussed in many policy and academic circles since the 1970s. Yet in August of 2002, at the World Summit on Sustainable Development in Johanesburg, South Africa, Israel's then Minister of Regional Cooperation, Roni Milo and the Jordanian Minister of Planning, Bassem Awadallah, announced that their two countries were reviving these plans [81].

The Project entails the transfer of seawater from the Red Sea to the Dead Sea via a 180 km conveyance system. Given that the Dead Sea is 400 meters below sea level the height difference can be utilized hydrostatically to desalinate the incoming seawater and supply Jordan, Israel and the Palestinians with freshwater. The main components of the project will therefore include: a) a seawater conveyance system, or "Peace Conduit", b) the reverse osmosis desalination plants that will take advantage of the inherited energy provided by the available head, and c) a conveyance system incorporating large pipelines and pumping stations to carry the desalinated water (Figure 6).

Figure 6. Red Sea-Dead Sea Peace Conduit and Desalination Project

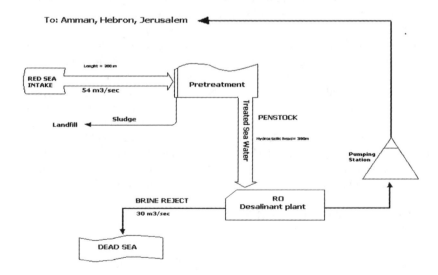

Source: Tahal Consulting Engineers Ltd., 2000. [77]

In the Project's first stage, 350 MCM/y (using 800 MCM/y of seawater) of freshwater will be produced. The plan is to eventually produce some 850 MCM/y of freshwater—using 1.9 billion cubic meters of seawater [78]. Initial investment needed for the Conduit comes to $800 million. This amount is to be solicited in the form of international grants. Private sector investment can then be solicited for the construction of the desalination plants [56]. Construction costs for the first and final stage are estimated to be $2,347 and $4,991 million (in 1996 prices), respectively. Annual operation costs are estimated to be $139 and $440 million, respectively (Table 10). The timetable for the commissioning of the first stage is estimated to be between 5-10 years.

Table 10. The Red Sea- Dead Sea Project construction costs and annual operating costs (US $ Millions)

Component	Estimated Total Construction Costs		Estimated Annual Costs	
	First Stage	Final Stage	First Stage	Final Stage
Peace Conduit-Conveyance Facilities	797	1662	10	44
Desalination/Power Plants	620	1332	23	77
Water Transmission to Jordan	656	1406	72	216
Water Transmission to Israel and the Palestinians	274	590	34	103
Total	2347	4991	139	440

Source: Tahal Consulting Engineers Ltd., 2001, 4 and 20. [78]

Despite the Project's water supply augmentation value, another selling point will be the potential cooperation between Israel and Jordan over the environmental sustainability of the Dead Sea area. In fact, the Dead Sea suffers from severe drought and evaporation. Sea level has been receding at between 0.5-1 meters per year and a net effect of –1000 MCM/y— is the result of the discrepancy between inflows and outflows [23, 56].[45] The Sea's level has already dropped 10 meters over the past 15 years and is expected to drop another 20 meters to 434 meters below sea level by the year 2020 if no seawater augmentation projects are pursued [78]. The decreasing level is causing fundamental changes in the region such as: sink holes with resulting damage to surrounding infrastructure, decline in the level of underground water, destruction of the natural landscapes and receding seashores. The above have had a large effect not only on the surrounding ecosystem but also on the tourism industry both Israel and Jordan depend on [56, 78]. The water conveyed from the Dead Sea via the Peace Conduit will, therefore, also be used to raise the Dead Sea to its original levels and replenish the annually evaporated water. Of the 1900 MCM/y of seawater that will be carried by the Conduit in the final stage of the Project, 1050 MCM/y, in resulting brine, will be discharged into the Dead Sea—reversing the deficit between inflow and outflow.

5.3. THE NEED FOR DONOR MONEY

Most of the supply side plans, currently under way or planned, will require large amounts of money. Israel and Jordan, for example, will not be able to shoulder the

---

[45] In the 1930s inflow to the Sea was 1.3 billion cubic meters/y. By 2000 inflow had dropped to just 300 MCM/y.

heavy financial burden of the initial investment in the Peace Conduit. The solution, besides encouraging the economic development of these entities and the generation of capital through reforms in the water sector, is outside assistance and subsidies. Financial plans for the Red Sea-Dead Sea Peace Conduit, as presented by Tahal, also highlight loans and contributions not only from donor countries but also fees from vacationing tourists and the benefiting tourism industry, in addition to royalties from potash and chemical factories, which profit from the mining and selling of Dead Sea minerals. The above can be used to cover the costs of the Peace Conduit and subsidize the price of desalinated water, at least in the first stage. With $90 million in annual fees paid by the Dead Sea tourism sector, the cost of one cubic meter of freshwater in Amman, for example, will be below one dollar [78].[46] Private sector involvement, similar to the Ashkelon desalination plant scenario, can also be instrumental in the construction and maintenance of the desalination plant.

Perhaps the majority of the capital for projects, in general, will have to come from donor countries. Industrialized nations and the oil rich Persian Gulf kingdoms, not to mention international financial institutions, are perhaps the only countries that have the ability to make large donations for cooperative projects between Israel and her neighbors and who have signaled their interest in a peaceful resolution to the Arab-Israeli conflict.

The concept of large sums of donor money invested in such projects is not at all far fetched. Naturally, investment will be more forthcoming if progress on the Middle East peace process track is seen and the political situation on the ground is conducive to negotiations. In fact, the issue of investment in desalination plants was already discussed in the context of the Israeli-Palestinian negotiations in Camp David in 2000. Out of the $35 billion, discussed as part of a financial package, that would be raised by the United States and other industrialized nations, $10 billion was to be used to develop water desalination plants to increase the usable water available to Israel, the Palestinians and Jordan. The Palestinians would be the principal beneficiaries of this development project [65]. When serious negotiations between Israel and the Palestinians resume, a similar package needs be arranged.

6. Conclusion

The rainy winter of 2002-2003 provided great respite both in Israel and Jordan from years of little rain and drought. As of July 1st 2003 the Israel Water Commission reported the level of the Sea of Galilee at 209.86 below sea level—only .96 meters away from its upper red line (208.9 meters below sea level)—where any amount above that will result in floods [47].[47] As of May 2003 total water reserves in Jordan's eight major dams reached 142 MCM—some 76 percent of the dams' total capacity of 186 MCM [91].[48] Yet a year of heavy rains is more the exception than the rule in the Middle East. With a UNESCO sponsored study projecting a period of prolonged drought in the Middle East [6, 34,

---

[46]The cost per cubic meter in Amman (after accounting for the desalination and conveyance to Amman) varies between $1-$1.5 according to a 6%-12% interest rate.

[47] Compare this to 214.07 below sea level in January 1, 2003 [47].

[48] Compare this to 46 percent capacity (86 MCM) [5], and 34 percent capacity (64 MCM) [13]; While winter recharging of aquifers will be calculated in September of 2003, some aquifer monitoring has shown that the water table rose between a centimeter and several meters [91].

68][49], betting on mother nature is risky. Immediate long-term planning can't be avoided. While conservation campaigns, pricing schemes, and reform of the water sector all need to be considered, supply side alternatives such as wastewater augmentation and desalination must also be rigorously factored in to the equation as serious options for the future.

It has been estimated by some sources that by 2050 Israel, Jordan and the Palestinians may reach a water deficit of 3000 MCM/y with total population expected to reach 30 million [51]. While this scenario is more likely a forecast of a region averse to sector reforms, cooperation and augmentation of existing water supplies with non-traditional methods, facts on the ground have demonstrated that the parties have gradually chosen the path of coordination and ingenuity to battle their water shortages. Yet despite implementing forward-looking strategies, both Jordan and Israel will undoubtedly still be contending with future water deficits. Likely financing problems combined with drought years are the main cause. Even if Jordan's impressive investment plan to provide an additional 400 MCM/y by 2011-2020 is completed as planned and on schedule[50] a deficit of between 360-437 MCM/y [18, 58, 88] still looms. To that end demand side reforms will have to play an important role in the country's augmentation strategy.

Great strides have already been made in reducing the agriculture sector's demands on freshwater. Israel has been a key player in this endeavor, introducing schemes such as drip-irrigation, computerized controlled irrigation and bio-engineered crops that can survive on either little water or very low quality water. Cooperation between Jordan, Israel and the United States has been key in enhancing these inventions and encouraging their use among them. Nonetheless, continued application and innovation of this technology continues to be necessary. At the same time, in both Jordan and Israel subsidies to the agricultural sector still exist. Because the price of water does not reflect its true value, especially in relation to farming, inefficiency on the side of farmers is still seen. Despite lower government subsidies and extraction tariffs, agriculture is still the main beneficiary [63].[51]

The transformation within the Jordanian agricultural sector towards more intensive production systems such as high value annual crops produced using drip irrigation in plastic houses should be noted and encouraged. Like in Israel, the raising of tariffs for irrigation water is also needed to raise efficiency. Perhaps the most water inefficient process unique to Jordan is the unaccounted-for water. The difference between water produced and water billed is at unsustainably high levels in all urban areas and reflects the inadequacies of system operation and maintenance. The efforts of the government to pursue private sector participation in rehabilitating the national water delivery system by repairing inadequate and inaccurate metering is, therefore, encouraging.

---

[49] It is important to note that precise time scales could not be inferred from the tools used in the study. But the range of the drought is within a few decades. According to Issar this is indicative enough, as this is the range of any future national plan [34]. Global warming scenarios point to continued decrease in rainfall in the Middle East. More precise conclusions will be ascertained in the coming year [6].

[50] As of February 2001 only 42 percent of the total volume of the investment program has been secured [88]; An updated review of Jordan's water investment program is currently being prepared by the Ministry of Water and Irrigation and will include a listing of the most recently secured funds [91].

[51] I am referring here to cited information from the Ministry of Finance, The State of Israel.

Despite the possibility of financial setbacks and time delays in the inauguration of supply side projects, there is no escaping such an alternative. Demand side reforms can just as well lag or run into political obstacles, thus necessitating a dual strategy. Where national coffers are stretched thin, private sector participation (including BOT), donor money and capital earned from better demand policies will have to be utilized.

The 1994 Treaty of Peace, which facilitated a cooperative scheme for managing the waters of the Jordan and Yarmouk Rivers and increased the amount of water allotted to the Kingdom, has already prompted the construction of several Jordanian dams necessary for stream diversions and catching unutilized floodwaters. Yet, assuming all the projects for providing Jordan with its allotted quantities are completed in the near future, even these set amounts will be either insufficient or precariously close for covering the current overdraft of water from renewable underground aquifers.[52]

Therefore, other supply side alternatives, such as wastewater reclamation, are indispensable. The agricultural sector will be the main beneficiary of this water, freeing up much needed freshwater for the domestic sector. Both Jordan and Israel are already engaging in such a program. However, a more efficient and rigorous campaign is needed to utilize the majority, if not all, of available effluents, especially in Jordan.

Seawater desalination provides perhaps the most serious and popular means to augment freshwater supplies in bulk. As the technologies for desalination are continuously being re-tested and improved the price of producing a cubic meter of freshwater is gradually decreasing. Israel has already inaugurated the construction of its first large-scale seawater desalination plant while Jordan has been considering a plant in Aqaba. Recently, both countries have resurrected plans for a joint desalination project. The majority of funds needed would most likely have to come from donor countries and include private sector participation. In an attempt to solicit the main funds, the project should be presented as a cooperative peace venture. Similarly, Jordan and the Palestinians would be the main beneficiaries. Given the potential of this massive project yet its long commissioning time, serious attention need be given immediately. Assuming the first stage of the project was to be completed between 2010-2015, the 350 MCM/y of freshwater produced could probably clear Jordan's water deficit and at the same time provide between 96-163 MCM/y of additional water for Israel and the Palestinians.[53] If the second stage of the project was to be completed by 2030, the 850 MCM/y of freshwater produced could probably reduce the estimated regional deficit[54] by about half.

---

[52] Assuming Jordan overdraws by 300 MCM/y [12] then even the 215 MCM/y allotted to Jordan from the Peace Treaty leaves about 100 MCM/y overdrawn; Considering the overdrawn water figure of 225 MCM/y [58], then the 215 MCM/y would also not cover the deficit; Given the overdrawn figure of 137 MCM/y [18] (and assuming this figure is sustainable), then the 215 MCM/y allotted to Jordan from the Peace Treaty will leave about 78 MCM/y available for other purposes.

[53] As calculated above, 187 MCM and 254 MCM are the estimated deficits for Jordan in 2010 and 2015, respectively.

[54] The regional deficit (about 1600-1800 MCM [56]) is probably calculated with little or no apparent reference to other water augmenting strategies such as Israeli, Palestinian or Jordanian desalination and wastewater reclamation campaigns. Therefore 850 MCM/y of freshwater produced could reduce the water deficit even further. It should also be noted that given Israel's own desalination campaign, Israel may forfeit its water allocations from the Red Sea-Dead Sea project to the Palestinians and Jordanians.

References

1. Al-Hadidi, M. (1999) Brackish water management and use in Jordan, *Desalination, 126.*
2. Allan, J.A. (2000) *The Middle East Water Question: Hydropolitics and the Global Economy.* I.B. Tauris, London and New York.
3. Water Ministry studies possibility of desalination plant in Aqaba, (2002), *Jordan Times*, April 17.
4. Aloul, S. (2003) Wihdeh Dam construction agreement to be signed today, *Jordan Times*, April 6.
5. Aloul, S. (2003) Zai water still diverted as Kingdom enjoys abundant rains, *Jordan Times*, February 16.
6. Alpert, P. (2003) (Professor of Atmospheric Sciences, Tel Aviv University, Tel-Aviv, and head of the Global Change in the Hydrological Cycle (GLOWA)-Jordan River), April 24; Personal correspondence.
7. Alpher, Y. (2002) Israel as a Regional Water Supplier, *Bittermellons.org, 29,* August 5, www.bitterlemons.org/previous/bl050802ed29.html
8. Arlosoroff, S. (2002) Integrated Approach for Efficient Water Use Case Study: Israel, paper presented at the World Food Prize International Symposium, *From the Middle East to the Middle West: Managing Freshwater Shortages and Regional Water Security*, October 24-25.
9. Bar-El, R. (1995) The Long-Term Water Balance East and West of the Jordan River, *National and Economic Planning Authority, Ministry of Economy and Planning, The State of Israel,* May, http://www.mfa.gov.il/mfa/go.asp?MFAH0ay90
10. Charkasi, D. (1999) Expert looks at desalination as alternative water source, *Jordan Times*, December 2.
11. Charkasi, D. (2000) Officials deny Israel stopped supplying Jordan with water, *Jordan Times*, June 6.
12. Committee on Sustainable Water Supplies (1999) *Water for the Future: the West Bank and Gaza Strip, Israel and Jordan.* National Academy Press, Washington, D.C.
13. Dalal, K. (2002) Kingdom's eight major dams holding 34 percent of their capacity, *Jordan Times*, December 22.
14. Deconinck, S. (2002) Israel Water Policy in a Regional Context of Conflict: Prospects for Sustainable Development for Israelis and Palestinians? http://waternet.ug.ac.be.waterpolicy1.htm
15. Drucker, R. (2002) *Harakiri-Ehud Barak: The Failure.* Yediot Achronot Books, Tel-Aviv. (in Hebrew).
16. Elmusa, S. (1995) The Jordan-Israel Water Agreement: A Model or an Exception, *Journal of Palestine Studies, 24, 3*, 63-73.
17. Elmusa, S. (1997) *Water Conflict: Economics, Politics, Law and the Palestinian-Israeli Water Resources.* Institute for Palestine Studies, Washington, D.C.
18. El-Naser, H. (2002) The Plan for the Response to Water Challenges, Ministry of Water and Irrigation, The Hashemite Kingdom of Jordan, 1-26.
19. Fedler. J. (1994) A problem that can't be wished or washed away, *The New Middle East Magazine, 1, 2,* http://archives.obs-us.com/obs/english/books/mem/n01a10.htm
20. Frisch, H. (2001) Water and Israel's National Security, paper presented in the conference on *Efficient Use of Limited Water Resources: Making Israel a Model State*, The Begin-Sadat Center for Strategic Studies, December 20, http://www.biu.ac.il/SOC/besa/waterarticle6.html
21. Ghadeer, T. and Khatib, A. (1998) Government Floats $700million Water Project, *Jordan Times*, October 18.
22. Global Water Intelligence (2002) Jordan Unveils Investment Plans, July.
23. Global Water Intelligence (2002) Red-Dead Sea Canal back on the Agenda, September.
24. Glueckstern, P. (2001) Desalination: Current Situation and Future Prospects, paper presented in the conference on *Efficient Use of Limited Water Resources: Making Israel a Model State*, The Begin-Sadat Center for Strategic Studies, December 20, http://www.biu.ac.il/SOC/besa/waterarticle1.html
25. Haddadin, M. (2000) Negotiated Resolution of the Jordan-Israel Water Conflict, in Dinar, S. and Dinar, A. (eds.) Negotiating in International Watercourses: Water Diplomacy, Conflict and Cooperation, *International Negotiation, 5, 2*, 263-288.
26. Haddadin, M. (2002) *Diplomacy on the Jordan: International Conflict and Negotiated Resolution*, Kluwer Academic Publishers, Dordecht.
27. Haddadin, M. (2003) (Former Minister of Water and Irrigation, The Hashemite Kingdom of Jordan and currently Visiting Professor at Oregon State University), January 4 and 6; Personal correspondence.
28. Harpaz, Y., Haddad, M. and Arlosoroff, S. (2001) Overview of the Mountain Aquifer: A Shared Israeli-Palestinian Resource, in Feitelson, E., Haddad, M. *Management of Shared Ground Water Resources: The Israeli-Palestinian Case with an International Perspective*, Kluwer Academic Publishers, Dordecht, 43-56.
29. Haruvy, N. (2002) (Institute of Soils, Water and Environmental Sciences, Agricultural Research Organization, Volcani Center, Bet Dagan, Israel), December 17; Personal correspondence.

30. Hillel, D. (1994) *Rivers of Eden: The Struggle for Water and the Quest for Peace in the Middle East*, Oxford University Press, New York.
31. IDE Technologies Ltd, www.ide-tech.com
32. Isaac, J. (2002) Chromo-hydropolitics in the Eastern Mediterranean, paper presented at the World Food Prize International Symposium, *From the Middle East to the Middle West: Managing Freshwater Shortages and Regional Water Security*, October 24-25.
33. Israel Radio (2002) Israel to provide Jordan with 50 MCM of water, June 17.
34. Issar, A. (2003) (Professor Emeritus, J. Blaustein Institute for Desert Research, Ben Gurion University, Beer-Sheba, Israel and head of the UNESCO sponsored study), April 5; Personal correspondence
35. Khatib, A. (1998) Israel reaffirms its commitment to water agreement, *Jordan Times*, December 15.
36. Khatib, A. (1999) Jordan seeks new water sharing terms with Israel, *Jordan Times*, August 17.
37. Khatib, A. (1999) Jordan signs JD66 million agreement for dam construction projects in the Southern Ghor Region, *Jordan Times*, January 16.
38. Khatib, A. (1998) Kingdom remains in search of definitive solutions to chronic water shortages, *Jordan Times*, August 15.
39. Khatib, A. (1999) Tensions ease in Jordan-Israel water dispute, *Jordan Times*, April 7.
40. Kislev, Y. (2001) The Water Economy of Israel, paper presented at the conference on *Water in the Jordan Valley: Technical Solutions and Regional Cooperation*, University of Oklahoma, International Programs Center, Center for Peace Studies, Norman, Oklahoma, November 13-14.
41. Kliot, N. (1994) *Water Resources and Conflict in the Middle East*, Routledge, London.
42. Landers, M. and Clarke, J. (1998) Overview of Middle East Water Resources, Water Resources of Palestinian, Israel and Jordanian Interest: Groundwater Basins, *compiled by the US Geological Survey for the Executive Action Team, Middle East Water Data Banks Project*, http://exact-me.org/overview/p11.htm
43. Libiszewski, S. (1997) Integrating Political and Technical Approaches: Lessons from the Israeli-Jordanian Water Treaty, in Gleditsch, N.P. (ed.) *Conflict and the Environment*, Kluwer Academic Publishers, Drodecht, 385-402.
44. Lithwick, H. (2000) Evaluating Water Balances in Israel, in Brooks, D. and Mehemt, O. (eds.) *Water Balances in the Eastern Mediterranean*, International Development Research Center, 29-58.
45. Mercado, A. (2001) Selected Groundwater Quality Parameters: Observed Trends and Possible Remedies, paper presented in the conference on *Efficient Use of Limited Water Resources: Making Israel a Model State*, The Begin-Sadat Center for Strategic Studies, December 20, http://www.biu.ac.il/SOC/besa/waterarticle2.html
46. Ministry of Foreign Affairs, The State of Israel (1999) Seawater Desalination Projects: The Challenge and the Options to Meet the Water Shortage, March, http://www.mfa.gov.il/mfa/go.asp?MFAH0eck0
47. Ministry of National Infrastructures, The State of Israel (2002/2003) Sea of Galilee Level Information, (in Hebrew), http://www.mni.gov.il/heb/kineret_screen.shtm
48. Ministry of National Infrastructures, The State of Israel (2001) The Water Sector: Water Resources and Water Availability, http://www.mni.gov.il/english/units/Water/WaterResourcesandWaterAvailability.shtml
49. Ministry of National Infrastructures, The State of Israel (2001) The Water Sector: Water Production and Consumption, http://www.mni.gov.il/english/units/Water/WaterProductionandConsumption.shtml
50. Ministry of National Infrastructures, The State of Israel (2001) The Water Sector: Non-Conventional Water Resources and Conservation, http://www.mni.gov.il/english/units/Water/NonconvetionalWaterResourcesandConservation.shtml
51. Ministry of National Infrastructures, The State of Israel (2001) The Water Sector: Regional Cooperation on Water Resources, http://www.mni.gov.il/english/units/Water/RegionalCooperationonWaterResources.shtml
52. Ministry of National Infrastructures, The State of Israel (2001) Wastewater Treatment and Reuse: Existing Wastewater Treatment and Reuse Projects, http://www.mni.gov.il/english/units/Water/ExistingWastewaterTreatmentandReuseProjects.shtml
53. Ministry of National Infrastructures, The State of Israel (2001) The Water Sector: Financing and Pricing of Water, http://www.mni.gov.il/english/units/Water/FinancingandPricingofWater.shtml
54. Ministry of National Infrastructures, Water Commission, The State of Israel (2002) Israel's Water Economy, August, www.mfa.gov.il/mfa/go.asp?MFAH0mb00
55. Ministry of National Infrastructures, The State of Israel (2001) Wastewater Treatment and Reuse: Ongoing Wastewater Treatment and Reuse Plan, http://www.mni.gov.il/english/units/Water/OngoingWastewaterTreatmentandReusePlan.shtml
56. Ministry of Regional Cooperation, The State of Israel (2002) Israel and Jordan Launch Global Campaign to Save the Dead-Sea, August, http://www.mfa.gov.il/mfa/go.asp?MFAH0mn70

57. Ministry of Water and Irrigation, The Hashemite Kingdom of Jordan,
    http://www.mwi.gov.jo/ministry_of_water_and_irrigationout.htm
58. Ministry of Water and Irrigation, The Hashemite Kingdom of Jordan, Slide show,
    www.jordanembassyus.org/new/index.shtml
59. Ministry of Water and Irrigation, The Hashemite Kingdom of Jordan, Proposed Investment Program 1997-
    2011, http://www.mwi.gov.jo/waj/WAJ%20Web%20Page/FUTURE.htm
60. Ministry of Water and Irrigation, The Hashemite Kingdom of Jordan (2002) Water Sector Planning &
    Associated Investment Program 2002-2011.
61. Orme, W. (2001) Israel Raises Its Glass to Desalination, *New York Times*, June 23, C1.
62. Platts Global Water Report (2002) Desalination Awards: Israel, August 12.
63. Plaut, S. (2000) Water Pricing in Israel, *Institute for Advanced Strategic and Political Studies, Policy
    Studies*, 47, July.
64. Rawashdeh, R. (1999) Ministry to award desalination project in Hisban to local firm, *Jordan Times*,
    November 29.
65. Reidel, B. (2002) Camp David: Two Years Later, *Bitterlemons.org*, July 15, 26,
    www.bitterlemons.org/previous/bl150702ed26extra.html
66. Rosenberg, M. (1999) Israel Taps the Mediterranean, *Utility Business*, October 1.
67. Rouyer, A. (2000) *Turning Water into Politics: The Water Issue in the Israel-Palestinian Conflict*, St.
    Martin's Press, New York.
68. Rudge, D. (2002) Prolonged Drought Projected for the Middle East, *The Jerusalem Post*, November 27.
69. Sabbah, W. and Isaac, J. (1995) Towards a Palestinian Water Policy, paper presented at the Seminar on
    Options and Strategies for Freshwater Development and utilization in Selected Arab Countries, Center for
    Environmental Development for the Arab Region and Europe (CEDARE), Amman, The Hashemite
    Kingdom of Jordan, June 26-28.
70. Schwarz, J. (2001) Water Resources Development and Management in Israel, paper presented in the
    conference on *Efficient Use of Limited Water Resources: Making Israel a Model State*, The Begin-Sadat
    Center for Strategic Studies, December 20, http://www.biu.ac.il/SOC/besa/waterarticle5.html
71. Shamir, U. (1998) Water Agreements Between Israel and its Neighbors, in Albert, J., Bernhardsson, M.,
    Kenna, R., Coppock, J. and Miller, J. (eds.) *Transformation of Middle Eastern Natural Environments:
    Legacies and Lessons, Bulletin Series, Yale School of Forestry and Environmental Studies*, 103, 274-296.
72. Shannag E. and Al-Adwan, Y. (2000) Evaluating Water Balances in Jordan, in Brooks, D. and Mehmet, O.
    (eds.) *Water Balances in the Eastern Mediterranean*, International Development Research Center, 85-93.
73. Sher, G. (2001) *Just Beyond Reach: Israeli-Palestinian Negotiations 1999-2001*, Yediot Achronot Books,
    Tel-Aviv.
74. Shuval, H. (1993) Approaches to Finding an Equitable Solution to the Water Resources Problems shared
    by Israelis and Palestinians in the Use of the Mountain Aquifer, in Baskin, G. (ed.) *Water: Conflict or
    Cooperation?* Israel-Palestine Center for Research and Information, Jerusalem, 37-84.
75. Sitton, D. (2000) Advanced Agriculture as a Tool Against Desertification, Applied Research Institutes,
    Ben-Gurion University of the Negev, October.
76. Soffer, A. (1999) *Rivers of Fire: The Conflict Over Water in the Middle East*, Rowman & Littlefield
    Publishers, Inc., Lanham.
77. Tahal Consulting Engineers Ltd. (2000) Red Sea-Dead Sea Peace Conduit, *Techno-Economic Report
    Prepared for the Ministry of Regional Cooperation, The State of Israel*, July.
78. Tahal Consulting Engineers Ltd. (2001) Red Sea-Dead Sea Peace Conduit, *Techno-Economic Report
    Prepared for the Ministry of Regional Cooperation, The State of Israel*, July 2000 (modified).
79. Tal, S. (2000) Long Term Tasks of the Israeli Water Sector, *Ministry of National Infrastructures, Water
    Commission, The State of Israel*, December.
80. Treaty of Peace Between The State of Israel and The Hashemite Kingdom of Jordan, October 26, 1994,
    www.mfa.gov.il/mfa/go.asp?MFAH00pa0
81. United Nations, Department of Public Information, News and Media Services Division (2002) Press
    Conference on Israel-Jordan Dead Sea Project, World Summit on Sustainable Development, September 1,
    http://www.un.org/events/wssd/pressconf/020901conf12.htm
82. Vengosh, A. and Weinthal E. (2003) The State of the Water Crisis in Israel and the Palestinian Authority:
    Integrating Science and Policy, seminar at the *Woodrow Wilson International Center for Scholars,
    Environmental Change and Security Project*, March 4.
83. Water Technology: The Website for the Water Industry, Eshkol, Nizzana and Ashkelon Desalination
    Plants: Israel, www.water-technology.net/projects/israel/index.html.

84. Water Technology: The Website for the Water Industry, Greater Amman Water Supply Project: Jordan, www.water-technology.net/projects/greater_amman/
85. Water Technology: The Website for the Water Industry, Tannur Dam: Jordan, www.water-technology.net/projects/tannur_dam/
86. Wolf, A. (1993) Water for Peace in the Jordan River Watershed, *Natural Resources Journal, 33,* 797-839.
87. Wollman, S, Environmental Hydrology Activities-Israel, Activity Report, *International Association for Environmental Hydrology (IAEH)*, http://www.hydroweb.com/israel-groundwater.html
88. World Bank (2001) The Hashemite Kingdom of Jordan: Water Sector Review Update, February 15.
89. World Bank (1996) The Hashemite Kingdom of Jordan: Water Sector Review, February 1.
90. Young, S. (1991) The Battle for Water: Storm Clouds Gathering, *Middle East International*, February 22.
91. Zoubi, M. (2003) (Director of Finance and Follow-up Projects, Ministry of Water and Irrigation, The Hashemite Kingdom of Jordan), April 27 and June 24; Personal correspondence.

Chapter 14

In the long term, will there be water shortage in Mediterranean Europe?

J. Margat
Plan Bleu-Regional Activity Centre – UNEP/MAP
15, rue L. Van Beethoven - Sophia Antipolis - 06560 VALBONNE

The European Mediterranean countries may seem to be less exposed to water scarcity risks in the future than their Southern and Eastern neighbours. But these risks cannot be excluded, so we should be concerned and make an appropriate assessment in order to be prepared to face them. This chapter tries to explore on a long term basis looking towards a 2050 horizon the possible changes in water resources and water demands of these countries in order to estimate the possible tensions and imbalances, their location and their degree of seriousness. Structural or cyclical, regional or local, water shortage demonstrates breaks in the balance between supply (linked to the resources and the state of their mobilisation) and the demand (of all utilisation sectors), whether these breaks occur from insufficiency or a breakdown in resources or from water production or too much demand. Therefore, the long-term forecast consists in understanding the occurrences of shortages –and envisaging where and when they could appear– on the basis of the examination and comparison of assumed future developments, according to various theories, supplies –first of all conventional water resources– and demands within the framework and timeframe considered, i.e. Mediterranean countries of Europe, the first half of the 21st century. Resources, like demand, are subject to change and variation, depending on dynamics that are often very different. But comparing them is only meaningful within a temporally and spatially relevant framework, i.e. a too extensive field may hide local imbalances; a too-restricted field may not be appropriate for the definition of resources and reduce the possibilities of transporting water. In the countries considered as the most extensive (France, Greece, Italy and Spain), the most appropriate fields are catchment areas or groups of catchment areas. The long-term forecast implies first of all analysing the initial conditions, i.e. the stress between mobilisable resources and demands which indeed exists right now in a few regions and overshadow the future. An exploration of future demands will be attempted, taking into account that, in the long term, it is easier to predict certain factors, in particular population, than total quantities directly. The effects of climatic changes on resources, despite the still considerable uncertainties that affect forecasting, cannot be avoided.

1. Imbalances between water demands and resources are already present locally

Regional comparisons between present demands and internal resources, even being limited to natural and average resources (therefore optimistic), show cases where demands are excessive or on a par with these resources, and are thus sensitive to the danger of a breakdown in dry years, i.e. catchment areas in south-east Spain (Jucar and

A. Marquina (ed.), Environmental Challenges in the Mediterranean 2000-2050, 233–244.
© 2004 *Kluwer Academic Publishers. Printed in the Netherlands.*

Segura), coastal catchment areas in Catalonia in Spain, Puglia and Sicily in Italy and Attica in Greece, not to mention Cyprus and Malta (figure 1).

Figure 1. Basins or regions of Europe's Mediterranean countries classified according to the link between real water demands (years 1995-2000) and internal average natural resources.

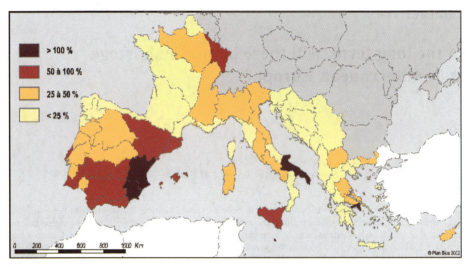

Source : Plan Bleu.

These situations are often associated with the proximity to large urban zones in the coastal areas with weak local resources, i.e. Barcelona, Valencia and Athens.

Comparing resources really used and those useable –less easy, as these are not so often evaluated on this scale, and the assessment criteria are more varied–could reveal more tenion and critical situations.

Structural or cyclical shortages already exist in a few regions where they have usually triggered calls for resources from surplus neighbouring catchment areas (transfers), i.e. Spain, Italy and Greece.

This analysis confirms another approach, which is indirect but more applicable in long-term forecasting, i.e. the comparison of resources to population by the "water resources (natural and average) per capita" indicator (or its opposite, the "competition index", i.e. the population per resource unit). The assessment is made in relation to meaningful water stress thresholds or generally acknowledged shortages of 1,000 and 500 cubic metres per capita (i.e. 1,000 and 2,000 inhabitants per cubic hm per year), which relate to the growing difficulties to satisfying water demands, for which the population is supposed to be the main determining factor.

According to this indicator and these thresholds, if we consider whole country units, only the two island countries are presently in situations of stress (Cyprus) or shortage (Malta). However, a more region-oriented analysis confirms that water depletion and potential shortage affect the same catchment areas or coastal regions in the Mediterranean Europe, which emphasises the discordance between the respective geographies of population and natural resources (figure 2).

Figure 2. Basins or regions of Europe's Mediterranean countries classified according to their annual internal average natural water resources per inhabitant (populations 1995).

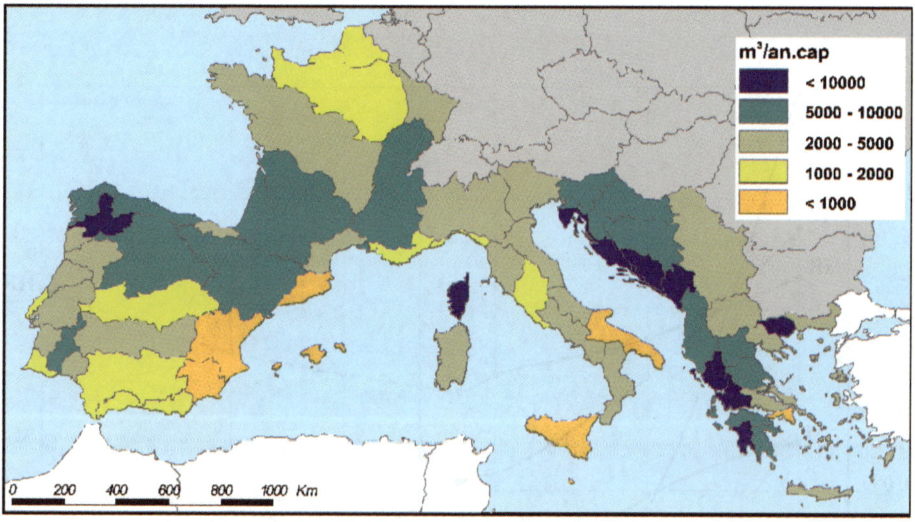

Source: Plan Bleu

2.How will water resources develop per capita?

To investigate the possible future increase in tension between water resources and demand, an initial traditional approach (but indirect and macroscopic) relies on the analysis of the forthcoming changes in the above-mentioned indicator, "water resources per capita". These changes will be the reflection of demographic projections in the hypothesis of constancy of the present average resources (in quantity) and comparing them to tolerable critical thresholds. This relies on the fact that, in the very long term (2050), population is the only relatively accessible variable and on the assumption that it will remain a basic factor in water demands.

According to a country-based overall view –the only one possible given the available demographic projections, that is a fall in population already underway (Italy) or foreseeable over the coming decades in most of the countries considered (the only exceptions being Albania, Cyprus and Macedonia)– prospects for the indicator up to about 2050 naturally show a general growth trend in average water resources per capita, thus averting the warning thresholds (figure 3). Only Cyprus and Malta would remain in their present situation.

This apparently comforting reflection nonetheless raises two notes of concern:

Relying only on the annual natural and average resources is not realistic. First of all it obscures the effects of the high year-to-year variability of resources inherent to the Mediterranean climate, e.g. in a dry year of a ten-year period, supply can fall as much as a third of the Spanish average. Cyclical regional risks of depletion and stress –if not shortage– are therefore not excluded.

Figure 3. Natural water resources growth (internal and external) per inhabitant, according to population changes, in Europe's Mediterranean countries. Trend 1950-2000 and forecast to horizon 2025, according to the United Nations 1998 medium variant projections.

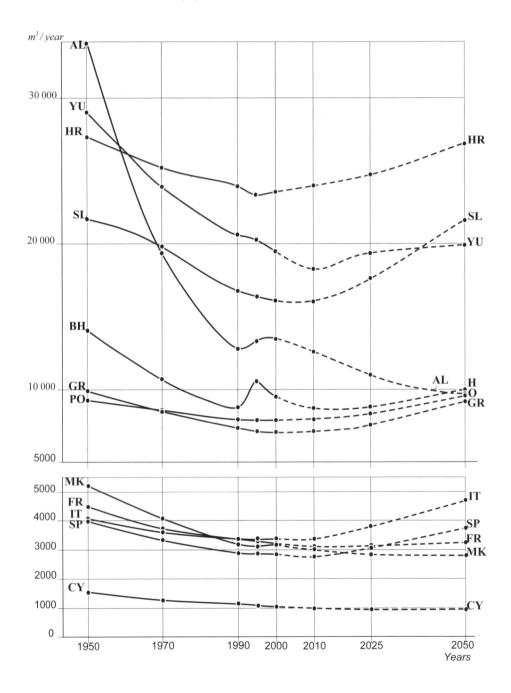

In the Spanish example, overall supply in a ten-year dry year only amounts to 880 cubic metres per year per capita taking into consideration the current population. It would only barely surpass 1,000 cubic metres per year (1,150) even in 2050.

Moreover, it would be better to take into account only those resources considered exploitable according to technological, economic, environmental and, even geopolitical criteria, belonging to each country. In the main Mediterranean countries considered, the share of average natural resources deemed exploitable or useable is approximately:

42 % in Spain
53 % in France
65 % in Italy
52 % in Slovenia
28 % in Croatia
30 % in Albania
41 % in Greece
69 % in Cyprus

Therefore, the computed resources per capita would consequently be reduced and sensitivity to cyclical droughts would increase, especially in Spain, France, Italy and Macedonia.

Moreover, these assessments of exploitable resources can themselves change and decrease if criteria, especially the environmental ones, become more severe.

As has already been emphasised above, a country-based framework to compare resources with population is not very appropriate in most cases. It hides big regional differences. Stress or shortages already present in a few regions, already mentioned (map in Fig. 2), do not appear in the initial conditions of forecasted development per entire country.

It is true that very long-term regionalised demographic prospects are difficult and not very available. Blue Plan trials up to about 2025 indicate more growth trends –or less diminishing ones– than the national averages for the coastal regions, yet less extensive trends than the catchment areas or regions which are considered above and less appropriate to comparisons with water resources that cannot be defined too locally.

The current contrasts between regions or catchment areas in the European Mediterranean countries will, in all probability, continue in the future. They are probably, even bound to increase through future internal population movements (effects of coastline attraction) and by developments in urbanisation and (tourist) activities in the coastal areas.

Thus the map of stressed or critical regions with difficulties of water supply by 2050 should probably not be very different from the current state of things.

3.Will water demand increase in the long-term?

Despite the not negligible uncertainties affecting the understanding of current water demands and their past variations –especially those of the generally dominant irrigation sector (except in France and in the northern Balkan countries)– a fairly general trend is discernible in their contemporary evolution and forecast in the short or medium term, i.e. a slow-down in growth, even a stabilisation or a drop –in Italy– (Fig. 4), rather similar to the trends in population changes.

Figure 4. Approximate growth 1970-2000 and trend projections at horizon 2025 of total fresh water demand in Europe's Mediterranean countries.

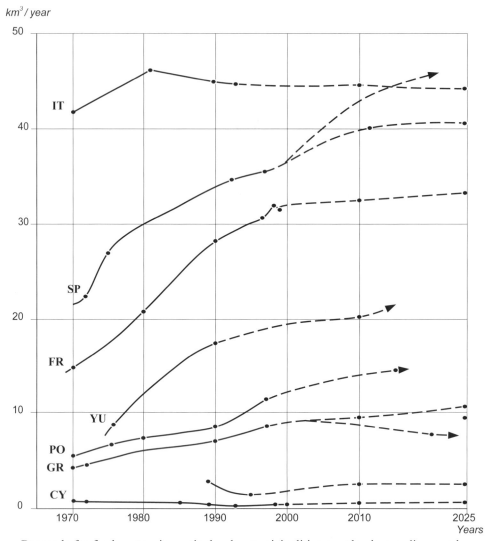

Demands for fresh water, in particular, by municipalities are slowly receding nearly everywhere [7].

The long-term outlook for demand has been made up to 2025. To do so, we have taken into account fairly varied scenarios in each sector of use (including the trend increase of the population, though less than in the 20th century and other factors like more anticipatory and environmentally active policies).

Breaks in long-term trends, the beginnings of which can be seen and which make it possible to predict slower and slower growth (even none), are unlikely.

In any case the demands of the most influential sector –irrigation– will change according to the economic context and the national and European agricultural policies and much more than in relation to the needs of food-industry production associated with the population. Predictions to about 2050 on this topic would be risky... These demands might nonetheless grow because of a climate change (more frequent droughts), discussed below (section 4).

Moreover, the growing determination to reduce the pressure on the environment and to protect nature in the European countries considered, should contribute to reduce certain uses by encouraging recycling and water saving, thus contributing to stabilising the demand or even, reducing it. This would be the case particularly in the energy sector in particular (the cooling of thermo-electric plants), by making closed circuits more common (France) or increasing the use of sea water.

On the other hand, the geographical distribution of demand in each country could change more by an increasing focus on coasts through progressive urbanisation and industrial or tertiary (tourism) activities, thus emphasising the imbalance between coastal and inland areas. This local growth is hard to quantify in the long term, but it will probably be more noticeable than the trends on a country-wide scale.

Faced with the resources –still in the hypothesis of the constancy of their current average state–, future demands should not notably modify the map classifying the catchment areas or regions according to the degree of pressure currently exercised on internal resources (Fig. 1).

With the total demand exceeding by half the average resources, a few additional catchment areas could nonetheless be affected, i.e. the Tagus in Spain and Portugal, the Sado-Melides in Portugal, the Romana-Marche in Italy and Thessaly in Greece, where current demands are already near this threshold.

4. What if the climate changes?

What would be the impact of a climate change attributable to additional greenhouse effects in the 21st century on the two variables in question, i.e. resources and water demand? The depletion of one and the increase of the other as predicted could increase the stress and thus the danger of shortages.

4.1. WILL THE FUTURE WATER RESOURCES BE THE SAME AS IN THE 20TH CENTURY OR MIGHT THEY BE DEPLETED OR DEGRADED IN THE 21ST CENTURY?

Natural water resources are strongly conditioned by the climatic variables which determine the inflows (precipitations, temperatures). They are therefore sensitive to changes in the averages used to define climates i.e. unchanged-average regime changes would probably have more effect on resources than changes of averages within the same regime.

Studies and research on climate change and its consequences have increased throughout the world over the past decade [10] and have begun to be regionalised, especially in the Mediterranean region [11,12,13,18]

The spatial resolution of overall climatic models still remains fairly vague and a long way from a relevant scale for assessing resources per catchment area. The estimates of change, for the most part, rely on those of the annual averages, and they describe changes of average condition without specifying the dynamic of change or the limits of probable occurrence.

Experts are much more loquacious about changes in temperature –first variable affected– than about changes in precipitation (translated only as a relative spread in relation to the current averages, uniformly applied to large regions independently of the rainfall gradients).There are even less details about the changes in hydrological variables (total runoff) on which the estimate of renewable natural water resources depends. These indications are of limited value for carrying out forecasting re-evaluations of resources that should be based on simulations of input regimes on, at least, a monthly basis, and preferably (in the Mediterranean region) every ten days or daily.

Despite these reservations, which call for a lot of caution, there is room to consider the results of the first forecasting trials for the European Mediterranean countries.

4.1.1.Precipitation

The scenarios for change in average rainfall do not predict differences in the same direction everywhere, e.g. an increase of winter rains in France (+15%), a fall in Spain (-5%, MIMAM 1998) and in Portugal, south of the Tagus (-10 to -15%, ESCAPE model, 1996-97). The Palutikof and Wigley projections (1996) for 2030-2100 foresee falls of 10 to 40 per cent in southern Spain, 10 per cent in central Spain, southern France and Greece but an increase of up to 20 per cent in central Italy.

Nonetheless experts for the most part agree about two predominant trends:

- A greater contrast of seasons, i.e. more rain in winter, less in summer;
- Sharper zonality, i.e. average increase in the north between the 40th and 50th parallels: France, Italy and in the northern Balkans; a fall would take place in the south: Spain, Portugal, southern Italy, Greece and Cyprus.

It is generally agreed that there will be an increase in the frequency and scale of cyclical droughts without, however, any details as to the extent.

4.1.2.Runoff.

The available results are less frequent and they mainly concern:

- Spain [7] With a 1 degree rise in the average temperature and a 5 percent drop in the average (uniform) rainfall, runoff would fall by an average of 20 per cent. However it would reach an increase of 25 to 50 per cent in the south east, in the catchment areas of the Ebre and the Guadiana (a detailed map of the differences in this hypothesis has been established).
- Portugal, [5] (ESCAPE model). With about a 1 degree rise in average temperature and a 10 to 15 per cent drop in the average rainfall south of the Tagus, runoff between 1990 and 2060 would fall an average of

0 to 5% in the Miňo (north) catchment area
5 to 10% in the centre
10 to 20% south of the Tagus

20 to 100% in the Guadiana and the Algarve

Local simulation by modelling has been tried in France for two sub-catchment areas of the Rhone: the Ain (Jura, oceanic regime) and the Eyrieux (on the edge of the Massif Central, Mediterranean regime), with the following hypotheses for 2050:

+20% in winter rain
-15% in summer rain   } Applied to daily intensity
+2% in average annual temperature
average annual runoff would increase by 9 per cent (Ain) and by 13 per cent (Eyrieux) with a 20 per cent increase in flooding and a significant drop of low water marks[1]

Increases in winter rainfall (the most effective) and decreases in summer rainfall would always increase the annual runoff, even though warming will diminish their effectiveness a little.

But whether supply on average rises or falls according to the scenarios or the regions, it is generally agreed that they would be more irregular (between seasons and years), though without being able to estimate to what degree. In any case that should make resources less exploitable while increasing the need to regulate. At the same time the runoff regime would increase erosion giving rise to the decrease of sediments and reducing reservoir capacity. Irregular water resource control would become in short both more necessary and more difficult.

As a result, regular water resources would be weakened due to the more prolonged and severe low water, above all where the already low ground water supply is small. This is mainly due to the reduction of the regulatory role of the snow cover and glaciers in the alpine catchment areas (France and Italy).

The sensitivity of lower water to pollution would consequently increase because of lesser dilution and the weakening of the self-purification capacities caused by the warming. Quality degradation would increase if purification and waste reduction were increased too.

4.1.3.Rise in sea level

The effects of the rise in sea level associated with climate change –predicted with a wide margin of uncertainty, i.e. an average of 45 to 50 cm between now and 2050–on water resources in the Mediterranean would be local but not negligible, especially in the coastal areas. In fact its importance is capital in the countries considered (their coastline should be kept in mind, i.e. nearly 35,000 km in the Mediterranean, 15,000 km of which are in Greece alone).

Relative levels will be more substantial in subsidence areas (the Camargue, Venice, etc.), of up to 1 m or 1.5 m in the 21st century, where the effects would be the worst.

Changes in the balance between fresh water and sea water in delta tributaries and coastal aquifers would diminish exploitable water resources and increase exposure to the dangers of overdraft (already present in several regions, e.g. Spain). Nonetheless water

---

[1] Personal communication, Leblois, Cemagref

supply to the coastal areas is much more reliant on inland resources. Rising sea levels would probably have other more serious impacts than that on local water sources, i.e. ground occupancy or lagoon ecosystems.

On the contrary, this rise could slightly reduce the under-sea fresh water discharge – not negligible in the Mediterranean– which benefits the coastal area springs.

4.1.4.On what water demands will climate change have an impact?
Among the various water demands most obviously conditioned by climatic factors, thus sensitive to climate change, is the demand for irrigation water, although this depends a lot on other factors (type of irrigation, type of ground, efficiency, type of crop, etc.). In the European Mediterranean countries, where irrigation is basically complementary, water needs could grow both because of less summer rainfall and because of rising temperatures, which increase potential evapotranspiration and reduce groundwater reserves. Yet local situations are too varied to make it possible to infer overall changes of average climatic variables or even, with everything else remaining stable, relative growth in irrigation water needs.

An exercise carried out on experimental plots in several Mediterranean countries (Spain, France and Greece) in cases of irrigation by pumping groundwater [22] concluded that irrigation water needs could grow by 500 to 1,500 cu. m per hectare per year.

In fact, as with resources, irrigation water needs would be a lot more sensitive to the changes of variability, both seasonal and year-to-year (especially the risks of drought) than to average changes of climatic variables.
Irrigation will most likely continue to be the major water-demand sector in the Mediterranean countries of Europe. Yet, among the factors of long-term evolution, climatic change is probably not the most decisive. Advances in efficiency could largely counterbalance the growth in water needs, not to mention the unknown possible developments in the future of irrigation agriculture.

5.Conclusions

In the Mediterranean countries of continental Europe the long-term structural water shortages should remain concentrated in a few regions or catchment areas where the present situations might deepen or worsen.

These situations would be due less to the increase in demand than to the depletion of resources:

- The future evolution of demand estimated for the first half of the 21st century will not reproduce the pattern of the latter half of the 20th. It will grow much less or even stabilise except for a few local effects from urbanisation (a factor which fosters the displacement of demand more than regional growth) and the effects of the probably more frequent and severe droughts.
- An average depletion of resources is to be feared in the most southern regions (the south of the Iberian peninsula, southern Italy, etc.) mainly due to the increasing risks of cyclical shortage (drought) and the difficulties of controlling water as a consequence of accumulated variability. Yet this depletion is hard to

quantify and predict for defined time periods. Experts agree only about the major uncertainties. They all predict a greater frequency of extreme events without measuring their magnitude, assessing their probability or setting deadlines. It is perhaps premature to interpret certain extreme contemporary phenomena, such as droughts, rainfall intensity and flooding, which are not without precedence but can be more destructive (because of increased modern awareness), like the first avowed symptoms of disturbance attributable to climate change.

Moreover, the reduction of residual availability, under the dual effect of withdrawals and final consumption, and the impacts on quality might contribute, as much as the depletion of supply (and the reduction of exploitability), to rarefying tributary supply sources of conventional resources in the most sensitive regions or catchment areas.

References

1. Attane, I., and Courbage, Y. (2001) La démographie en Méditerranée. Situation et projection, *Economica-Les Fascicules du Plan Bleu 11*, Paris.
2. Benblidia, M., Margat, J. and Vallee, D. (1998) Pénuries d'eau prochaines en Méditerranée?, *Futuribles 223, juillet-août*, 5-29.
3. Drain, M. (1996) Les conflits pour l'eau en Europe méditerranéenne, *Espace rural, 36*, 65.
4. EC/Montgomery, W. (1997) Spain/Portugal Hydrological Appraisal, *Final Report. EC, DGXVI-Cohesion Fund*, 54.
5. EEA (1996) Water resources problems in Southern Europe – An overview report, Estrella, T., Marcuello C., and Iglesias, A. *Topic report 15, Inland Waters*.
6. Elliniki, E. (1996*) Long-range study on water supply and demand in Europe Study at country level: Greece*, The Hellenic Society for the Protection of Environment and the Cultural Heritage, Athens.
7. Estrella, T., Marcuello, C., and Dimas, M. (2000) *Las aguas continentales en los países mediterráneos de la Unión Europea*, Ministerio de Fomento y Medio Ambiente, Centro de Estudios y Experimentación de Obras Publicas, Madrid
8. EUROSTAT (1998) *Water in Europe. Part 1. Renewable Water Resources*, Office for Official Publication of the European Communities, Luxembourg.
9. Falkenmark, M. (1986) Fresh water – Time for a modified approach, *Ambio, 15, 4*, 92-200.
10. Houghton, J.T., Meira Filho, L.G., Callander, B.A., Harris, N., Kattenberg, A. and Maskell, K., (eds.) (1996) *Climate Change 1995: The Science of Climate Change.* Contribution of Working Group I to the Second Assessment Report of the Intergovernmental Panel on Climate Change, Cambridge University Press, Cambridge.
11. Jeftic, L., Milliman, J.D. and Sestini, G. (eds.) (1992) *Climatic Change and the Mediterranean*. Edward Arnold, London.
12. Jeftic, L., Keskes, S., and Pemetta, J.C. (eds) (1996) *Climatic Change and the Mediterranean*. Edward Arnold, London.
13. Karas, J. (1997) Climate Change and the Mediterranean Region, *Greenpeace International, 34*.
14. Leblois, E., and Margat, J. (2000) Effets possibles sur les écoulements superficiels et les eaux souterraines in Premier Ministre, Mission Interministérielle de l'Effet de Serre *Impacts potentiels du changement climatique en France au XXIe siècle*, M.A.T.E., Paris.
15. Margat, J. (1998) *Sécheresse et ressources en eau en Méditerranée*. Rapport à la Conférence sur la politique de l'eau en Méditerranée, Réseau méditerranéen de l'eau, Valencia, 16-18 avril, session Gestion des sécheresses
16. Margat, J. (1999) *Ressources en eau, sécheresse et risques de pénurie en Méditerranée*. International Conference Euromed-Safe, Naples, 27-29 octobre 1999, 11
17. Margat, J. and Vallee, D. (2000) *Vision Méditerranéenne sur l'eau, la population et l'environnement au XXIe siècle/Mediterranean Vision for Water, Populations and the environment in the 21st Century*. Plan Bleu. MEDTAC, document pour le Forum mondial de La Haye, Global Water Partnership, Conseil Mondial de l'Eau, Sophia Antipolis, 62.

18. Medias-France (2001) *The present states of knowledge of global climate change, its regional aspects and impacts in the Mediterranean Region,* Plan Bleu, Rapport de MEDIAS-FRANCE. WWW.PLANBLEU.ORG
19. OIE (1996). *Long Range study on water supply and demand in Europe. Study at country level: France.* Office International de l'Eau, Limoge-Orléans.
20. Palutikof, J.P.,Guo X.,Wigley,T.M.L.,and Gregory, J.M. (1992) *Regional Changes in Climate in the Mediterranean Basin due to Global Greenhouse Gas Warming.* MAP Technical Report Series, UNEP, Athens.
21. Palutikof, J.P., and Wigley, T.M. (1996) Developing climate change scenarios for the Mediterranean Region, in Jeftic, L., Keskes, S., and Pernetta, J.C. (eds) *Climate Change in the Mediterranean,* Edward Arnold, London.
22. Vachaud, G., Bourani, F., and Chen,T. (1997) *Evolution des ressources en eau souterraines en zone de culture irriguée du bassin méditerranéen en cas de changements climatiques,* Colloques Agriculture et développement durable en Méditerranée. Montpellier, (internal publication).
23. WRI (1997) *Long – Range study on water supply and demand in Europe. Level A: Studies at country level* Water Research Institute (NRC) Rome.

Section V

Population Growth

Chapter 15

Implications of Population Decline for the European Union (2000-2050)

J. Díez-Nicolás
Universidad Complutense
Facultad de CC. Políticas, Campus de Somosaguas, 28223 Madrid

Population prospects for the European Union seem to coincide in forecasting a general population decline for this area, due to a persistent low level of fertility (below replacement level) and an increasing ageing of the population [9]. Population trends in the European Union countries are very similar at present, both when the present 15 or the 10 candidate countries are considered. They all have the lowest rates of growth, the lowest fertility rates, and the highest life expectancies, which combined are producing the most aged populations in the world. If these trends continue the population of the European Union will not only decline, but will become even more aged, and will have a smaller population force. Some of the measures that have been proposed to avoid this prospect are to increase fertility and to increase immigration, under the assumption that they will increase population growth, that they will make the population younger and that they will increase the population force. This paper intends to demonstrate that higher fertility and immigration are not the only possible answers to these processes, and that some benefits may also be derived from non growing and ageing populations, provided that societies adopt the necessary changes to adapt to this new demographic situation.

1. The population of the European Union within the context of world population

When the four regions of Europe (according to UN definitions) are compared with other world regions, it is evident that they show the lowest projected rates of increase for the periods 2002-2025 and 2025-2050. More specifically, the four rates of increase are below 0.5% for the first period, and Eastern Europe even shows negative growth. In the second period three of the rates are negative, while Northern Europe shows a zero-population growth, but only East Asia (outside Europe) is expected to experience negative growth between 2025 and 2050 (due to expected negative growth both in Japan and China). All other world regions show positive growth rates for the two periods (apart from the already mentioned exception regarding East Asia), though the rates for the second period are expected to be lower in all regions, without any exception. As a matter of fact, all non-European regions, except North America and East Asia, will eventually grow over 1% per year between 2002 and 2025, but only two regions, Sub Saharan Africa and Western Asia might grow over 1% between 2025 and 2050. Population

A. Marquina (ed.), Environmental Challenges in the Mediterranean 2000-2050, 247–263.
© 2004 *Kluwer Academic Publishers. Printed in the Netherlands.*

growth will therefore decline throughout the next two twenty-five year periods only in Eastern Europe and in the European Union, and it will also decline, but only between 2025 and 2050, in Western and Southern Europe, so that probably only Western Europe will grow between 2000 and 2025.

The four European regions and the European Union as a whole show at present the lowest total fertility rates, all five of them well below the generally accepted replacement level of 2.1 children per woman [10], while the world average is at 2.8. And the range varies from a low 1.7 in East Asia to a high 5.6 in Sub Saharan Africa. All other non-European regions, with the only exception of East Asia, have a total fertility rate that is above replacement level, though there seems to be a general trend towards lower fertility rates in all regions of the world.

Table 1. Demographic Indicators and Projections for World Regions, 2002-2050

	Population mid 2002 (Millions)	Projected population (Millions)		Projected annual rate of increase (%) (a)		Total fertility rate	Life expectancy at birth (both sexes)	Percent of population of age		Projected percent of population in 2050 (b)	
		2025	2050	2002-2025	2025-2050			< 15	65 +	65 +	80 +
World	**6,215**	**7,859**	**9,104**	**1.15**	**0.63**	**2.8**	**67**	**30**	**7**	**15.6**	**4.1**
Northern Africa	180	249	302	1.67	0.85	3.5	66	36	4	14.1	2.6
Sub-Saharan Africa	693	1,081	1,606	2.43	1.94	5.6	49	44	3	6.9	1.1
North America	319	382	450	0.86	0.71	2.1	77	21	13	21.4	7.7
Latin America & Caribbean	531	697	815	1.36	0.68	2.7	71	32	6	16.9	4.1
Western Asia	197	298	404	2.23	1.42	3.9	68	36	5	11.3	2.4
South Central Asia	1,521	2,047	2,474	1.50	0.83	3.3	63	37	4	13.2	2.6
Southeast Asia	536	706	811	1.38	0.60	2.7	67	32	5	16.1	3.5
East Asia	1,512	1,690	1,608	0.51	-0.19	1.7	72	22	8	23.6	7.4
Northern Europe	96	103	103	0.32	0.0	1.6	78	19	15	27.2	10.4
Western Europe	184	187	178	0.07	-0.19	1.5	78	17	16	29.0	11.9
Eastern Europe	301	279	231	-0.32	-0.69	1.2	68	18	13	27.9	7.3
Southern Europe	147	149	139	0.06	-0.27	1.3	78	16	17	33.5	11.9
Oceania	32	40	46	1.09	0.60	2.5	75	25	10	18.0	5.6
European Union-15	**381.2**	**391.7**	**374.1**	**0.12**	**-0.18**	**1.6**	**79**	**17**	**16**	**30.8**	**12.0**

Source: Population Reference Bureau, *World Population 2002.*
(a) Calculated by the author.
(b) United Nations, *World Population Ageing, 1950-2050* [11].

Three of the four European regions (all but Eastern Europe) and the European Union show at present the highest total life expectancy at birth, 78 and 79 years. The exception of Eastern Europe is mainly due to the recent increase in mortality rates in Russia and some other republics of the former Soviet Union, such as Moldova, Ukraine and Belarus. This fact is quite exceptional (only comparable to the rise of mortality rates in many African countries due to AIDS). And it seems to be a result of social disorganisation processes that have taken place in some of those countries as a consequence of the economic change from a state planned economy to a free market economy.

All four European regions and the European Union show the lowest proportions of population of less than 15 years of age (below 20% in all cases), as well as the highest proportions of population of 65 years and over (above 13% in all cases). And the four European regions are expected to reach the highest proportions of population of over 65 and over 80 years of age in 2050. The proportions will be above 27% and above 7% respectively in the four regions and in the European Union, though East Asia will have a similar proportion to Eastern Europe regarding those of 80 years and over.

To summarise, there seems to be a common European demographic pattern, from Portugal to Russia, which definitely contrasts with all other regions of the world. Only North America and East Asia, and less so also Oceania (because of Australia and New Zealand), resemble the European pattern in some respects. And, certainly, there are also differences within Europe, with Eastern Europe being the region that more frequently departs from the common pattern shown by the other three regions. As a consequence of this pattern, the total European population, which amounted to 728 million in 2002, is expected to decrease to 718 million in 2025 and to 651 million in 2050. As for the enlarged European Union of 25 countries, its population may increase from 456 million in 2002 to 465 million in 2025, but will decrease to 439 million in 2050 if the United Nations forecasts are fulfilled. The main factors that explain those prospects are the maintenance of fertility below replacement level and a very low level of mortality (high life expectancy) that will produce an increasingly ageing population [14].

In general, many politicians and social scientists evaluate these prospects as non desirable. They argue that negative population growth, below replacement fertility level and increasingly ageing populations are all undesirable, because they will produce a demographic structure that may jeopardise the payment of retirement pensions and, ultimately, endanger the whole Social Security system and the Welfare State that have characterised free market European societies in the past century. The European pattern just described seems to reflect very accurately the situation of the present fifteen countries of the European Union, that account for 381.2 million out of the 728 million that make up the total population of Europe. The remaining 346.8 million correspond to Eastern Europe (301 million) and to countries in Northern, Western and Southern Europe that do not yet belong to the European Union (45.8 million). Therefore, the population of the European Union, in 2002, represents 52% of the total population of Europe. And the prospects estimate that the EU-15 population will be 391.7 million in 2025 and only 374.1 million in 2050, representing 55% and 58% respectively of the total European population on those two dates. The projected population for the European Union in 2050 is smaller than in 2002, so that its weight over the total European population will grow from the present 52% to a projected 58%. This can only imply that the European population outside the European Union as defined today will have an even lower (and more negative) rate of population growth over the next fifty years.

2. Demographic trends in member and candidate countries of the European Union

The demographic trends that have been described for Europe as a whole, compared with those of other world regions, are also characteristic of the present fifteen member countries of the European Union. As has been said, the total population of these countries represents a little more than half of the total European population. But, because of its low growth rates, the total projected population for the fifteen countries in 2025 will be only slightly higher than that of 2002. This may be due to the fact that four countries (Portugal, Greece, Germany and Italy) will experience negative growth rates during that period, while only two of the remaining eleven countries are expected to grow over 0.5% per year (Luxembourg and Ireland). However, it is estimated that the European Union will have a negative population growth between 2025 and 2050, when only six countries may have positive growth rates (Denmark, Sweden, Belgium, the Netherlands, France and the United Kingdom) and two additional ones are expected to have zero-population growth (Ireland and Luxembourg). The total European Union population will have a more important loss between 2025 and 2050, but considering the whole period 2002-2050, the total population of the European Union will decline only slightly. This may be a result of the very significant decrease expected in Germany and to some extent in Finland, Greece, Italy and Portugal, that will be compensated by annual average positive growth rates in the remaining twelve countries. Therefore, it seems that most countries in the European Union will grow positively for the next twenty-three years, but most of them too will have negative growth rates between 2025 and 2050. A second conclusion seems to be that the total population of the European Union may decline over the next fifty years, but only because of a population decline in five countries, since the other ten countries will grow, slowly, but positively.

The prospects for slow and even negative population growth rates in the European Union are due to the fact that the average number of children per woman is at present below replacement level in all countries (almost half that level in Spain, for example). The combined effect of such low fertility with very low mortality (that results in very high life expectancy) is a very pronounced process of demographic ageing. In fact, the proportion of the population of 65 years and over is greater than the proportion that is less than 15 years of age already in three countries (Italy, Greece and Spain), and equal to it in two additional countries (Germany and Portugal). That proportion is at present between 15-20% in most countries, though still lower in Ireland, Luxembourg and the Netherlands, but it will be greater than 20% in all countries by 2050, and even greater than 30% in six countries (Spain, Italy, Greece, Austria, Germany and Sweden). In that same year, population prospects estimate that the proportion of the population of 80 years and over will be higher than 10% in all but three countries (Ireland, Luxembourg and Portugal). Population ageing, according to United Nations projections, will accelerate in all European Union countries over the next 50 years, but especially in Southern European ones [8].

When discussing the future population of the European Union the ten countries that will become members in 2004 must be included. The total population for the ten applying countries was 74.9 million in 2002 (19.6% of the present total population of the European Union, or 16.4% of the total population of the enlarged EU-25, which in 2002 will amount to 456.1 million).

Table 2. Demographic Indicators and Projections for the Member and Candidate Countries of the European Union, 2002-2050

	Population mid 2002 (Millions)	Projected population (Millions)		Projected annual rate of increase (%) (a)		Total Fertility Rate	Life expectancy at birth (both sexes)	Percent of population of age		Projected percent of population in 2050 (b)	
		2025	2050	2002-2025	2025-2050			< 15	65 +	65 +	80 +
Members: (a)	**381.2**	**391.7**	**374.1**	**0.12**	**-0.18**	**1.6**	**79**	**17**	**16**	**30.8**	**12.0**
Austria	8.1	8.4	8.2	0.16	-0.10	1.3	78	17	15	34.0	14.5
Belgium	10.3	10.8	11.0	0.21	0.07	1.7	78	18	17	29.0	11.5
Denmark	5.4	5.9	6.4	0.40	0.34	1.7	77	19	15	25.9	9.7
Finland	5.2	5.3	4.8	0.08	-0.38	1.7	78	18	15	27.9	10.6
France	59.5	64.2	65.1	0.34	0.06	1.9	79	19	16	26.7	10.4
Germany	82.4	78.1	67.7	-0.22	-0.53	1.3	78	16	16	31.0	13.2
Greece	11.0	10.4	9.7	-0.24	-0.27	1.3	78	15	17	34.1	11.8
Ireland	3.8	4.5	4.5	0.80	0.00	1.9	77	21	11	22.0	5.9
Italy	58.1	57.5	52.2	-0.04	-0.37	1.3	80	14	19	35.9	14.1
Luxembourg	0.5	0.6	0.6	0.87	0.00	1.8	78	19	14	19.7	6.9
Netherlands	16.1	17.7	18.0	0.43	0.07	1.7	78	19	14	26.5	10.1
Portugal	10.4	9.7	8.6	-0.29	-0.45	1.5	76	16	16	29.8	9.1
Spain	41.3	44.3	42.1	0.32	-0.20	1.2	79	15	17	37.6	13.4
Sweden	8.9	9.5	9.8	0.29	0.13	1.6	80	18	17	30.4	12.2
United Kingdom	60.2	64.8	65.4	0.33	0.04	1.6	78	19	16	27.3	10.8
Candidates: (a)	**74.9**	**73.6**	**65.0**	**-0.08**	**-0.47**	**1.3**	**74**	**18**	**13**	**28.9**	**7.8**
Cyprus	0.9	1.0	1.0	0.48	0.00	1.7	77	22	10	23.2	7.5
Czech Rep.	10.3	10.3	9.4	0.00	-0.38	1.1	75	16	14	32.7	9.5
Estonia	1.4	1.2	0.9	-0.62	-1.00	1.3	71	18	15	26.9	7.3
Hungary	10.1	9.2	8.1	-0.39	-0.48	1.3	72	17	15	29.0	7.7
Latvia	2.3	2.2	1.8	-0.19	-0.73	1.2	71	17	15	28.2	8.2
Lithuania	3.5	3.5	3.1	0.00	-0.46	1.3	73	19	14	28.8	9.4
Malta	0.4	0.4	0.4	0.00	0.00	1.7	77	20	12	26.9	9.0
Poland	38.6	38.6	33.9	0.00	-0.49	1.3	74	19	12	27.9	7.4
Slovakia	5.4	5.2	4.7	-0.16	-0.38	1.2	73	19	11	28.9	
Slovenia	2.0	2.0	1.7	0.00	-0.60	1.3	76	16	14	34.8	11.8

Source: Population Reference Bureau, *World Population 2002* [2].
(a) Calculated by the author.
(b) United Nations, *World Population Ageing, 1950-2050* [4].

It must be underlined that all the trends that were ascribed to the European Union countries are even more pronounced in the group of ten candidate countries. In fact, it is estimated that their total population will decrease not only between 2025 and 2050 (as in the present EU-15), but also between 2002 and 2025 (contrary to the present European

Union countries). Growth rates, however, will be even more negative in the second period than in the first one (similarly to what was observed for the present EU-15). Only one country is expected to have a positive growth between 2002 and 2025 (Cyprus) and none at all between 2025 and 2050 (while there were ten and five countries, respectively, that were estimated to achieve positive growth rates in the same two time periods). Fertility levels in the present EU-15 range from 1.2 children per woman in Spain to 1.9 in France and Ireland, while among the ten candidate countries the range varies from 1.1 in the Czech Republic to 1.7 in Cyprus and Malta. Only five out of the fifteen present member countries of the European Union have a fertility rate below 1.5 compared to eight out of ten among the candidate countries. Life expectancy at birth varies from 76 to 80 among the EU-15 countries, and it varies from 71 to 77 among the applying countries, which implies still higher mortality (lower life expectancy) among the candidate countries.

As a consequence, the population of the candidate countries is at present a little younger than that of the present members. In fact, the proportion of the population which is 65 years and older is not greater or equal to the proportion which is less than 15 years of age in any of the candidate countries, while it is larger or equal to it in five of the fifteen member countries. And the population of the candidate countries in 2050 will continue to be a little younger than that of the present members, though their age distributions will be more homogeneous. Thus, of the present fifteen members, two will have less than 25% of their population aged 65 or more (Ireland and Luxembourg), while six will have more than 30% (Spain, Italy, Greece, Austria, Germany and Sweden). And of the ten candidate countries only one will have less than 25% of its population aged 65 or more (Cyprus), while two will have more than 30% (Slovenia and Czech Republic).

Taking into account the enlarged EU-25, its present population is 456.1 million, and according to United Nations it will increase to 465.3 million in 2025 but will decrease to 439.1 million in 2050, with annual rates of growth of 0.09% during the first period and - 0.22% during the second period. The total fertility rate for this enlarged European Union would be, at present, 1.4 children per woman (lower than the present average for the fifteen member countries), and its life expectancy at birth would be 78 years. The age structure of the resulting enlarged population is, at present, a little younger than the present fifteen member-European Union since, although the proportion of the population under 15 years of age is 17%, the proportion of 65 years and older is only 14%. But the future projected population for the enlarged European Union in 2050 will be almost as old as if only the present fifteen countries were projected. Thus, 30.5% of the population will be 65 years or older and 11.4% will be 80 and older, both proportions being slightly lower than those projected for the present fifteen countries, because of the younger projected population for the candidate countries.

3. The components of population growth

The data that have been presented seem to demonstrate that the future of the European Union population will not be very different whether one considers only its present composition of fifteen countries or the enlarged composition of twenty five member countries. All European populations, even those outside the twenty five, show very similar patterns and processes, characterised by very low fertility and very low mortality levels (high life expectancy at birth) that in combination lead to small or even negative population growth rates and to increasingly ageing population distributions.

This is the present situation and the forecasted future, provided that the present trends continue for the next twenty-three and the following twenty-five years. But it must be underlined that the projections require that present trends continue for such a long period of time, and they also require the assumption of a closed population, without migration flows. Both assumptions, as will be now discussed, are only a demographic exercise, and do not necessarily have to be accepted as the most likely outcomes. And, besides, there are other variables that are not demographic and should be considered as potential intervening variables with very important consequences.

First of all, the assumption that present trends will continue is certainly the most conservative hypothesis, and to challenge it evidence must be presented to the contrary. Thus, it does not seem reasonable to expect a rise in mortality levels, unless the European Union suffers great natural or human catastrophes (wars that produce massive casualties, huge economic crises that produce mass starvation, etc.). Quite on the contrary it seems more natural to expect still more important gains in life expectancy during the next fifty years, due to the great advances that are expected in biology, medicine, genetics and other related fields.

For similar reasons, a rise in fertility is not expected. All European Union countries, both present members and candidates, show fertility levels that are, for the most part, significantly below replacement. Under what conditions could a rise in fertility be expected? Marriage patterns, new roles for men and women, incorporation of women to the labour force, new life styles and, more important, rising expectations in standards of living and consumption patterns, all seem to go in a direction that is not favourable to rising fertility.

In fact, if it were accepted that most of these variables are responsible for the recent fall of fertility since the decade of the eighties in European countries, in order to expect a change in fertility it would first be necessary to expect changes in those explanatory variables.

But, on the contrary, research on attitudes towards marriage, the family and childbearing across most European countries during the past twenty years does not seem to guarantee rising fertility that might have any significant consequences on the structure of their populations, at least within the near future [5].

The second assumption, based on the model of a closed population, must also be rejected, not only because migration flows must always be considered when discussing population growth, but because they represent at present the most important component of population growth in the majority of European Union countries.

Until a few decades ago, net migration (the difference between immigrants and emigrants) represented a small proportion of the total demographic growth in any European Union country, since natural growth (the difference between births and deaths) contributed to total growth more than net migration. Certainly, births were more numerous than deaths, and the opposite was an exception until very recently, due to the significant fall in fertility. Net migration was positive in most Central and Western European countries (the more developed ones), but it was negative in Southern European countries (economically less developed) during the decade of the seventies, and positive

Table 3. Total Population Growth, Natural Growth and Net Migration Rates per 1,000 inhabitants in Member and Candidate Countries of the European Union, 1980 and 2000

	Population mid 2002 (Millions)	1980			2000		
		Total growth	Natural growth	Net Migration	Total growth	Natural growth	Net migration
Members: (a)	**381.2**						
Austria	8.1	0.10	-0.02	0.12	0.23	0.02	0.21
Belgium	10.3	0.08	0.10	-0.02	0.24	0.11	0.12
Denmark	5.4	0.03	0.03	0.01	0.36	0.17	0.19
Finland	5.2	0.34	0.39	-0.05	0.19	0.14	0.05
France	59.5	0.55	0.47	0.08	0.50	0.41	0.09
Germany	82.4	0.28	-0.11	0.39	0.04	-0.09	0.12
Greece	11.0	1.15	0.63	0.52	0.21	-0.02	0.23
Ireland	3.8	1.17	1.19	-0.02	0.11	0.61	-0.51
Italy	58.1	0.16	0.15	0.01	0.28	-0.04	0.31
Luxembourg	0.5	0.38	0.02	0.37	1.28	0.45	0.83
Netherlands	16.1	0.83	0.47	0.36	0.77	0.42	0.36
Portugal	10.4	1.08	0.65	0.43	0.63	0.14	0.49
Spain	41.3	1.05	0.75	0.30	0.97	0.06	0.91
Sweden	8.9	0.18	0.06	0.11	0.24	-0.03	0.28
United Kingdom	60.2	0.10	0.17	-0.07	0.40	0.12	0.28
Candidates: (a)	**74.9**						
Cyprus	0.9	1.14	1.11	0.04	0.57	0.46	0.11
Czech Rep.	10.3	0.20	0.18	0.02	-0.11	-0.18	0.06
Estonia	1.4	0.68	0.27	0.41	-0.37	-0.39	0.02
Hungary	10.1	-0.04	0.03	-0.07	-0.38	-0.38	0.00
Latvia	2.3	0.23	0.14	0.10	-0.58	-0.50	-0.08
Lithuania	3.5	0.53	0.47	0.06	-0.16	-0.13	-0.03
Malta	0.4	0.98	0.74	0.25	0.61	0.34	0.27
Poland	38.6	0.90	0.96	-0.06	-0.02	0.03	-0.05
Slovakia	5.4	0.83	0.89	-0.06	0.07	0.04	0.03
Slovenia	2.0	0.87	0.58	0.29	0.12	-0.02	0.14

Source: Council of Europe, *Recent Demographic Developments in Europe 2001* [1]
(a) Calculated by the author.

but relatively small during the decade of the eighties. In any case, net migration, whether positive or negative, had in general less weight on the total growth of a country's population than natural growth. Seven countries had negative net migration in 1980, and only one of them had, in addition, negative natural growth: Hungary. Net migration represented more than half of the total growth only in five of the eighteen countries that had positive net migration (Germany, Austria, Luxembourg, Sweden and Estonia) [1]. It must be also pointed out that the relatively high positive migration rate in some Southern European countries in 1980 was partly due to the return of former emigrants who, during the decades of the fifties and sixties, had migrated to more developed European countries searching for better jobs. This would be the case of Spain, Portugal and Greece. The situation has changed significantly during the past two decades. First, total population growth has generally been lower in 2000 than it was in 1980, and this has been particularly true in all the candidate countries, though seven of the fifteen member countries show higher rates of growth in 2000. Besides, while only one country had a negative rate of growth in 1980 (Hungary), six of them (all candidate countries) show negative population growth in 2000. Second, of the eighteen countries that had positive net migration in 2000, in nine of them it represented more than half of their total growth for that year (Germany, Sweden, Slovenia, Spain, Austria, Portugal, the United Kingdom, Luxembourg, and Denmark). Besides, comparing the data for 2000 with those of 1980, it may be seen that total population growth has decreased in eighteen of the twenty-five countries, natural growth has diminished in nineteen countries, but net migration has increased in sixteen countries.

Some general conclusions can be proposed at this time. First, population growth has decreased in most European Union countries, both members and candidates, over the past twenty years. Second, natural growth was the main component of population growth two decades ago, but net migration is at present the more important component of growth. Third, the rate of immigration more and more often compensates for the low or even negative rate of natural increase, as seems to be the case with respect to Austria, Germany, Greece, Italy, Spain and Sweden. Immigration is compensating or supplementing low or negative natural population growth [10], not only through the direct effect of immigrants themselves, but also through their contribution to fertility, since most immigrants are young adults in their childbearing years who generally come from countries where fertility is higher. Therefore, population projections that assume the continuation of present demographic trends in the European Union may be greatly underestimating the population that will be reached in 2025 and 2050, because they do not take immigration into account. Though significant increases in fertility rates are not to be expected in the native receiving populations, the double contribution of immigration, through the increase of population numbers themselves, and through their contribution to fertility, may result in larger populations than those that are usually forecasted.

4. Implications of present and expected population trends in the European Union

Population projections seem, therefore, to be underestimating future growth of the European Union member and candidate countries, mainly because they have generally not taken into account the role of migration in-flows [6]. But, even if it were true that European populations will not grow or that they will decrease within the next fifty years, it is not clear why that should be a matter of concern. There is no empirical evidence that population size or population growth rates are positively correlated with economic wealth (GNP) or social well being (HDI). In fact, one could even argue in favour of the opposite relationship, that is, that high economic and social development leads to low population growth rates [13].

Besides, from a global perspective, the real problem in world population today does not seem to be the lack of growth, but rather the maintenance of still high population growth rates that have characterised world population since the end of World War II. Even with the present growth rate (1.3% per year) world population would double within 70 years, though with great differences between the more and the less developed regions. The latter are growing sixteen times as rapidly as the former, since the more developed countries (which include the European Union) have reached almost zero population growth. One should not forget that world population took approximately sixteen centuries and a half to double since the year 0 of the Christian era, but doubled again in only two hundred years, and again in one hundred years, and has trebled since 1950 to the present. Therefore, decreasing world population growth rates may even be beneficial, since population pressure on natural resources will be lessened. And, as has been happening since the beginning of the agricultural and the industrial revolutions, the more developed countries are usually the ones that first initiate new demographic trends, which are followed afterwards by the less developed countries. Thus, the second demographic transition, characterised by below replacement fertility levels and close to zero or even negative population growth, first started in the European countries during the decade of the eighties, but is being followed more or less intensively by practically all countries in the world at present. The total fertility rate in less developed countries is still above 3 children per woman, but it used to be 5 or even more children per woman only a few decades ago, and consequently population growth rates have also decreased dramatically in most less developed countries. It must not be forgotten, in this respect, that for the past fifty years the less developed countries have benefited from a very rapid decline in mortality rates that would have caused higher growth rates had it not been for the decline in fertility.

If it is accepted that low, zero or even negative population growth rates are preferable to high rates of growth, the conclusion must be that this can only be achieved through low fertility, because rising mortality levels are not a desirable social goal in any society. The problem today seems to be that while the developed countries (in particular European countries) have almost achieved zero population growth, the less developed countries still have a high rate of growth that will lead them to double their population in only thirty five to forty years. All signs point in the direction that population growth is falling in most of these countries due to falling fertility rates that seem to be adjusting themselves to the falling mortality rates of previous and future decades. On the other hand, most of the less developed countries are also reducing their population growth through negative net migration.

Some politicians and social scientists express a second area of concern regarding the possibility of population decline in the European population within the next fifty years that refers to fertility. Before 1970 no country in the European Union had a fertility level below replacement. Even in 1970 only four countries had already decreased below that level (Luxembourg, Denmark, Sweden and Finland), but by 1985 fertility in all fifteen countries except Ireland was already below the replacement level of 2.1 children per woman. And that situation, including Ireland since 1995, has been maintained since then, that is, no European Union country has returned to the level of replacement fertility. As a matter of fact, it is not possible to identify a trend of increasing fertility in recent years, as some would like to argue, because the lowest level of fertility has been reached in some countries precisely in very recent years. Thus, if one examines fertility rates since 1970, when the highest rates were attained, it may be noticed that the lowest fertility rates were reached in 1998 or later

Table 4. Total Fertility Rates in European Union Member and Candidate Countries, 1970-2000

	Total Fertility Rate (number of children per woman)										
	1970	1975	1980	1985	1990	1995	1996	1997	1998	1999	2000
Members:											
Austria	2.29	1.83	1.65	1.47	1.45	1.40	1.42	1.37	1.34	1.32	1.34
Belgium	2.25	1.74	1.68	1.51	1.62	1.55	1.59	1.59	1.59	1.61	1.66
Denmark	1.95	1.92	1.55	1.45	1.67	1.80	1.75	1.75	1.72	1.73	1.77
Finland	1.83	1.68	1.63	1.64	1.78	1.81	1.76	1.75	1.70	1.74	1.73
France	2.47	1.93	1.95	1.81	1.78	1.71	1.72	1.71	1.76	1.79	1.89
Germany	2.03	1.48	1.56	1.37	1.45	1.25	1.32	1.37	1.36	1.36	1.36
Greece	2.43	2.32	2.22	1.67	1.39	1.32	1.30	1.31	1.29	1.28	1.29
Ireland	3.97	3.43	3.24	2.48	2.11	1.84	1.89	1.92	1.93	1.88	1.89
Italy	2.38	2.17	1.64	1.42	1.33	1.20	1.19	1.22	1.20	1.23	1.23
Luxembourg	1.98	1.55	1.49	1.38	1.60	1.69	1.76	1.71	1.68	1.73	1.79
Netherlands	2.57	1.66	1.60	1.51	1.62	1.53	1.53	1.56	1.63	1.65	1.72
Portugal	2.84	2.63	2.20	1.72	1.57	1.40	1.44	1.46	1.46	1.49	1.52
Spain	2.86	2.79	2.20	1.64	1.36	1.18	1.17	1.18	1.16	1.20	1.24
Sweden	1.92	1.77	1.68	1.74	2.13	1.73	1.60	1.52	1.50	1.50	1.54
United Kingdom	2.43	1.81	1.89	1.79	1.83	1.71	1.72	1.72	1.71	1.68	1.65
Candidates:											
Cyprus	2.54	2.01	2.46	2.38	2.42	2.13	2.08	2.00	1.92	1.83	1.83
Czech Rep.	1.90	2.40	2.10	1.96	1.90	1.28	1.18	1.17	1.16	1.13	1.14
Estonia	2.16	2.04	2.02	2.12	2.04	1.32	1.30	1.24	1.21	1.24	1.39
Hungary	1.98	2.35	1.91	1.85	1.87	1.57	1.46	1.38	1.33	1.29	1.32
Latvia	2.02	1.96	1.90	2.09	2.01	1.26	1.16	1.11	1.10	1.18	1.24
Lithuania	2.39	2.18	1.99	2.09	2.02	1,49	1.42	1.39	1.36	1.35	1.27
Malta	...	2.17	1.98	1.99	2.05	1.83	2.10	1.95	1.81	1.72	1.67
Poland	2.26	2.26	2.26	2.32	2.05	1.62	1.58	1.51	1.44	1.37	1.34
Slovakia	2.41	2.53	2.31	2.26	2.09	1.52	1.47	1.43	1.38	1.33	1.29
Slovenia	2.12	2.17	2.10	1.71	1.46	1.29	1.28	1.25	1.23	1.21	1.26

Source: Council of Europe, *Recent Demographic Developments in Europe 2001* [1].

in eight of the fifteen member countries and in all of the candidate countries. (Two of these countries may be considered exceptions, Denmark and Finland, both of which reached an all period-low rate in 1985 and 1980 respectively, but attained another peak in 1995 and another lowest rate in 1998, with increasing rates again afterwards. Ireland is another exception, in that its rates fell till 1995, rose until 1998 and fell again afterwards.) Seven member countries, then, reached their lowest rates around 1995 or earlier, and have experienced increasing rates since then. When fertility rates in 2000 are compared with the most recent lowest rates throughout the whole thirty year period, it is observed that the increase in fertility has been under 5% in seven of the member countries, 5% to 10% in four countries, and over 10% in only three countries. Only the United Kingdom reached its lowest rate ever in 2000. And, among the candidate countries, the increase has been less than 5% in three countriesand more than 10% in two countries (Estonia and Latvia), but all other five countries reached their lowest rate precisely in 2000.

It may be concluded that no European Union country, member or candidate has returned to replacement fertility levels. The highest rates in 2000 are those of Ireland and Cyprus (1.89 and 1.83 respectively), both of which may be considered as rates that are still on their way down, and France (1.89). France is usually taken as an example of recovered fertility, but it must be noticed that its rate has never been lower that 1.71 children per woman. The only real examples of recovered fertility are those of Luxembourg and the Netherlands, both of which had their lowest rates in 1985 and their highest rates since then in 2000, with increases of 30% and 14% respectively. But Denmark, as has been said, though reaching an all time lowest rate in 1985, rose again to another highest of 1.80 in 1995 which has not been matched since then. And the case of Sweden, often cited too as an example of recovered fertility, is certainly not an example. Though it reached a low rate of 1.68 in 1980, it rose to 2.13 in 1990 but fell again afterwards, reaching its lowest rates of 1.50 in 1998 and in 1999 exceeded only by .04 points in 2000. Therefore, there seems to be no ground for claiming a clear recovery of fertility in European countries, since all rates in 2000 were not only below 2.1, but only ten out of twenty five were above 1.6 children per woman.

Unemployment among young men and women, increased women's participation in the labour force and access to housing facilities, have often been cited as the main causes for the fall of fertility since 1970. However, empirical evidence has not supported these supposed relationships at the country or the individual levels of analysis with sufficient reliability, nor where either a cross-sectional or a longitudinal analysis is performed. Rather, it seems that new cultural values and life styles are responsible for retarding the emancipation of youngsters, the age at which they construct couples and families (if at all), the age at which they have children and, consequently, the number and spacing of their children (if any) [3]. Therefore, unless the values and life styles that have caused the fall of fertility change, it seems difficult to foresee fertility increasing significantly in the near future. It does not seem casual that some of the lowest fertility rates are found in Southern and Eastern European countries, that is those that have arrived later to mass consumption and have more recently experienced great economic, social and political change in a short period of time.

Fertility decline, combined with a very low mortality, is responsible for the ageing of population, a process that has taken place first in the countries of the European Union and other developed countries, but which is being followed by all other countries in the world. Therefore, population ageing does not seem to be a temporary process, but one that will probably last for a long and unpredictable period of time.

The first thing that should be said is that population ageing must not be considered a social problem. It would be ironic to label as a problem one of the most important successes in the history of mankind: making it possible for the great majority of human beings to postpone death until what continues to be the culmination of human life, one hundred years. Life expectancy at birth was around thirty-five years for most populations in the world until the beginning of the last century, and in more developed countries until the middle of the nineteenth century. Today there is almost no country in the world with such a low life expectancy at birth. On the contrary, more than 80% of each cohort in developed countries can expect to survive till the age of 70, and more than half can expect to reach the age of 80. This great and unprecedented historic success cannot be turned into a "problem". The real problem is that society has not yet assimilated this new fact, and has not produced the social reorganisation necessary to cope with this new situation.

Interactions between population, environment (resources), technology (material culture) and social organisation (non-material culture), including belief and value systems, have been explained by the social ecosystem paradigm [4]. According to this theoretical model, human populations adapt to (and survive in) their environment through culture, something that makes this adaptation process totally different from what is found in other biotic but non-human populations (plants and animals). Significant changes in any of the four elements of the social ecosystem may have repercussions in the other three elements, so that change is an immanent trait of human ecosystems. The equilibrium among the four elements is always unstable, never complete, so that tensions and conflicts in their interactive processes make conflict and change ever present to a greater or lesser degree. Changes in value systems and life styles (non-material culture), as has been argued above, may probably explain the low fertility of more developed countries and the present process of falling fertility in developing countries. For the same reason, it is argued here, changes in the population structure must bring changes in the social organisation of more developed societies [2].

The only possible ways to avoid population ageing in the European Union, and in effect, in any country in the world, are three: to increase mortality, to increase fertility, or to increase positive net migration. When considering the total world population, needless to say, the third alternative is not viable. Certainly, if mortality increases, the proportion of each cohort arriving to high ages would decrease, and would produce a pyramidal shape of the population, as in pre-industrial times. But it does not seem plausible that any government should want to implement a policy to increase mortality. On the contrary, efforts to increase life expectancy even beyond the traditional threshold of 100 years of age will continue. Therefore, at world level there is no other alternative than to increase fertility, so that new in-coming cohorts will be larger than the preceding ones. In this way, though most of the population in each cohort will reach older ages, the numbers in younger cohorts will always be larger than in the older ones, thus producing a pyramidal distribution. But, as it may be easily demonstrated, this alternative requires that the number of births in each cohort should continue to increase indefinitely, because whenever the number of births in two successive cohorts is the same, a rectangular rather than a pyramidal shape will result. Besides, according to the analysis presented above, there is no evidence whatsoever that fertility may return to replacement levels in the more developed countries, and more specifically in European Union member or candidate countries. On the contrary, empirical evidence seems to support the hypothesis that fertility will continue to be below 2.1 children per woman and, furthermore, that all countries will tend to drop more or less rapidly towards fertility levels that stay around or below replacement. As regards positive net migration, it may be considered as a temporary alternative for those countries that have initiated their ageing process earlier. Thus, more developed countries, including members and candidates of the European Union, may reduce the process of ageing of their populations through net immigration, as has been discussed above, because migrants tend to be young adults. The flow of migrants towards the more developed countries, though contributing to retard population ageing in the receiving countries, will contribute to the ageing of the populations of origin, since the weight of the young and old age groups will increase at the expense of diminishing the middle age groups. The temporary character of net immigration as an alternative to ageing is also due to the fact that immigrants will eventually also age, so that if population ageing is to be avoided, a continuous current of net immigration must be maintained indefinitely. (Needless to say, net immigration occurs for reasons other than avoiding ageing of the population in developed countries, though some politicians often use that explanation. Net immigration is more a consequence of economic, than purely demographic, factors).

In other words, to avoid the process of population ageing European Union countries should increase their fertility rates indefinitely and/or increase their net immigration also indefinitely. If increases are only temporary, the process of ageing will recur. Besides, increases in fertility will produce higher population growth rates, a scenario that is not free of criticisms and problems, especially at world level. Consequently, and not considering the alternative of rising mortality, the only other alternative is to admit that populations will tend to age. This implies making the necessary changes in social organisation to take into account this new demographic situation, characterised by low fertility, low mortality, low or even negative growth and an age distribution that will be more similar in shape to a rectangle than to a pyramid.

The age distribution of populations has changed through time as a consequence of the levels of mortality and fertility, though migration, as has been mentioned above, has only had a more significant weight in specific populations and dates in present times and with respect to European Union countries. When the population is divided into three major age groups: the young (less than 15 years of age), the adult and potentially active (15 to 64 years) and the old (65 and over), it is observed that in pre-industrial populations, the proportions are 30-35, 60-65 and 3-5 per cent respectively. In industrializing populations the proportions were 40-50, 40-55 and 5-7 per cent. In industrial populations the distribution is 25-35, 60-65 and 10-15 per cent. And in post-industrial populations the proportions may be 15-20, 55-65 and 20-30 per cent. The threat of present demographic trends on the potentially active population is not really such a threat, because that segment of the population will not diminish significantly. The greater changes are taking place, and will take place, when comparing the young and the old age groups. But even the apparent reduction of the potentially active population may be solved by intelligent social responses, that is, adaptive changes in social organisation.

Most of the fears attached to low fertility refer to the assumption that smaller cohorts will imply that the number of contributors to Social Security will be less than the number of retired people entitled to receive a pension, and that this situation will lead to a bankrupt Social Security. This argument has led many social scientists and politicians to support rising fertility rates. However, this response misplaces the real goal that is desired. The goal is to have more contributors to Social Security than pensioners, and therefore one must promote measures to increase contributors, not necessarily to increase births. Increasing births will only affect the number of contributors to Social Security after about thirty years, when the newborn reach the age to enter the labour force, and therefore is a long-term answer. But there are other short-term measures that may help to increase the number of contributors, such as policies to reduce unemployment among the young, to facilitate women's access to the labour force (through family support policies), or to increase the number of immigrants that are admitted into the country. More specifically, there is one measure that may have immediate and profound effects:

retarding the age of retirement, or even better, to make retirement voluntary. Besides, retirement pensions should be dependent on the total time that a person has contributed to Social Security (something that is compatible with a minimum pension for everybody, regardless of having contributed or not to Social Security).

When life expectancy at birth was 60 years (only a few decades ago), it seemed reasonable to establish retirement age at 65 years, because only about 5-7 per cent of the population survived to that age or even surpassed it. In those times, people entered the labour force at around 20 years of age and retired at 65, so that in a 60 year long average life span, individuals were self-sufficient for about three-quarters of their life. At present, youngsters enter the labour force at around 30 years, and due to early retirement or long-term unemployment many individuals retire at 55 years. But, since the average life expectancy is 80 years, they are self-sufficient only during one third of their life, the rest being dependent on their families or on society. Besides, being compulsorily retired for more than 30 years of one's life is certainly not a very desirable future. It seems as if society parks these individuals in a waiting hall until they die. Certainly, individuals have a right to retire from active work, and to receive a pension that will help them to live decently, but a right cannot and should not be transformed into an obligation. As individuals reach older ages in much better physical and mental conditions than ever before, they should have the right to decide when they want to retire. Retirement means, no matter how high pensions are, a loss of income, a loss of social prestige and self-esteem, and a loss of power. On the contrary, if retirement age is postponed the same number of years as the average age of entry into the labour force, around ten years, the proportion of the population over 75 years (about 25% in developed countries) will be similar to the proportion over 65 years at present. It is a question of adjusting retirement age to the age of entry into the labour force, and of accommodating it to the new life expectancy. In the theoretical example of a stationary population there is a permanent in-flow of 100,000 thousand births every year. If nobody died till age 100, there would be exactly 100,000 persons in each of the one hundred age groups and there would be zero population growth (because there would be 100,000 deaths per year, due to a cohort of 100,000 reaching the upper limit of 100 years of age). And consequently there would be a fixed age distribution with 15 per cent of the population under 15 years of age, 25 per cent over 75 years of age, and 60% in the potentially active population age group. This theoretical example provides a way to visualise the impact of changing fertility or mortality levels in the shape and volume of the total population.

What is needed, then, is that societies make the necessary changes in their organisation to give individuals over 60 years a social role that cannot be that of waiting patiently to die, a role that must be coherent with the one they had until they reached the age of 60, not better nor worse. This is the real challenge of European Union societies today, and the challenge that developing societies will have to face in a non-distant future. A challenge that implies changes in the social organisation, and one that does not require imperative changes in fertility, mortality or even migration.

On the contrary, these changes should be compatible with individuals' decisions about the number of children that they have, with new gains in life expectancy, and they should not require massive population redistribution due to scarcity of living opportunities in some places and the abundance of opportunities in other places.

5. Conclusions

This paper argues that population trends in the enlarged European Union will continue to be characterised by high and even increasing life expectancy, below replacement fertility, close to zero population growth and, consequently, increasing ageing of the population. In order to reverse these trends, if that was considered necessary, and discarding a rise in mortality, only a significant increase in fertility would produce a younger population. This would have the cost of high rates of population growth, which might not be convenient from a world perspective due to its impact on natural resources and on the environment.

For the past years the European Union has received an increasing number of immigrants, and it seems likely that it will continue to do so for the next few decades. However, this demographic input has not had, and is not likely to have, a significant impact on the structure of the receiving population, though it may increase slightly the younger adult age-groups and the labour force, and even the rate of fertility. But immigration flows will not alter significantly the fact that European populations will continue to age (among other things because immigrants will also age).

Therefore, unless European societies are prepared to accept and/or promote a rise in mortality (something that is unthinkable), their populations will continue to have a close to zero population growth, and they will continue to age. The shape of their age distributions will approximate a rectangle rather than a pyramid meaning that most of the individuals in each cohort will survive till 100 years of age. But, as individuals reach higher ages in much better physical and mental conditions, European societies should probably consider the need to make structural changes that postpone retirement age till 75 years of age or, even better, to make retirement voluntary. This measure would reduce the weight of the dependent older population to a proportion similar to the present one with retirement age at 65 years.

In brief, social rather than demographic changes are needed to adapt to the present and expected population trends in the European Union. As for net immigration flows, they will probably continue and even increase in the future, but their demographic impact on fertility or the age structure will not alter significantly present and expected population trends.

References

1. Council of Europe (2002) *Recent Demographic Developments in Europe 2001*, Strasbourg.
2. Díez-Nicolás, J. and Inglehart, R. (eds.) (1993) *Tendencias mundiales de cambio en los valores sociales y políticos*, Fundesco, Madrid.
3. Díez-Nicolás, J. (2001) Causas y consecuencias del reciente descenso de la fecundidad en España, in *Demografía y Cambio Social*, Consejería de Servicios Sociales. Comunidad de Madrid, Madrid.
4. Hawley, A.H. (1986) *Human Ecology. A theoretical essay*, The University of Chicago Press. Chicago.
5. Kaa, D.J.van de (1993) *The second demographic transition revisited: theories and expectations, in Population and Family in the Low Countries: Late Fertility and other Current Issues*, NIDI-CBGS pub. 30, The Hague.
6. Martin, Ph. And Widgren, J. (2002) International migration: facing the challenge, *Population Bulletin of the Population Reference Bureau, 1*, 1-40.
7. Population Reference Bureau (2002) *World Population Data Sheet*, PRB, Washington.

8. Shroots, J.J., Fernández-Ballesteros, R. and Rudinger, G. (eds.) (1999) *Ageing in Europe*, IOS Press, Amsterdam.
9. United Nations (1999) *Population Ageing*, Population Division, Department of Economic and Social Affairs, New York.
10. United Nations (2000a) Below Replacement Fertility, *Population Bulletin of the United Nations, Special Issue Nos. 40/41*, New York.
11. United Nations (2000b) *World Population Ageing 1950-2050*, Population Division, Department of Economic and Social Affairs, New York.
12. United Nations (2001a) *Replacement Migration*, Population Division, Department of Economic and Social Affairs, New York.
13. United Nations (2001b) *Population, Environment and Development, The Concise Report*, Population Division, Department of Economic and Social Affairs, New York.
14. United Nations (2002) *World Population Prospects: The 2000 Revision*, Population Division, Department of Economic and Social Affairs, New York.

Chapter 16

Population Growth in The Euro-Mediterranean Region

H. Makhlouf
Cairo Demographic Center, (CDC)78 (St. No. 4) El-Hadaba-Elolya, Mokattam
11571 Cairo

The main objectives of this chapter are the presentation of: vital events that characterize member countries of the Euro-Mediterranean region and to what extent they affect their rate of population growth and the required number of years for doubling in size. The chapter will try to determine the major factors accounting for the fertility differential between countries in the region and the contribution of each to its persistence; assess the negative effect of population growth on the socioeconomic and demographic development of the region; and project the population size for each country and region as a whole by the years 2025 and 2050.

1. Introduction

Undoubtedly, the world is witnessing today crucial demographic changes manifested in a substantial differentiation in rates of population growth that certainly affects the future size of global population. A close review of the early demographic history of the world indicates that very few countries in Europe and the United States have experienced, since the middle of the 17th century, a significant decline in mortality followed, after an elapse of a considerable period of time, by a corresponding trend in fertility. The duration of the transitional period required for a convergence of the vital rates at low levels in these countries had stretched over more than 150 years. In the meantime, almost all the other countries of the world were still distinguished by high levels of both fertility and mortality. This situation has changed remarkably since the end of the Second World War. The application of Western medical discoveries in various fields of medicine, the advancement of knowledge in the realm of health and sanitation, and above all national and international campaigns for vaccination against infant and child diseases became widely adopted by many developing countries. Consequently, mortality in these countries dropped at a faster rate than that experienced earlier by Western societies while fertility remained for some time at its traditional level. This tendency led to a widening of the demographic gap that produced an unprecedented population growth rate of 3%, if not even more, particularly in many African countries.

Today, the efforts of both formal and informal organisations in the area of population planning and control has begun to show a concomitant effect in the downward trend of

A. Marquina (ed.), Environmental Challenges in the Mediterranean 2000-2050, 265–280.
© *2004 Kluwer Academic Publishers. Printed in the Netherlands.*

the level of fertility that has become evident in many parts of the world, particularly in North Africa, Latin America, Europe and South Eastern Asia. One of the interesting regions that presents an excellent example of mixed stages of a population transition is that of the Euro-Mediterranean which contains countries located on the continents of Africa, Asia and Europe. Therefore, current research is designed to examine in detail the differences that may exist in the levels of population growth and the stages of a demographic transition that distinguish member countries belonging to this region, and to project their expected population size by the year 2025 and 2050.

Geographically, this region consists of several sub-regions located in Africa, Asia and Europe. It comprises at present 29 member countries distributed unequally among these sub-regions. According to the UN regions' classification, they fall in 5 sub-regions of which one is located in Africa, another one in Asia and the remaining three in Europe. The African sub-region located in the Northern part of the continent includes 5 countries namely Algeria, Egypt, Libya, Morocco and Tunisia. On the other hand, 7 member countries are in the Western region of Asia (Middle Eastern countries). These are Cyprus, Israel, Jordan, Lebanon, Palestine, Syria and Turkey. The remaining three sub-regions fall within the European continent of which the Northern one comprises Denmark, Finland, Ireland, Sweden and the United Kingdom. The Southern one includes Albania, Greece, Italy, Malta, Portugal and Spain, while the Western sub-region comprises Austria, Belgium, France, Germany, Luxembourg and the Netherlands.

2. Population Size

The demographic situation of the Euro-Mediterranean countries is viewed in terms of population size and occupied area as well as in the magnitude of vital events that determine the turn rates of natural growth. In the meantime, efforts have also been made to display the demographic and territorial variations that may exist between the constituent countries of the region. Table 1 shows the percentage distribution of population, area and density (persons per square mile) of the Euro-Mediterranean member countries by sub-regions in 2001.

It appears from the data presented in table 1 that the population size of the region amounts to 630.2 million people or about 10.3% of the total world population in 2001. It occupies an area of 3.908.104 square miles that represents 7.5% of that of the world as a whole. The relationship between population size and area indicates that a number of countries have less than their equitable share as reflected by the high density. Consequently, the average population pressure per unit of land in this region is about 161 persons per square mile compared to 118 for the world as a whole.

However, a close examination of the population size and area reveals remarkable differences in density, between as well as within, each sub-region. It is obvious from the table that the Western Europe sub-region shows a great disparity in its population size compared to other areas to the extent of having a population density that exceeds by more than 3 times that of the world as a whole. It has a total population of 176.2 million people occupying an area of 411,696 square miles giving a density of 428 persons per unit of land area. On the other hand, the Northern Africa sub-region is inhabited by 144.9 million people but has a land area of 2,221,193 square miles.

Table 1. Percent Distribution of Population and Area of Euro-Mediterranean Member Countries by Sub-regions and their Population Density per square mile in 2001.

Sub-region & Country	Population in million	Area sq. mile	Percent Popu lation.	Percent Area	Density per sq. mile
Northern Africa					
Algeria	31.0	919,591	21.4	41.4	34
Egypt	69.8	386,660	48.2	17.4	181
Libya	5.2	679359	3.6	30.6	8
Morocco	29.2	172,413	20.2	7.8	169
Tunisia	9.7	63,170	6.6	2.8	154
Total	144.9	2,221,193	22.5	47.9	65
Middle and Near East				.	247
Cyprus	.9	3,571	.9	.8	791
Israel	6.4	8,131	6.2	1.0	150
Jordan	5.2	34,444	5.0	8.1	1061
Lebanon	4.3	4,015	4.2	.9	1365
Palestine	3.3	2,417	3.2	.6	231
Syria	17.1	71,498	16.4	16.9	221
Turkey	66.3	299,158	64.1	70.8	
Total	103.5	423,234	16.4	10.8	245
Northern Europe					
Denmark	5.4	16,637	6.5	3.8	322
Finland	5.2	130,560	6.2	29.5	40
Ireland	3.8	27,135	4.6	6.1	142
Sweden	8.9	173,730	10.7	39.2	51
United Kingdom	60.0	945,487	72.0	21.4	635
Total	83.3	442,610	13.2	11.4	188
Southern Europe					
Albania	3.4	11,100	2.8	2.7	310
Greece	10.9	50,950	8.9	12.4	214
Italy	57.8	116,320	47.3	28.4	497
Malta	.4	124	.3	--	3157
Portugal	10.0	35,514	8.2	8.7	282
Spain	39.8	195,363	32.5	47.8	204
Total	122.3	409,371	19.4	10.5	299
Western Europe					
Austria	8.1	32,378	4.6	7.9	251
Belgium	10.3	11,787	5.8	2.9	872
France	59.2	212,934	33.6	51.7	278
Germany	82.2	137,830	46.7	33.5	597
Luxembourg	.4	999	.2	.2	446
Netherlands	16.0	15,768	9.1	3.8	1018
Total	176.2	411696	28.0	10.5	428
Grand Total	630,2	3,908,104	100.0	100.0	193

Source: 2001 World Population Data Sheet [7].

This means that about 23% of the total region population occupies 56.8% of its total area. The disparity between its population share and that of the total area accounts for lowering the density to a level below that of other sub-regions as well as the world as a whole. Between these two levels falls the other sub-regions with Southern Europe third in terms of population size but fourth in terms of density. In contrast, the Northern Europe sub-region stands second in low level of density but fifth in population size.

Another distinction that can be noted from the data is the lack of an even distribution of both population and area within each sub-region, which leads to a further increase in its heterogeneous characteristics. Within each sub-region, only one country seems to capture more than 45% of its total population, which indicates a high index of population concentration. Egypt's share reaches nearly 48% of the total population of its Northern Africa sub-region. This same tendency applies equally to Turkey in the Middle and Near East sub-region that accounts for about 64% of its total population. The United Kingdom with 72%, Germany with 47% and Italy with 49% tend to have the highest percentage of population in their respective sub-regions. In contrast, these countries do not have an equitable share of their area which leads to fantastic differences in population pressure per square mile. Algeria has about 41.4% of its sub-region total area but with only 31.0 million inhabitants which reduces its density level to 34 peoplerson per square mile. Density in Libya is even much lower than that. With 30.6% of the total area of the Northern Africa sub-region, and a population share of only 3.6%, its density is the lowest not only compared to its neighboring countries but also to all those constituting the Euro-Mediterranean region.

The only highly populated countries that have more than their equitable land share in each sub-region are Turkey, Spain and France where the former accounts for 64.1% of its sub-region population but with a corresponding 70.8% of the area. This reduced its population density to 221.0 people per square mile compared to an average of 244.5 for the sub-region as a whole. The same relationship applies to Spain and France. In contrast, countries like Egypt, Morocco, the United Kingdom, Italy, Belgium, Germany, and the Netherlands are in a critical position with respect to the population pressure on their occupied land area. Within each of their sub-regions, they display an extremely high density compared to that of the general average. Perhaps the level of density may have a significant bearing on the rate of population growth and strong motives for early control of fertility.

3. Rate of Population Growth

In the present analysis, population growth is measured by the rate of natural increase that represents the difference between Crude Birth Rate (CBR) and Crude Death Rate (CDR). Unlike the commonly stated growth rate, it excludes the effect of net migration. And since many of these countries vary in terms of the net migration rate, it is necessary to use the rate of natural increase as an indicator of growth to measure the changes in population size attributed solely to the balance between births and deaths in the Euro-Mediterranean member countries.

On the basis of this index, these countries manifest, as shown by the data presented in table 2, marked differences in the rate of population growth that ranges between -0.1% and 3.7%. The ascending ranking order of these countries indicates that Germany, Italy and Sweden have a growth rate of -0.1%. In other words, these three countries have reached a stage not anticipated by the classical theory of demographic transition. Nevertheless, these countries are not located in the same sub-region of the European continent. This suggests that the demographic transition on the European continent might not have started at the same time in these countries though they currently share this unique feature of depopulation. In this respect, Sweden is well known to be the country that experienced the phenomenon of demographic transition earlier than any other in the world. This will be verified later by the level and trend of its TFR (Total Fertility Rate) in the second half of the 20th century.

On the other hand, Greece is the only country in the Southern Europe sub-region that has a stationary population of a zero rate of growth. It is followed by 7 countries having an annual growth rate between 0.1% -0.2% but distributed in several sub-regions of the European continent of which three fall in the Northern sub-region, two in the Western and two in the Southern. Denmark, Finland and the UK fall in the first, Austria and Belgium in the second and Portugal and Spain in the third respectively. An annual rate of natural increase between 0.3-0.5% is noted in 6 countries of which Malta has the lowest with 0.3% followed by France, Luxembourg, and the Netherlands with 0.4% then the Netherlands and Cyprus with a relatively higher rate of 0.5%. The highest rate of population growth in the European sub-regions is found in Albania that reaches annually 1.5%. In short, with the exception of Albania, none of the European member countries of the region had an annual growth rate exceeding 0.5% in 2001. However, 7 of them still display a demographic gap between 0.2 and 0.5%.

In contrast, 5 non-European countries showed a rate of natural growth ranging between 1.3% and 1.9%. Two of them fall in the Northern Africa sub-region while 3 in the Middle and Near East one. The first area includes Tunisia and Algeria and the second Turkey, Israel and Lebanon. These countries are progressing through the third stage of demographic transition characterized by starting a decline in fertility but still far from completing it successfully. Out of the remaining 6 member countries, five have an annual growth rate between 2.0% and 2.5% of which three are located in the Northern sub-region of Africa and the other two in the Middle and Near East sub-region in Asia.

The first sub-region comprises Egypt, Libya and Morocco while the second Jordan and Syria. These countries are still at an early stage of the demographic transition and some of them have just experienced recently a slow decline in fertility. The highest growth rate is found in Palestine that falls in the Middle and Near East sub-region.

However, similar rates of population growth may result from different rates of vital events. On the basis of this assumption, marked differences in rates of population growth between Euro-Mediterranean countries might have evolved in response to concomitant differences in birth rather than death rates.

Therefore, analysis of the available data pertaining to births and deaths provides an answer to this and other related questions. Data presented in table 2 show levels of crude birth rate in the member countries of the Euro-Mediterranean region.

Although it may be argued that the crude birth rate is not an accurate measure of fertility, particularly in comparisons between countries that vary markedly in terms of age and sex composition, the answer is that standardisation for these variations yields the same results. This can be seen from the data presented in table 2 that show the TFR of Euro-Mediterranean member countries as a standardized measure of fertility not affected by age or sex composition.

Table 2. Crude Birth Rate (CBR), Crude Death Rate (CDR), Rate of Natural Increase and Total Fertility Rate (TFR) of Euro-Mediterranean Countries and their Ranking Order by Rate of Natural Increase (RNI) in 2001

NO.	Country	Rate Natural Increase	Crude Birth	Crude Death	Total Fertility
1	Germany	-0.1	9	10	1.3
2	Italy	-0.1	9	10	1.3
3	Sweden	-0.1	10	11	1.5
4	Greece	0.0	10	10	1.3
5	Austria	0.1	10	9	1.3
6	Belgium	0.1	11	10	1.6
7	Finland	0.1	11	10	1.7
8	Portugal	0.1	12	11	1.5
9	Spain	0.1	10	9	1.2
10	UK	0.1	12	11	1.7
11	Denmark	0.2	13	11	1.7
12	Malta	0.3	11	8	1.7
13	France	0.4	13	9	1.9
14	Luxembourg	0.4	13	9	1.7
15	Netherlands	0.4	13	9	1.7
16	Cyprus	0.5	13	8	1.8
17	Ireland	0.5	14	9	1.9
18	Albania	1.2	17	5	2.8
19	Tunisia	1.3	19	6	2.3
20	Turkey	1.5	22	7	2.5
21	Israel	1.6	22	6	3.0
22	Lebanon	1.6	23	7	2.5
23	Algeria	1.9	25	6	3.1
24	Morocco	2.0	26	6	3.4
25	Egypt	2.1	28	7	3.5
26	Jordan	2.2	27	5	3.6
27	Libya	2.4	28	4	3.9
28	Syria	2.5	31	6	4.1
29	Palestine	3.7	42	5	5.9

Source: 2001 World Population Data Sheet [7]

Comparison between the fertility measures cited in table 2 does not show marked differences in the ranking position of member countries according to each in the region. In the meantime, the correlation coefficient between RNI and TFR is .9755** and with CBR is.9948**. Both values are highly significant at .001 level. This is substantiated further by the value of coefficient association between CBR and TFR that reaches .9845**.

Accordingly, the CBRs presented in table 2 can be used to classify the Euro-Mediterranean member countries into homogeneous sub-groups that vary markedly in terms of fertility level. At first glance, the range of rate difference extends from as low as annual births of 9 per 1000 population in Germany and Italy to 42 in Palestine. Out of the other 17 European countries, these all six with an annual birth rate of 9-10 per 1000 population, five with 11-12, six with 13-14 and finally one with 17 births.

Outside the European continent, only Tunisia has a CBR of 19 per 1000 population. Israel, Turkey and Lebanon follow it with an annual birth rate of 22-23 per 1000 population. Four countries that include 3 from the Northern Africa sub-region and one from the Middle East have an annual birth rate that varies between 25-28. Finally, Syria and Palestine each stand in a separate birth category. The first has a crude birth rate of 31 per 1000 population while the second 42. With stabilized mortality rate at a level of 5-6 deaths per 1000 population, it takes the first country 28 years to double its population and only 19 years for the second.

Unlike the case of fertility, Euro-Mediterranean member countries show a narrow range of differential mortality as can be seen from the data cited in table 2. They reveal a unique situation where the rates of fertility and mortality are not positively associated. This is attributed to the fact that the crude death rate in this case is much more distorted than the crude birth rate by the age-sex composition of the population. It is apparent from the data that the maximum range of difference in crude death rates among Euro-Mediterranean countries is only 7 persons per 1000 population but surprisingly lower among high fertility countries.

The lowest mortality rate of 4 deaths per 1000 population is found in Libya followed by Jordan and Palestine with 5 deaths each. Eight countries of mixed locations have a death rate of 6-7 per 1000 population. None of the European countries is represented in this category of low CDR which coincides with those currently characterized by a declining high crude birth rate. All countries of Northern Africa and the Middle East sub-regions fall in this low mortality level. In contrast, all European member countries of the Euro-Mediterranean region display a higher level of mortality, which is explainable by the aging structure of their population. Cyprus and Malta have an annual death rate of 8 per 1000 population. Three levels of a relatively higher mortality rate distinguish the remaining countries. The majority of European member countries fall in the category of annual deaths of 9 per 1000 population. Four of these countries belong to the Western sub-region, and the Northern and Southern sub-regions share the remaining two equally. These are followed in ascending order by 5 countries of mixed geographical location having a death rate of 10 persons per 1000 population. Two countries from each of the Western and Southern sub-regions are represented in this mortality level and the remaining one belongs to the Northern area. The highest level of crude death rate of 11 per 1000 population is found in 4 countries of which 3 are located in the Northern sub-region of Europe and the fourth in the Southern one. In fact, it is evident that countries with 9+ crude death rate are those have which successfully completed the final stage of demographic transition that evolved over almost 180 years in UK. This will be illustrated by the historical data pertaining to the early demographic situation in this country. They demonstrate levels, trends and duration of each stage of the demographic transition that most probably have taken place in all Europe during the past 3 centuries.

In short, the data suggest an inverse association between the CDR and each of RNI, CBR and TFR. Coefficients of association are -.8986**, -8419** and -.8163** respectively. This pattern of association should have been positive and still noticeable in many parts of the world. Accurate assessment of the health and sanitation standard in this group of countries can be better measured by the mortality rate among infants, children below 5 years of age and mothers giving birth. Mortality rates of infants and children below 5 years of age show unexpectedly an inverse association of -. 7469** and -.7379** with the CDR respectively. The comparable figures are .7654** and .7847** with NRI. Coefficient of association between the two indices of children's mortality approaches unity. It reaches .9612**. Unlike the CDR, each has a positive coefficient of .7476** and .7730** with the CBR, respectively. With the TFR, the coefficient is .6640** with the IMR and .6929** with mortality rate of children below 5 years of age. Table 3 presents a summary of the coefficient of association between all these demographic variables.

Table 3. Coefficient of Association between 9 Demographic Variables of Euro-Mediterranean countries in 2001.

Variable	X1	X2	X3	X4	X5
X1 Rate of natural increase	--				
X2 Crude birth rate	.9948**	--			
X3 Crude death rate X3	-.8986**	-.8491**	--		
X4 Total fertility rate	.9755**	.9845**	-.8163**	--	
X5 Infant mortality rate	.7654**	.7476**	-.7469**	.6640**	--
X6 Mortality rate –5 years	.7847**	.7730**	-.7379**	.6962**	.9621**

** Significant at .001 level.

It is obvious from the table that the highest positive coefficient of .9948** exists between the RNI and the CBR. The comparable value with total fertility rate is .9755. In the meantime, is unlikely to be witnessed in high fertility countries a negative association between the CDR and the CBR that reaches -. 8986** which is peculiar in the present region under investigation.

4. Stages of Demographic Transition

Perhaps the most interesting model that describes the early demographic transition that occurred in most European countries can be displayed by the available data for the UK. Data for Sweden or France which started their transition about or a little earlier than UK can also serve the same purpose.

The UK historical experience is used to demonstrate in an empirical manner what is meant by demographic transition, its successive stages, duration and the characteristics of each as they occurred in almost all low fertility member countries of the Euro-Mediterranean region. Table 4 presents the relevant data that explain this model.

Table 4. Crude Birth Rate, Crude Death Rate and Rate of Natural Increase for England and Wales between 1741 to 1961.

Period	Rate		
	Birth	Death	Natural Increase
First stage: Stabilizing high fertility and mortality			
1741-50	36.9	33.0	3.9
Second stage: Declining mortality			
1751-60	36.9	30.3	6.6
1771-80	37.5	31.1	6.4
1791-1800	37.3	26.9	10.4
1810-20	36.6	21.1	15.5
1831-40	36.6	23.4	13.2
1851-60	34.1	22.2	11.9
1871-80	35.4	21.4	14.0
Third stage: Declining fertility with mortality			
1881-90	32.4	19.1	13.3
1891-1900	29.9	18.2	11.7
1901-09	27.2	15.4	11.8
1910-19	21.8	14.4	7.4
1920-29	18.3	12.1	6.2
Fourth stage: Stabilizing low fertility and mortality			
1930-34	15.0	12.0	3.0
1935-39	14.7	12.5	2.2
1940-44	15.9	13.2	2.7
1945-49	18.0	11.5	6.5
1950-54	15.5	11.6	3.9
1955-59	17.4	11.8	5.6
1960-61	15.9	11.6	4.3

Sources: C. P. Blaker and D. V. Glass, The Future of Our Population (Population Investigation Committee, London) 1937, 7.
The Registrar General's Annual Statistical Review of England and Wales.
Donald J. Bogue, Principles of Demography (John Wiley and Sons, Inc., New York) [1].

The data presented in table 4 furnish valuable information that helps to visualize the process of demographic transition that took place in most European countries from 1741 onward. Due to limited space, brief discussion of the table is presented. It is obvious that mortality started declining earlier than fertility by a considerable period of time. It took the UK about 140 years to reduce its mortality rate from 33.0 per 1000 population in 1741-50 to 21.4 in 1871-80 without a concomitant change in fertility. It retained its traditional level of 36.9 and 35.4 respectively during this same period.

Beginning in 1881-90, the birth rate started manifesting a significant decline to reach 18.3 per 1000 population after 50 years. During the next 30 years, it stabilized at its lowest level of 15 to 17 marking the end of the 4th or final stage. With the exception of the baby boom period of 1945-49, little fluctuation is witnessed in the crude birth rate. In the meantime, the CDR shared with the CBR at this stage a concomitant decline. It dropped from 19.1 in 1881-90 to 12.1 in 1920-29. By 1930, it reached its lowest level for more than 40 years of about 12 per 1000 population. Consequently, it is believed that the pattern of demographic transition of the UK has been strongly duplicated in most European countries in the Euro-Mediterranean region as verified by the levels and trends of their TFR during the past three decades. This can be judged from the data presented in table 5.

Table 5. Total Fertility Rates of Euro-Mediterranean Member Countries between 1970-2000.

Country	Total Fertility Rate						
	1970	1975	1980	1985	1990	1995	2000
Austria	2.29	1.83	1.65	1.47	1.45	1.40	1.34
Belgium	2.25	1.74	1.68	1.51	1.62	1.55	1.66
Cyprus	2.54	2.01	2.46	2.38	2.42	2.13	1.83
Denmark	1.95	1.92	1.55	1.45	1.67	1.80	1.77
Finland	1.83	1.68	1.63	1.64	1.78	1.81	1.73
France	2.47	1.93	1.95	1.81	1.78	1.71	1.89
Germany	2.03	1.48	1.56	1.37	1.45	1.25	1.36
Greece	2.43	2.32	2.22	1.67	1.39	1.32	1.29
Ireland	3.97	3.43	3.24	2.48	2.11	1.84	1.89
Italy	2.38	2.17	1.64	1.42	1.33	1.20	1.23
Luxembourg	1.98	1.55	1.49	1.38	1.60	1.69	1.79
Malta	....	2.17	1.98	1.99	2.05	1.83	1.67
Netherlands	2.57	1.66	1.60	1.51	1.62	1.53	1.72
Portugal	2.84	2.63	2.20	1.72	1.57	1.40	1.52
Spain	2.86	2.79	2.20	1.64	1.36	1.18	1.24
Sweden	1.92	1.77	1.68	1.74	2.13	1.73	1.54
United Kingdom	2.43	1.81	1.89	1.79	1.83	1.71	1.65

Source: Council of Europe, Recent Demographic Development in Europe, 2001.

It is interesting to note from the table that Sweden continued, due to its early demographic transition, to maintain the lowest TFR until 1970. This situation has changed today and many European countries have sustained a fertility decline below the Swedish and UK levels. The case of Austria, Germany, Greece, Italy and Spain testifies to this fact. Perhaps the prerequisites that triggered the transition in the early years of the 17th century in Sweden and the UK have spread later in most European countries. This can be revealed by a brief illustration of the historical events that set the stage for demographic transition in Europe.

Historically, since 1650 Europe has experienced a number of social and economic changes that have been responsible for providing somewhat more favorable conditions for a population transition than those which prevailed in the previous centuries. Poverty, disease and unsettled political conditions that kept death rates at a level almost as high as that of birth rates. Three major technical revolutions have taken place in Europe that fostered an early drop in mortality. There have been crucial discoveries in the fields of medicine and sanitation, agriculture and industry. From the population point of view, the medical and sanitary revolution is the most important one. It has totally eliminated many epidemic diseases that used to claim the life of millions of people, and it has, on the other hand, reduced the incidence of many others. For example, Europe did not experience during the 17th century a repetition of the deadly plague, the Black Death, that spread between 1348-1350. Even its occurrence in London in 1665-1666 was not severe compared with those of three centuries before. In the meantime, the introduction of vaccination led to a rapid decline in infant mortality from as high as 200 per 1000 births in the early years of the 17th century to about 100 by the end of the 18th century. By 1898/1902, it dropped in Sweden to 98, in England and Wales to 152 and Austria to 220. By 1956/60, infant mortality declined to reach 17, 23 and 41 in these countries respectively.

On the other hand, the agricultural revolution resulted from a steady increase in scientific knowledge in the fields of agronomy and animal husbandry, the use of fertilizers, crop rotation, methods of cultivation and mechanisation. By application of this technological know-how in agriculture, yields of major field crops have scored substantial increases. This led to an elimination of famines and the improvement of the nutritional conditions of the population to resist major diseases.

The industrial revolution brought close on its heels a further decline in mortality. It provided better and faster transportation facilities to move food from areas of abundant production to those suffering from a severe shortage and hence eliminated the incidence of famine and poor health. It made the mechanisation of agricultural land possible which increased the output of major agricultural crops substantially. Finally, it made an exchange of food for manufactured goods possible.

One may add to these a fourth revolution that resulted in the discovery of modern contraceptive methods that accelerated the pace of birth control. Without the widespread use of these methods, one would have expected severe starvation and disaster in many countries facing shortage of food and agricultural resources. The practice of traditional forms of contraceptive methods in the second half of the 18th century was responsible for an early decline in fertility in many European countries.

5. Determinant of Population Growth

The following analysis aims at singling out the most significant factors that have had a significant effect on the rate of population growth in the Euro-Mediterranean member countries. The application of the stepwise regression model indicated that the crude birth rate (excluding total fertility rate) is the only variable determinant in the rate of population growth with $R^2 = .98958$ and Beta .99477. This means that about 99% of the differences in the rate of population growth between Euro-Mediterranean member countries are attributed solely to corresponding differences in the crude birth rate and that its decline by one unit leads to a similar trend of .995 unit in the growth rate. This relation can be expressed in the following regression equation:

$$X1= -11.691017 \ (constant) + .994776 \ X2$$
Where $X1$ = *rate of natural increase*
$X2$ = *crude birth rate*

Now, the question that arises is, what accounts for noted differences in the CBR of the member countries? The answer to this question is explained by two indicators represented by a human development index and the mortality rate of children below 5 years of age. The coefficient of determination with the first is .79875 and Beta -.893730 and a constant of 48.361654. This means that 80% of the differential in the CBR among member countries of this region is attributed to the compound measure of health, education and income that constitute the index of human development. On introducing the second variable, the coefficient increased to .86255 with Beta -. 677617 for the human development index and .332422 for the mortality of children below 5 years of age. The constant for the regression equation of this function was 37.952675. This means that 86% of the differences in the CBR between countries of this region is explained by the human development index and the mortality of children below 5 years of age. A change of a unit of the first variable leads to a corresponding change of -.68 unit in the CBR while that of the second to .33. The final equation of this function can be expressed as follows:

$$X2 = 37.952675 - .677617 X15 + .332422 X6$$

However, excluding the index of human development from the regression analysis, it shows that differentials in the CBR between Euro-Mediterranean countries are determined by 4 variables in successive order of importance. They are the mortality rate of children below 5 years of age (X6), density per square mile (X7), percent of urban population (X9) and percent of females employed in industry (X12). The coefficient of determination increased from 59.746 with X6 to 71.391 by introducing X7 to 80.027 with X9 and 83.817 with X12. This function can be expressed in the following equation:

$$X2 = 17.84703 + .744654 X6 + .301777 X7 - 277085 X9 - .209098 X12$$

6. Implications of Population Growth

Even in countries with abundant natural resources, a population growth may also be a hindering factor for development. Its devastating effect is not only limited to the depletion of scarce natural resources, but also of capital earning crucially needed for the modernisation of the national economy.

The creation of a sufficient number of suitable job opportunities represents another problem to economies with a rapidly growing population even in countries with abundant natural resources. The recruitment of the required number of qualified specialists in the fields of education, health and sanitation, marketing, finance, communication, etc., to meet the increasing demands of an annually added new cohort of population is an important task of modern government systems. They can hardly function properly in the presence of a high rate of population growth. In the meantime, a rapid population growth also creates an age-sex structure in a population having a high dependency ratio that represents a heavy economic burden on both governments and families. To what extend the differential RNI affects the quality and quantity of education, health, sanitation, per capita income and economic services of the population is shown in the analysis of the data presented in table 6.

The table shows a high significant association between the rate of population growth and maternal mortality rate (.5988**), male and female expectation of life at birth (-.7551** and -.8338** respectively), net enrollment ratio in secondary education

(-.5612*), per capita expenditure on health (-.6638**), per capita purchasing parity on health (.6593**), number of TV sets per 1000 population (-.6898**), number of daily papers per 1000 population (-.5773**) and number of hospital beds per 1000 population (− .5640*),. This means that all public services related to health, education and culture deteriorate as a result of a high rate of population growth.

Table 6. Correlation Coefficient between Rate of Population Growth and 20 Indicators of Socioeconomic Development of Euro-Mediterranean Member Countries in 2000.

Indicator	Rate of Population Growth
1-Marital mortality rate	.5988**
-Male life expectancy at birth	-.7551**
3-Female life expectancy at birth	-.8338**
4-Net enrollment in Secondary school	-.5612*
5-Per capita health expenditure	-.6638**
6-PPP health expenditure	-.6593**
7-TV per 1000 population	-.6898**
8-Daily papers per 1000 population	.5773**
9-N° of hospital beds/1000 population	-.5640*
10-Age dependency ratio	.1412
11-% of male labor force in agriculture	.3040
12-% of females labor force in agriculture	.4135
13-% of unemployed males	-.1248
14-% of unemployed females	-.1322
15-Net enrollment ratio in primary	-.3920
16-N° of physicians /1000 population	-.4150
17-% from GNP for health expenditure	-.4292
18-% from GNP for education expenditure	-.2054
19-% of population having access to safe water	-.3603
20-% of population having access to sanitation	-.0979

** Significant at .001 level.
*Significant at .01 level.

7. Population Projection

Based on the current level and trend of population growth for each member country in the Euro-Mediterranean region, a projection was made to estimate their size by the years 2025 and 2050. The result of this projection is reported in table 7.

By 2025, the Northern Africa sub-region replaces Western Europe as having the highest share of the total population in the region. It increases from 22.5% in 2001 to 27.2% in 2025. On the contrary, the Western Europe sub-region drops to second position in terms of population size; from 28.3% to 25.1%. Due to its continuous depopulation, it is expected to occupy the third position by year 2050 with a share of 22.7%, while the Northern Africa sub-region continues to have the highest share of 30.9%.

With regard to the Middle East sub-region, it occupies the third position by having a share of 20.1% of the region's total population in 2025 then the second with a percentage share of 23.0% in 2050. In the meantime, the Southern Europe sub-region drops from the third position in 2001 to the 4th in 2025 with a population share of 15.2%.

Table 7. Projected Number of Population by Country and Sub-region of Euro-Mediterranean area for 2025 and 2050.

Country and Region	Population in million			Percent		
	2001	2015	2050	2001	2015	2050
Northern Africa						
Algeria	31.0	46.6	57.7	22.2	23.8	24.5
Egypt	69.8	97.4	117.1	50.0	49.6	49.6
Morocco	29.2	39.3	46.2	20.9	20.0	19.5
Tunisia	9.7	12.9	15.0	6.9	6.6	6.3
Total	139.7	196.2	235.9	22.5	27.2	30.9
Middle East						
Cyprus	.9	1.0	1.1	.9	.7	.6
Israel	9.7	8.3	9.4	6.2	5.0	5.3
Jordan	5.2	8.8	12.0	5.0	6.1	6.8
Lebanon'	4.3	5.6	6.5	4.2	3.9	3.7
Palestine	3.3	7.4	11.2	3.2	5.1	6.4
Syria	17.1	26.9	35.3	16.4	18.6	20.0
Turkey	66.3	880	100.7	64.1	60.6	57.1
Total	103.5	145.0	1762	16.7	20.1	23.0
Northern Europe						
Denmark	5.4	5.8	6.1	6.5	6.5	6.9
Finland	5.2	5.3	4.8	6.	6.0	58.4
Ireland	3.8	4.5	4.5	4.6	5.1	5.1
Sweden	8.9	9.3	9.2	10.7	10.4	10.4
United Kingdom.	62.2	64.1	64.2	72.0	72.0	72.3
Total	83.3	89.0	88.8	13.4	12.4	11.5
Southern Europe						
Greece	10.9	10.4	9.7	9.2	9.5	10.8
Italy	57.8	52.4	41.9	48.6	48.0	46.0
Malta	.4	.4	.4	.3	.4	.4
Portugal	10.0	9.3	8.2	8.4	8.5	9.0
Spain	39.8	36.7	30.8	33.5	33.6	33.8
Total	118.9	109.2	91.0	19.1	15.2	11.9
Western Europe						
Austria	8.1	8.1	7.7	4.6	4.5	4.4
Belgium	10.3	10.3	10.0	5.8	5.7	5.8
France	59.2	64.2	65.1	33.6	35.5	37.4
Germany	82.2	80.2	73.3	46.7	44.4	42.2
Luxembourg	.4	.6	.6	.2	.3	.3
Netherlands	160	17.3	17.2	9.1	9.3	9.9
Total	180.7	180.7	173.9	28.3	25.1	22.7
Grand Total	621.6	720.1	765.0	100.0	100.0	100.0

Source: Projection from PRB population data sheet of 2000 [6].

It occupies that same position with a share of 11.9% in 2050. Though the population share of the Northern Europe sub-region is expected to drop from 13.4% in 2001 to 12.4% in 2025 then to 11.5% in 2050, its ranking position in terms of population size among the sub-regions remains the 5th during the 50 years of the projection period.

The table also shows that the projected population size of the Northern Africa sub-region will increase by 40.4% up to 2025 and by 68.9% up to 2050. Similarly, the Middle East sub-region will experience a comparable population growth to add another 40.1% to its present size by 2025 and 70.2% by 2050. On the other hand, the population growth in the three sub-regions of the European continent undertakes a different path. The population of the Northern Europe sub-region is expected to increase by 6.8% between 2001 and 2025 then experience a drop to have a total growth of 5.6% by the year 2050. Similarly, the Western Europe sub-region is expected to continue its population growth to add 2.6% by 2025 then experience a drop due to depopulation of 1.3% in the next 25 years. In contrast, the Southern sub-region is unique in the trend of its population growth during the next 50 years. Unlike the rest of the Euro-Mediterranean sub-regions, it is expected to lose through depopulation 8.2% by 2025 and increase this loss to reach 23.5 by 2050.

8. Conclusions

This study was carried out to reveal the differences that may exist in the rate of population growth among Euro-Mediterranean member countries and their stages of demographic transition. Data from different international sources were used to provide a satisfactory answer to the questions raised by the objectives of this research. These questions were related to the current status of differentiation in the rate of population growth among countries of the Euro-Mediterranean region, factors affecting it, implications of this growth and their projected population by the years 2025 and 2050.

Analysis of the available data reveal remarkable differences between these countries in the rate of population growth (natural increase) that extends from as low as -0.1% to 3.7%. Only one country has zero growth where population is just replacing itself. A growth of 0.5% or less was noted in 9 countries. On the other hand, five countries showed a growth rate between 1.0% and 1.9% while 4 between 2.0%-3.0%. Only one country has a growth rate that reached 3.7%. This means that 17 member countries located in the European continent have successfully completed all stages of the demographic transition and maintain a growth rate that will take them several centuries to double their population. With the exception of 3 countries having a TFR approximating 2.1 children per woman, the rest of the other countries are far from a successful completion of the demographic transition in the near future.

With regard to the factors of differential rates of population growth between member countries of the Euro-Mediterranean region, the CBR proved by a stepwise regression analysis to be the only explanatory variable. On the other hand, two variables explained noted differences in the CBR of these countries. They are the human development index followed by the mortality of children below 5 years of age. Excluding the human development index, the regression analysis shows in order of importance that the mortality of children below 5 years of age, percentage of urbanisation, density per square mile and percentage of females in industry explain 83.8% of the variations in the CBR between the Euro-Mediterranean member countries.

Assessment of the implications of population growth on socioeconomic and demographic development revealed an adverse effect on maternal mortality rate, male and female expectation of life at birth, net enrollment ratio of children in secondary education, per capita expenditure on health, number of hospital beds per 1000 population, PPP of health expenditure, number of TV sets and number of daily papers per 1000 population.

In the light of the present findings, it seems obvious that a high emphasis should be placed on helping countries distinguished by a high rate of population growth to complete, as promptly as possible, the third stage of the demographic transition. Improvements of the component criteria of the human development index which requires a high standard of health and sanitary services to eradicate mortality among infants and children below 5 years of age, increasing per capita purchasing power for a large segment of the population and raising cultural and educational status particularly of females through prevention of children dropout after completion of primary education, should receive first priority in plans for development. With the adoption of a proper population policy and effective cooperation, rich countries of this region can contribute financially, technically and socially to achieve these desirable goals. Socioeconomic and demographic differences between Euro-Mediterranean member countries should be reduced in order to create a homogeneous region that represents a model of genuine cooperation and meaningful coordination between countries of the world.

References

1. Bogue, D. J. (1969) *Principles of Demography*, John Wiley & Sons, New York, Ch. 3.
2. Council of Europe (2002) *Recent Population Development in Europe*, Council of Europe, Strasbourg.
3. Hassan, S.S. (1999) Effect of Demographic Disparities on Socioeconomic Development, *CDC Research Monograph 29*.
4. Hassan, S. and Koraa, A. M. (2000) Factors Affecting Fertility and Mortality in Egypt, *CDC Research Monograph 30*.
5. Political and Economic Planning (1962) *World Population and Resources*, George Allen & Unwin Ltd, London.
6. Population Reference Bureau (2000) *World Population Data Sheet*, Population Reference Bureau, Washington D.C.
7. Population Reference Bureau (2001) *World Population Data Sheet*, Population Reference Bureau, Washington D.C.
8. United Nations (1996) *UN Demographic Yearbook*, UN, New York.
9. World Bank Report (1998) *World Development Indicators*, The World Bank, Washington D.C.
10. World Bank Report (2000) *World Development Indicators*, The World Bank, Washington D.C.

Section VI

Food Prospects

Chapter 17

Food Trade of the Mediterranean Countries in a World Context: Prospects to 2030

N. Alexandratos[1]
Consultant to FAO and former Chief, Global Perspective Studies, FAO, Rome

This chapter presents food projections to 2030 of the Mediterranean countries in the context of possible developments in world cereal trade[2]. This first section gives a brief review of the food consumption in the different countries and the dependence of such consumption and the overall agriculture on international trade. Section 2 presents projections to 2030 and section 3 discusses briefly the land and water endowments of the countries. The final section draws some conclusions.

1. Dependence of the Mediterranean Countries on Food Trade

The food and agriculture sectors and the food security of the Mediterranean countries depend in varying, but generally significant, degrees on world markets: some of them depend on imported food for meeting large parts of their consumption, others on exports for selling large parts of their production. For example, the non-EU countries (outside Turkey and Syria) had in the three-year average 1997/99 net cereal imports of 29 million tons. Such imports accounted for 54 percent of their consumption. At the other extreme, France's cereals sector depends greatly on world markets for selling its output. The country has net exports of some 30 million tons, or 45-50% of its production. Its sugar sector has a similar degree of dependence on export markets, while smaller but still significant dependence on exports are encountered in the country's livestock sector. Many countries are also large exporters of the typical

---

[1] This chapter is largely based on the author's work for Alexandratos [1] FAO [21] and Bruinsma [13]. The views expressed are the author's and do not express those of FAO. All historical data are from FAO: *Faostat data base* (http://faostat.fao.org) and the projections are from the working files for the preparation of FAO [12] and Bruinsma [13] unless otherwise indicated.
[2] Historical developments with particular reference to trade in the typical Mediterranean agricultural products are analyzed in Alexandratos [7] and (revised) in Alexandratos [8]. The following 15 countries are covered in this paper: the Mediterranean countries who are members of the EU, namely France, Greece, Italy, Portugal, Spain, and the non-EU Mediterranean countries of the Middle East: Israel, Jordan, Lebanon, Syria, Turkey, and North Africa: Algeria, Egypt, Libya, Morocco and Tunisia, referred to as the "non-EU countries" group.

A. Marquina (ed.), Environmental Challenges in the Mediterranean 2000-2050, 283–300.
© 2004 *Kluwer Academic Publishers. Printed in the Netherlands.*

Mediterranean products, e.g. fruit, vegetables, olive oil and wine. Such dependence is particularly marked in some countries, e.g. Spain (fruit and vegetables, olive oil), France (wine), Greece (olive oil).

There is a great divide between the EU countries of the Mediterranean and the non-EU ones as regards the forces that drive demand and production of food. On the demand side, the major determinant - population growth - is still fairly high in the non-EU countries while it is close to zero in the EU ones. In addition, the non-EU Mediterranean countries have greater scope than the EU-countries for increasing food consumption per capita, mainly through shifts of consumption away from staples like cereals, which is high[3] – particularly in the countries of North Africa– towards higher value products like those of the livestock sector (Figure 1). In contrast, per capita food consumption[4] is high in the EU countries with high components of livestock products in the diet. This leaves little scope for further increases, though expenditure on food may keep rising because of the growing share going to processing, packaging, transport and distribution. These two variables together (differences in demographic growth and in the scope for increases in per capita consumption) contribute to the aggregate demand for food and agriculture products growing fairly fast in the non-EU countries and very little in those of the EU (Table 1).

On the production side, the divide is even more significant: in the EU group, the decelerating demand is confronted with surplus agricultural production potential (land, water, technology, management, supportive policies) mainly on account of France which is a world scale agricultural power. The opposite holds for the non-EU countries: with the possible exception of Turkey, their land and water resources are scarce and/or of low agroecological suitability for rainfed production, being predominantly semi-arid. By and large, cereal production in the non-EU group during the last three decades barely kept pace with population growth, with per capita production fluctuating in the range 240-280 Kg. Over the same period, consumption per person (both for food and feed) grew from 325 kg to 385 kg. The difference covered by growing imports.

France, Spain and Turkey are net exporters of agricultural products (primary and processed, food and non-food), while all the others are net importers. Relevant balance of payments data are shown in Appendix I. Countries with large deficits in their balance of payments and high dependence of food consumption on imported food can be vulnerable in their food security if a combination of adverse conditions in world food markets (e.g. scarcities and rising prices) coincided with similar conditions in the balance of payments, e.g. servicing foreign debt. Among the Mediterranean countries, Syria and Jordan are classified by the World Bank in the *severely indebted category* and Algeria, Morocco, Tunisia, Lebanon and Turkey are classified as *moderately indebted*[5].

---

[3] Note that the per capita consumption of cereals for all food and non-food uses (mainly for animal feed) is higher in the EU countries than in the non-EU ones. In the latter, the per capita food consumption of cereals will decline in the future but that for all uses will increase because of the growth of feed use (see Table 2).

[4]The food consumption data used here are derived in the framework of the national Food Balance Sheets (FBS) constructed by FAO on the basis of countries' reports on their production and trade of food commodities, after estimates and/or allowances are made for non-food uses and for losses. Therefore, the more correct term for this variable would be "national apparent food consumption". The term "food consumption" is used in this sense in this paper.

[5]World Bank, *World Development Indicators 2002*, CD ROM, Quick Reference Tables. The European countries and Israel are not classified by the World Bank for indebtedness

2. Prospects to 2030

Population growth will continue to be fairly high in the non-EU countries and fuel the demand for food. By 2030, it will have added another 110 million persons to the group's present population of 240 million, notwithstanding the rather significant deceleration in terms of growth rates (from 2.4% p.a. in the preceding three decades to almost half that in the coming three decades – Table 1). By contrast, there will be no growth, and indeed a marginal decline, in the population of the EU Mediterranean countries. The demographic weights of the two groups will continue to change in favour of the non-EU group just as they have done in the past. Thus thirty years ago (in 1970), the non-EU countries had a smaller population (120 million) than the EU ones (156 million). By the mid-1980s, the two groups had equal population (170 million each). By 2030, the non-EU group will have reached 350 million, while the EU countries will still have 170 million (population data and projections are from UN [27]).

Figure 1. Food Consumption (Kcal/person/day and kg/person/year), Average 1998/2000

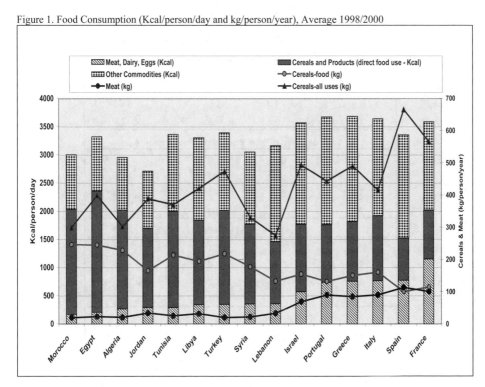

Concerning incomes, the other major determinant of food demand growth, the latest World Bank *economic growth* outlook for the period 2000-15 has the whole Middle East and North Africa[6] region lagging behind the other developing regions, with the

---

[6]This World Bank region excludes Turkey and Israel and includes, in addition to the other non-EU Mediterranean countries covered here, also Iran, Iraq, Saudi Arabia, Yemen, Bahrain, Oman, Djibouti and West Bank/Gaza.

growth rate of per capita income (GDP) projected at 1.3 percent p.a., the same as that for sub-Saharan Africa, compared with rates of 5.3 percent, 3.8 percent and 1.8 percent for East Asia, South Asia and Latin America, [30] (Table 1.). The very recent (2002) *Arab Human Development Report* of the UNDP lists what the authors consider to be the major impediments to development in the Arab countries, among them lack of participatory government, gender disparity and persistence of low educational levels and human skills. Of the narrower economic factors, the price of petroleum, on which many of the economies depend, is expected to decline in real terms (Figure 2).

Table 1. Aggregate Agriculture (all crops & livestock) & Cereals

	69-99	79-99	89-99	91-2001	97/9-2030
	Population growth rates % p.a.				
EU Medit	0.4	0.3	0.3		-0.1
N. Africa	2.4	2.3	1.9		1.3
M. East	2.4	2.3	2.2		1.3
	Aggregate Agriculture: Demand growth rates % p.a.				
EU Medit	1.1	0.7	0.7		*
All EU	*0.7*	*0.4*	*0.4*		*0.2*
N. Africa	4.0	3.4	3.0		2.0
M. East	3.0	2.6	2.4		1.7
	Aggregate Agriculture: Production growth rates % p.a.				
EU Medit	1.3	0.9	1.0		*
All EU	*1.1*	*0.6*	*0.5*		*0.2*
N. Africa	3.0	3.6	3.3		2.0
M. East	2.6	2.0	2.0		1.6
	Cereals: Demand growth rates % p.a.				
EU Medit	0.8	0.4	2.4		*
All EU	*0.3*	*0.1*	*2.0*		*0.3*
N. Africa	3.8	3.2	2.2		1.7
M. East	2.4	1.7	1.7		1.4
	Cereals: Production growth rates % p.a.				
EU Medit	1.8	1.2	1.2	1.2	*
All EU	*1.6*	*1.1*	*1.2*	*1.6*	*0.6*
N. Africa	2.5	3.6	2.0	0.9	1.8
M. East	2.2	1.3	1.6	-0.5	1.1

* Projections not available for the individual EU countries. The projection for all EU is shown instead.

Given the fairly high levels of food consumption already achieved in most countries (Figure 1), further growth in *food demand* will be determined primarily by population growth and structural change in diets towards more high-value products, essentially livestock. On both these counts, it is the non-EU countries that offer growth potential. However, their demand growth rate will be much lower than in the past (Table 1), given the lower population growth rate and the much higher consumption levels from which they now start compared with the past. In addition, the slow growth of incomes will restrain the thrust towards a structural change in the diets. Thus, per capita food consumption (in terms of Kcal/person/day) will increase only modestly while the rate of *structural change in diets* away from the direct consumption of cereals towards other commodities, particularly meat, will also be slow (Table 2).

The differences in the diet structures between the EU and the non-EU countries will likely persist. In particular, the gap in meat consumption between the EU countries and the rest may in 30 years be as large as it is today (Table 2). While factors other than income (e.g. cultural, religious, climatic) explain in part the lower livestock consumption levels in the non-EU countries of the Mediterranean, in the long term these other factors tend to be overwhelmed by the spread of diets high in livestock content when incomes rise sufficiently. It has happened in several high-income countries of the Gulf - Kuwait, the UAE and to some extent, Saudi Arabia. Further evidence is provided by the EU Mediterranean countries themselves which, with the exception of France which had traditionally high livestock consumption levels have made the transition from low to mid-high consumption of meat in the last four decades, following the rise in incomes.

On the production side, the constraints confronting most countries in the non-EU group will continue to bind, though, under improved policies, there is scope for better performance than that of the stagnant 1990s. An increase in total cereal production of the non-EU countries from the present 58 million tons (average 1999-2001) to 93 million tons by 2030 is considered feasible (see next section for land-irrigation-yield combinations underlying these estimates). This is an improvement over the stagnation of production observed during the last ten years (production was also 58 million tons at the beginning of the 1990s).

The import needs for basic foods (cereals, livestock, sugar, vegetable oils) will continue to grow. Net cereal imports may increase from the 30 million tons currently to 53 million tons in 2030, a less than doubling in 30 years (Tables 2, 3). This is a much slower growth than that of the past: net imports were only 5.5 million tons at the beginning of the 1970s and they increased almost six-fold in the next three decades to the present 30 million tons, with much of the increase having occurred during the oil boom period. The above-mentioned slowdown in the growth of demand and some recovery of production from the stagnation of the 1990s explain the projected slowdown in the growth of the region's cereal deficits.

An additional factor is that growing meat imports have been substituting for feed cereal imports and will continue to do so (see Table 2), though domestic meat production will continue to fuel the demand for feed import [29] and, as noted (footnote 3), will contribute to raising the per capita consumption of cereals for all uses, no matter that per capita food consumption will decline.

On the EU side, we have no projections for the individual EU countries, but only for the EU as a whole. They indicate that the long term trend towards a decline in the growth rate of aggregate agricultural production in the EU will continue because of (a) little growth in domestic demand, and (b) limited scope for further expansion of net exports with

Figure 2: World Market Price Indices, 1960-2001 and Projections to 2015 (1990=100)

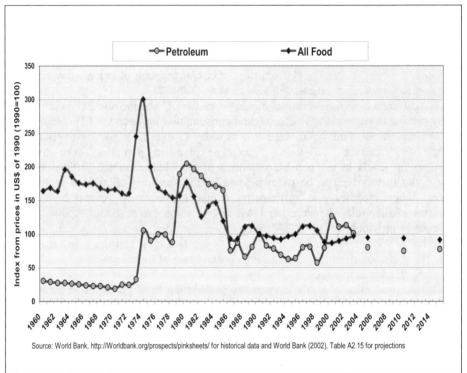

Source: World Bank, http://Worldbank.org/prospects/pinksheets/ for historical data and World Bank (2002), Table A2.15 for projections

Source: World Bank, http://Worldbank.org/prospects/pinksheets/ for historical data and World Bank (2002), Table A2.15 for projections

the aid of subsidies, given the WTO limits for subsidized exports and the additional declines in domestic support and export subsidies that may be applied in the context of further trade liberalisation. These possible developments may imply outright declines in the production of some commodities, e.g. sugar, rice [24]. The growth of exports without subsidies will also be constrained by the competition of other exporters in a world market that will not be growing very fast. An additional factor that may restrain the growth of production is the rising opposition to further intensification of agriculture and the emphasis on environmental conservation. However, even with lower production growth rates, the potential exportable surpluses of EU cereals will continue to grow, if present policies under the Common Agricultural Policy (including reforms already agreed upon as at present - autumn 2002) were to continue. Net exports exceeding about 17 million tons will materialize as actual exports only if they can be exported without subsidies. This will depend on a number of variables, including the policy prices of the EU, the prices in world markets and the exchange rate €/US$[7].

---

[7]Exports of cereals without subsidies over and above the subsidized exports are visualized in the Commission's latest projections study which projects net exports of 24 million tons in 2009 [15] Table 1.9). The WTO limits for EU exports with subsidies are, roughly, 25 million tons. These limits refer to gross

There seems to be a fair degree of consensus on this matter, i.e. that the EU will be exporting its cereal surpluses (or parts of them) without subsidies above the WTO limits. Thus, the European Commission's latest projections to 2009 conclude that "over the medium term, they [cereal exports] are projected to recover substantially thanks to greater availability and to exceed the annual limit for subsidised exports set by the URAA limits [i.e. 25.4 million tons for total cereals, including 0.5 million tons of food aid] as durum wheat, some common wheat and barley/malt would be exported without subsidies" [15]. Other projection studies seem to concur. The 2002 projections to 2011 of the US Dept of Agriculture (USDA) [28] concludes that "due to the declines in intervention prices and the weak euro, projected domestic and world prices indicate that EU wheat and barley can be exported without subsidies throughout the baseline period" It projects net EU exports of wheat and coarse grains of some 35 million tons for the year 2011 [28] (Tables 36-40). The projections of the Iowa State University's Food and Agricultural Policy Research Institute give net exports of 24 million tons for the year 2012 [17]. The results of the longer term study to 2020 of the International Food Policy Research Institute (IFPRI) point in the same direction [25] (Table D.10). The FAO projections [13] on which this paper is based indicate net EU exports of 54 million tons in 2030 (Table 3).

As noted, the food security of countries with heavy dependence on imported food is vulnerable to the vicissitudes of world markets, the eventual development of scarcities and disruptions of the normal flows of trade and, of course, to events that may affect the import capacity of the country itself. On this latter aspect, we cannot foresee how things may develop, except perhaps to underline the above mentioned prospect that the price of petroleum may decline in real terms (Figure 1), which does not augur well for the import capacity and overall economy of several non-EU Mediterranean counties. We can be somewhat more concrete about the longer term evolution of the world food markets. Overall, a number of recent projection studies [13, 21, 25] suggest that the cereal surpluses of the exporting countries together with the eventual emergence of Eastern Europe and the countries of the former USSR as net exporters, could easily meet the foreseeable increases of import demand on the part of the importing countries. The future positions of the major importing and exporting country groups are shown in Table 3. It can be seen that the historical trend for the non-EU Mediterranean countries to account for a rising share of world import demand may be reversed in the future, as other countries and regions will be playing a growing role in the growth of world imports. Both China and India, presently net exporters, will probably revert to being net importers. Structural scarcities on a world scale that would drive up food prices permanently should not develop. Price projections to 2015 (Figure 2) suggest that the index of food prices in international markets may in 2015 be a little below that of the current year 2002 [30] (Table A2.15). This, of course, is not to exclude short term price spikes under extraordinary circumstances

---

exports, while the Commission study projects imports to reach about 8.0 million tons, including those under the WTO commitments. It follows that any projection study showing net EU exports over 17 million tons must assume (implicitly or explicitly) that in the future the right combination of domestic and foreign prices and exchange rates will prevail.

Table 2. Food consumption, net trade, self-sufficiency

				64/6	74/6	84/6	97/9	2030
EU Medi	Consumption	Calories	Kcal/person /day	3013	3248	3352	3530	*
	"	Cereal-Food	kg/person /year	151	141	139	140	*
	"	Cereal all uses	kg/person /year	376	441	478	513	*
	"	Meat	kg/person /year	46	64	77	92	*
	Net Trade	Cereals	Th. tons	-3053	651	**17507**	**18413**	*
	"	Meat	Th. tons	-694	-1029	-1392	-741	*
	"	Dairy (milk equiv.)	Th. tons	342	-1458	-4213	-3201	*
	Self-Sufficiency	Cereals	%	96	100	122	123	*
All EU	Consumption	Calories	Kcal/person /day	3066	3196	3321	3431	3510
	"	Cereal-Food	kg/person /year	138	132	134	132	127
	"	Cereals all uses	kg/person /year	409	463	477	480	546
	"	Meat	kg/person /year	56	70	80	86	96
	Net Trade	Cereals	Th. tons	- 24852	-21406	15118	23741	54000
	"	Meat	Th. tons	-1381	-468	1048	2444	2000
	"	Dairy (milk equiv.)	Th. tons	-396	5846	11821	10408	11000
	Self-Sufficiency	Cereal	%	80	87	111	115	127
N. Africa	Consumption	Calories	Kcal/person /day	2168	2499	2957	3177	3350
	"	Cereals-Food	kg/person /year	172	192	220	242	230
	"	Cereals all uses	kg/person /year	234	269	321	345	386
	"	Meat	kg/person /year	10	11	16	20	34
	Net Trade	Cereals	Th. tons	-2764	-7696	-17402	-23198	-36500
	"	Meat	Th. tons	-46	-78	-426	-266	-480
	"	Dairy (milk equiv.)	Th. tons	-337	-902	-2519	-2537	-5000
	Self-Sufficiency	Cereal	%	76	72	51	53	55
M. East	Consumption	Calories	Kcal/person /day	2802	3031	3341	3423	3500
	"	Cereals-Food	kg/person /year	189	197	216	212	197
	"	Cereal all uses	kg/person /year	462	451	493	441	460
	"	Meat	kg/person /year	17	19	24	24	38
	Net Trade	Cereals	Th. tons	-1363	-3039	-4975	-6452	-16700
	"	Meat	Th. tons	-43	-46	-84	-196	-450
	"	Dairy (milk equiv.)	Th. tons	-238	-340	-686	-670	-1200
	Self-Sufficiency	Cereals	%	87	97	87	83	74

* Projections not available for the individual EU countries. The projection for all EU is shown instead.

Table 3. World Cereal Balances (million tons)

	69/71	79/81	89/91	99/01	2030
non-EU Mediterranean Countries	-5.5	-15.1	-23.7	-29.4	-53
(share in aggregate of all importers, %)	12.5	10.1	14.0	17.8	16
(Greece, Italy, Portugal, Spain)	-9.1	-14.3	-6.3	-15.8	
Other Importers					
Other Middle East	-0.8	-8.4	-14.9	-22.8	-67
Sub-Saharan Africa	-2.4	-7.3	-8.0	-15.0	-40
Latin America, excl. Argentina, Uruguay	-5.8	-22.4	-20.8	-40.2	-62
South Asia	-5.5	-1.8	-3.1	-0.4	-22
East Asia, excl. Thailand, Vietnam	-9.9	-28.6	-33.7	-28.9	-89
Industrial Country Importers	-16.4	-24.0	-28.6	-31.3	-34
East Europe and former Soviet Union	1.9	-41.6	-36.8	3.2	25
Total Importers	-44.5	-149.2	-169.7	-164.7	-343
Exporters					
Argentina, Uruguay., Thailand, Vietnam	10.5	18.1	16.9	31.8	65
EU	-22.3	-10.9	25.8	18.8	54
(of which France)	10.4	17.9	28.8	30.4	
North America, Australia	58.0	143.0	130.1	122.1	231
Total Exporters	46.1	150.3	172.8	172.7	351

Source: historical data FAO Faostat (accessed 1-12-02); projections Bruinsma, J. (ed.) (2003), *World Agriculture: Towards 2015/30, an FAO Perspective,* Earthscan, London and FAO, Rome.

3. Resource and Technology Potential for Future Production

The differences between the EU and the non-EU countries of the Mediterranean in the balance between population and food production resources will be further accentuated in the future, particularly if Turkey is excluded from the non-EU group. The Agroecological zones (AEZ) study of FAO and the International Institute of Applied Systems Analysis (IIASA, [22]) provides an evaluation of the land and climatic resources of each country and their suitability for producing a given crop or a combination of several crops under rainfed conditions and alternative technologies. For example, considering wheat production potential only, France has some 33 million ha of land classified as "good" for producing rainfed wheat at an average yield of 5.5 tons/ha under intermediate technology. About two thirds of it is prime wheat land (class Very Suitable and Suitable in the AEZ evaluation, VS, S in Table 4, Col. 3) in which the obtainable yield is 6.6 tons/ha. In principle, if all this land were used to produce only wheat with intermediate technology, France would be capable of producing some 3 tons per person in its 59 million population. In practice, France produces 650 kg per person using only 5.2 million ha of land for wheat at a yield of 7.3 tons/ha, using technology that is more advanced than the intermediate one of the AEZ evaluation variant reported in Table 4. At the other extreme is, of course, Egypt with virtually no land suitable for rainfed production. Jordan follows closely behind Egypt, with a rainfed wheat production potential of only 40 kg/person and actual production of 5-10 kg/person in recent years, some of it from irrigated land. Estimates of such potentials for other countries are given in Table 4, considering agroecological suitability for growing rainfed wheat only (Columns 1-2) or, more realistically, a mix of several crops (Columns 5-6).

These estimates, being theoretical potentials under very restrictive assumptions, are of some use for inter-country comparisons, but in no case indicate potential production of wheat that could actually be achieved, given the uses of land for other crops and purposes, including land that must remain under forest cover (see Appendix II for discussion of issues pertaining to the evaluation of land suitability for agricultural use). In addition, production potential of a country can be greatly increased by irrigation, including irrigation of desert land with no inherent agroecological suitability for rainfed cropping. The last three columns of Table 4 indicate the presently irrigated land and the extent to which water resources and land suitability provide potential[8] for further expansion of irrigation and associated production. By and large, there still exists potential for irrigation expansion in the non-EU Mediterranean countries, concentrated mainly in Turkey and to a smaller extent Egypt. It is expected that some of this potential will be used for expanding irrigation over the projection period to 2030. The estimates of such expansion are given in the last column of Table 4. Part of the water required to support the expanded irrigation should come from improved water use efficiency[9]. The latter could increase from an estimated 40 percent at present to about 53 percent in 2030 [13] (Table 4.10). Some increase in the cropping intensities[10] of irrigated areas should continue to take place so that the total harvested irrigated land may increase by more than implied by the projected numbers in Table 4.

In general, crops other than cereals will be taking a growing share of total irrigation. Even so, the estimates of future cereal production in the non-EU Mediterranean countries also depend crucially on expanding the irrigated area devoted to cereals, as the scope for sustainable production increases in the rainfed areas would not be sufficient to deliver even the relatively low growth rates of production of Table 1.

---

[8]Irrigation potential is defined as "...area of land suitable for irrigation development. It includes land already under irrigation. Assumptions made in assessing irrigation potential vary country to country. In most cases, it was computed on the basis of available land and water resources, but economic and environmental considerations may also have been taken into account. Some countries include the possible use of non-conventional sources of water for irrigation. Except in a few cases, no consideration is given to the possible double counting of shared water resources between riparian countries. Wetland and floodplains are usually, but not systematically, included in irrigation potential". [19]

[9]Irrigation efficiency: ratio between consumptive water use in crop production (the amount of water needed to compensate for the deficit between potential evapotranspiration and effective precipitation during the growing period of the crop) and water withdrawals from the reservoirs. For details see [13]

[10]Cropping intensity: the ratio of harvested area to physical area. The ratio rises when fallow periods are reduced and/or more than one crop is grown on the same piece of land in a year. Historically, cropping intensities have been rising everywhere under population pressure and the need to produce more food on land growing scarcer with time. Irrigated land offers much more scope for raising cropping intensities than rainfed land.

Table 4. Land and Water Resources for Agriculture

	Potential land suitable for rainfed wheat production under intermediate technology				Potential land with rainfed potential for a combination of 22 crops and up to 3 technologies				Irrigated (th.ha)		
							In Rainfed Use		Potential	In Use	
	th. ha	%VS orS	Aver. Yield (kg/ha)	Potential Production -kg/perso	th. ha	%Vsor S.	Current	2030		Current	2030
	1	2	3	4	5	6	7	8	9	10	11
France	33105	64	5472	3080	38787	71					
Greece	5420	30	4329	2221	6654	55					
Italy	13924	47	4484	1086	16101	64					
Portugal	3493	58	3955	1385	4415	62					
Spain	12170	13	3406	1040	27567	43					
Israel	470	51	3724	303	719	50					
M. East (excl. Israel)	**31147**	**16**			**42925**	**37**	**28006**	**27500**	**10013**	**5712**	**7600**
Jordan	78	5	2402	40	577	30	309		85	75	
Lebanon	137	25	3118	126	235	52	143		178	88	
Syria	3646	16	3008	713	6550	50	4313		1250	1189	
Turkey	27286	16	4006	1691	35563	34	23241		8500	4360	
N. Africa	**14318**	**24**			**30961**	**55**	**19650**	**19540**	**8131**	**5992**	**7680**
Algeria	3798	14	2818	366	12882	55	7623		730	560	
Egypt	16	0	3163	1	121	13	0		4435	3300	
Libya	1569	9	2538	786	2406	46	553		750	470	
Morocco	7127	33	3328	824	12303	56	8490		1653	1282	
Tunisia	1808	23	3147	614	3249	57	2984		563	380	

Sources: Columns 1-6, Fischer, G., Van Velthuizen, H., and Nachtergaele, F. (2000) *Global Agro-Ecological Zones Assessment: Methodology and Results*, Interim Report, IR-00-064 IIASA FAO, Luxemburg, Austria and Rome. Col. 7-8, 10-11, Bruinsma, J. (ed.) (2003). *World Agriculture: Towards 2015/30, an FAO Perspective*, Earthscan, London and FAO, Rome.; Col. 9, FAO (1997) Irrigation in the Near East in Figures, *FAO Water Report 9*, Rome.- Potentially irrigable land but not yet irrigated is included in the land with rainfed potential, except if it has no such potential, e.g. if it is desert that can be irrigated..

combinations of irrigated areas and yields that would underlie the cereal production growth in the non-EU Mediterranean countries from the present 53 million tons to the projected 93 million tons in 2030 are plotted in Figure 3.

Figure 3. Irrigated Land Use (Cereals, Other Crops) & Cereal Yields (irrigated, Rainfed, Total)

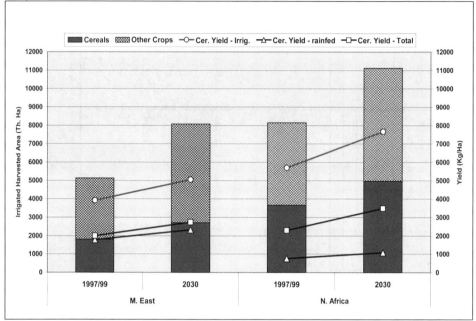

4. Conclusions

The dependence of the Mediterranean countries on world markets for importing considerable parts of their food needs (most non-EU counties) and for exporting considerable parts of their production of both basic foods (mainly France) and the typical Mediterranean products (mainly the EU countries but also several non-EU ones – Turkey, Morocco, etc) would continue. Therefore, developments in these markets (factors influencing demand and supply of the different commodities and the institutions and rules governing the conduct of international trade - WTO, etc) are of prime interest to all countries.

In the Mediterranean area, negotiating power on these matters is skewed in favour of the EU which is a major world trading power and can influence negotiations in ways that it considers beneficial to its agriculture. The non-EU countries have much less negotiating power, although their welfare depends on world agricultural markets much more than that of the EU countries: the food security of their population depends on continued access to imported food at terms and prices they can afford; and access to export markets for their horticultural products (predominantly to the EU) influences the living standards of parts of their rural population. The continuation under the Doha

Round of the multilateral trade negotiations to, *inter alia*, further liberalize agricultural trade as well as the related efforts towards further CAP reform represent an opportunity for improving access to the EU markets (see chapter 19). Yet, the debate on future reform of the EU's Common Agricultural Policy (CAP) in the context of the Doha round seems to concentrate on products and policies (reductions of tariffs, export subsidies and domestic support) that do not address directly the trade barriers (e.g. minimum entry prices) facing the main export products of the non-EU Mediterranean countries, viz. fruit and vegetables [16, 23]. In time honored fashion, such reform efforts seem to be dominated by the agricultural and trade concerns of the major exporters of the traditional temperate zone commodities – cereals, livestock, oilseeds.

In this context, attention needs to be drawn to the validity of the much touted benefits to the developing countries of further trade liberalisation. If such a liberalisation is going to concentrate again, as is likely, on the cereals-livestock sector, it will certainly benefit the few developing countries (Argentina, etc) that are actual or potential major exporters of these commodities. However, most developing countries are net importers of these products. If the removal of subsidies on the part of the major OECD countries were to raise world market prices, countries like those of the majority of the non-EU Mediterranean that are structurally dependent on imported basic foods (cereals. livestock, sugar) would be harmed. Therefore, everything is not clear-cut and straightforward when it comes to evaluating the pros and cons of further agricultural trade liberalisation from the standpoint of food security and the more general welfare of particular countries. Impacts will vary, and can be positive or negative, depending on country profiles, commodities affected and the modalities of liberalisation.

As another example of fairly significant positive and negative outcomes for different countries we can think of the liberalisation of the sugar sector which is heavily protected in most OECD countries (the EU, Japan, the USA). Its liberalisation will benefit countries like Brazil, Thailand, Cuba, Australia but will harm the poor ACP countries that have preferential access for exports to the heavily protected EU market, as well as those enjoying preferential access to the equally highly protected USA market. In parallel, the many net importing countries will probably have to pay higher prices for their sugar imports. In conclusion, the structural characteristics of the food and agriculture sectors of the non-EU Mediterranean countries require that extreme care be exercised when speaking of the possible benefits accruing to them from more liberal agricultural trade.

Given the dependence on imported food, the balance of payments constraints affecting the ability of countries to import food would continue to condition their food security. On this count, uncertainties will continue to prevail in several non-EU countries given that their overall economic prospects and the outlook for the petroleum sector are not particularly bright, while servicing foreign debt will continue to tax their external payments. In contrast, the probability is low that threats to food security could emerge from the possible development of structural shortages of food in world markets[11]. As noted, this is the conclusion of several medium and long-term projection studies which do not foresee significant rises in the prices of the major internationally traded food commodities. These findings reflect essentially two factors: (a) the

---

[11]This is not to exclude the possibility of temporary shortages due to weather fluctuations and possible abrupt policy changes in the main world market players.

continuous slowdown in the growth of the demand for food on a world scale, a result of the slowdown in population growth and the progressive approaching of near saturation in per capita food consumption in several countries, and (b) the existence of still significant potential for further production increases in several countries of the world, including the major traditional exporters as well as new ones that may emerge, e.g. the countries of Eastern Europe and the former USSR[12].

Naturally, such potential for growth in food production at the global level does not automatically translate into improved prospects for the food security of poor population groups in the different countries. Food insecurity and poverty problems in several countries, particularly in those in sub-Saharan Africa, will likely continue to prevail side by side with excess food production potential in other countries, just as they do at present [13, 21]. Countries with such problems usually have the majority of their population in agriculture and depend for their food supplies predominantly on what they themselves can produce and, on the demand side, on the incomes generated in agriculture and the rural economy. If their natural resources, infrastructure, etc, are too poor for accelerated agricultural development, they will find it difficult to solve their food security problems within a reasonable time horizon. None of the countries in the Mediterranean region fall in that class. Although several of them face severe natural resource constraints, none of them has widespread poverty associated with high percentages of the population being undernourished [20]. Naturally, this is not necessarily a permanent status. Economic collapses pushing significant parts of the population below the poverty line do occur occasionally, particularly for severely indebted countries – witness Argentina.

In the longer term, and given the structural characteristics of most non-EU Mediterranean countries (growing population, paucity of natural resources), enduring food security will increasingly depend on the development of robust and resilient non-agricultural sectors. The role of international trade and the more general international economic relations in this direction is as important, and perhaps more so, for their food security as the more narrow issues pertaining to agricultural trade.

References

1. Alexandratos, N. (ed.) (1995) *World Agriculture: Towards 2010. An FAO Study,* Wiley, Chichester and FAO, Rome; in French: Polytechnica, Paris; in Spanish: Mundi-Prensa, Madrid and Mexico.
2. Alexandratos, N. (1996a) Book Review of Lester Brown: Who Will Feed China?, *European Review of Agricultural Economics, 23, 4*, 511-513.
3. Alexandratos, N. (1996b) China's Cereals Deficits in a World Context, *Agricultural Economics, 5, 1*, 1-16 (in Chinese, in: *Chinese Rural Economy, 4, 1996*, 13-20).
4. Alexandratos, N. (1997a) China's Consumption of Cereals and Capacity of the Rest of the World to Increase Exports, *Food Policy, 22, 3*, 253-267
5. Alexandratos, N. (1997b) The World Food Outlook: A Review Essay, *Population and Development Review, 24, 4*, 877-888.

---

[12] There is a lively literature on the uncertainties concerning the extent to which the earth has sufficient potential to produce, in ways that do not threaten sustainability, enough food for the ever growing global population. For pessimistic views, see Ehrlich and [14]. For more balanced assessments, see [2, 3, 4, 5, 6, 10, 11, 12, 13, 21, 26]

6. Alexandratos, N. (1999) World Food and Agriculture: the Outlook for the Medium and Longer Term, *Proceedings of the US National Academy of Sciences*, *96*, 5908-5914.
7. Alexandratos, N. (2001) Mediterranean Countries and World Markets: Basic Foods and Mediterranean Products in B. Hervieu, N., Maraveyas, S., Vizantinopoulos (eds.) *Mediterranean Conference for Agricultural Research Cooperation – The Mediterranean Nutritional Model and Cooperation to Promote the International Trade of Mediterranean Products,* Papazissis Publishers, Athens.
8. Alexandratos, N. (2003) Mediterranean Countries and World Markets: Basic Foods and Mediterranean Products, in Brauch, H.G., Liotta, H.P., Marquina, A., Rogers, P., Selim, E.M. (eds.) *Security and the Environment in the Mediterranean - Conceptualising Security and Environmental Conflicts,* Springer, Heidelberg.
9. Alexandratos, N., and Bruinsma, J. (1999) Land use and land potentials for future world food security, in Fu-chen, Lo., Matshusita, J., and Takagi, H. *The Sustainable Future of the Global System,* UN University, Tokyo.
10. Alexandratos, N., Bruinsma, J., and Schmidhuber, J. (2000) China's Food and the World, in: *Agrarwirtschaft, Special Issue on Meeting the Food Challenge of the 21st Century, 9,10,* 327-335.
11. Brown, L., and Kane, H. (1995) *Full House, Reassessing the Earth's Population Carrying Capacity* Earthscan, London.
12. Brown, L. (1995) *Who will Feed China? Wake-up Call for a Small Planet* New York: W.W. Norton.
13. Bruinsma, J. (ed.) (2003) *World Agriculture: Towards 2015/30, a FAO Perspective,* Earthscan, London and FAO, Rome.
14. Ehrlich, P., and Ehrlich, A. (1990) *The Population Explosion,* Simon and Schuster, New York.
15. European Commission (2002a) *Prospects for Agricultural Markets, 2002-2009,* EU Commission, Directorate General for Agriculture, Brussels.
16. European Commission (2002b) *WTO and agriculture: European Commission proposes more market opening, less trade distorting support and a radically better deal for developing countries,* EU Document IP/02/1992, EU Commission, Brussels.
17. FAPRI (2002) *November 2002 Preliminary Baseline Projections,* Iowa State University and University of Missouri-Columbia (http://www.fapri.missouri.edu/BaselineReview.htm)
18. FAO (1981) Report on the Agro-Ecological Zones Project, *World Soil Resources Report 48,3,* FAO, Rome.
19. FAO (1997) Irrigation in the Near East in Figures, *FAO Water Report, 9,* Rome.
20. FAO (2001) *The State of Food Insecurity in the World 2001* FAO, Rome.
21. FAO (2002) *World Agriculture: Towards 2015/30,* Summary Report, FAO, Rome. http://www.fao.org/docrep/004/y3557e/y3557e00.htm
22. Fischer, G., Van Velthuizen, H., and Nachtergaele, F. (2000) *Global Agro-Ecological Zones Assessment: Methodology and Results,* Interim Report, IR-00-064 IIASA, Luxemburg, Austria and FAO, Rome.
23. Fischler, F. (2002) *European Agricultural Model in the Global Economy,* speech delivered at the Second International Conference on Globalisation, University of Leuven, (EU document DN: SPEECH/02/590, EU Commission, Brussels).
24. Oxfam (2002) *The Great EU Sugar Scam: How Europe's Sugar Regime is Devastating Livelihoods in the Developing World,* Oxfam, London.
25. Rosegrant, M., Paisner, M., Meijer, S., and Witcover, J. (2001) *Global Food Projections to 2020,* IFPRI, Washington, D.C.
26. Smil, V. (2000) *Feeding the World, A Challenge for the 21st Century,* MIT Press, Cambridge.
27. United Nations (2001) *World Population Prospects: The 2000 Revision,* UN Population Division, New York.
28. USDA (2002a) *Agricultural Baseline Projections to 2011,* Staff Report WAOB-2002-1, Washington DC
29. USDA (2002b) Middle East/North Africa Region: A Major Market for U.S. Feeds, *Agricultural Outlook,* Washington DC.
30. World Bank (2002) *Global Economic Prospects and the Developing Countries 2003,* World Bank, Washington DC.

Appendix I. Agriculture in the Balance of Payments: Net Balances, average 1998-2000, US$ million

Country	Aggregate current account[1]	All Agricultural Products (Crops and Livestock, Primary and Processed) Total	Food, excluding Fish Total	Cereals & Prepar.	Live Anim., Meat, dairy, eggs	Oilseeds, Oils, Fats	Sugar	Fruit& Vegetables	Other Food	Wine	Non-Food Feedingstuffs[2]	Other Non-Food[3]
EU Countries	**9066**	**4069**	**5174**	**3154**	**-3148**	**-928**	**629**	**5261**	**206**	**9141**	**-2000**	**-8246**
France	31056	11147	7280	4205	3911	28	1082	-2127	180	5187	-102	-1219
Greece	-8557	-669	-650	-245	-1143	197	-27	762	-195	51	-136	66
Italy	7480	-6578	-3607	448	-5261	-687	-125	1601	417	2317	-776	-4512
Portugal	-9527	-2619	-2049	-594	-512	-254	-96	-357	-237	385	-286	-669
Spain	-11384	2788	4201	-660	-143	-213	-206	5382	40	1200	-700	-1912
Middle East	**-8023**	**-1343**	**-365**	**-1233**	**-501**	**-975**	**-320**	**2690**	**-27**	**8**	**-318**	**-667**
Israel	-1966	-728	-677	-458	-132	-210	-127	332	-82	1	-53	1
Jordan	161	-529	-488	-261	-114	-9	-45	45	-105	-1	-45	4
Lebanon	-3594	-977	-699	-135	-308	-86	-26	-78	-68	1	-28	-251
Syria	440	-183	-370	-383	51	-200	-223	463	-78	0	-38	226
Turkey	-3065	1075	1870	3	2	-470	102	1927	305	7	-154	-647
North Africa		**-7519**	**-6061**	**-3203**	**-1292**	**-977**	**-743**	**329**	**-176**	**11**	**-528**	**-940**
Algeria		-2608	-2259	-1009	-519	-235	-238	-144	-115	3	-75	-277
Egypt	-1786	-2966	-2467	-1049	-513	-534	-253	-65	-54	0	-253	-245
Libya	797	-845	-698	-311	-147	-105	-39	-79	-17	0	-62	-86
Morocco	-262	-776	-477	-610	-80	-230	-142	576	11	3	-81	-221
Tunisia	-646	-323	-161	-225	-33	127	-70	42	-1	5	-57	-111
Total		**-4792**	**-1252**	**-1282**	**-4940**	**-2879**	**-433**	**8281**	**2**	**9160**	**-2846**	**-9853**
exc.France		*-15939*	*-8532*	*-5488*	*-8851*	*-2907*	*-1516*	*10408*	*-178*	*3973*	*-2744*	*-8635*

[1] Current account balance is the sum of net exports of goods, services, net income, and net current transfers. Data not available for Algeria; Greece average 99-00; Libya average 98-99. [2] Cereals for feed are included in Cereals, not in feedingstuffs. [3] Includes agr. raw materials (cotton, wool, rubber, hides/skins, etc)

Source: FAO, Faostat data base (http://faostat.fao.org/, latest update 4 July 02), except for current account balances which are from World Bank, *World Development Indicators 2002* - CD ROM.

Appendix II. Issues in estimating the land potential for rainfed agriculture[13]

The evaluation of land potential undertaken in the Agro-Ecological Zones (AEZ) study starts by taking stock of (a) the biophysical characteristics of the resource (soil, terrain, climate), and (b) the growing requirements of crops (solar radiation, temperature, humidity, etc.). The data in the former set are interfaced with those in the second set and conclusions are drawn on the amount of land that may be classified as suitable for producing each one of the crops tested [22].

The two data sets mentioned above are not immutable over time. Climate change, land degradation or, conversely, land improvements, together with the permanent conversion of land to non-agricultural uses, all contribute to change the extent and characteristics of the resource. Such potential changes are of particular importance if the purpose of the study is to draw inferences about the adequacy of land resources in the longer term. In parallel, the growth of scientific knowledge and the development of technology contribute to modifying the growing requirements of the different crops for achieving any given yield level. For example, in the present round of AEZ work the maximum attainable yield for rainfed wheat in sub-tropical and temperate environments is put at about 12 tons/ha in high input farming and about 4.8 tons/ha in low input farming. Some 25 years ago, when the first FAO Agro-Ecological Zones study was carried out [18] these yields were put at only 4.9 and 1.2 tons/ha, respectively. Likewise, land suitable for growing wheat at, say, 5 tons/ha in 30 years time may be quite different from that prevailing today, if scientific advances make it possible to obtain by that time such yields where only 2 tons/ha can be achieved today. A likely possibility would be through the development of varieties better able to withstand stresses such as drought, soil toxicity and pest attacks. Scientific knowledge and its application obviously will have an impact on whether or not any given piece of land will be classified as suitable for producing a given crop.

Land suitability is crop-specific. To take an extreme example, more than 50 percent of the land area in Congo DR is suitable for growing cassava but less than 3 percent is suitable for growing wheat. Therefore, before statements can be made about the adequacy or otherwise of land resources to grow food for an increasing population, the information about land suitability needs to be interfaced with information about expected demand patterns - volume and commodity composition of both domestic and foreign demand. For example, Congo DR's ample land resources suitable for growing cassava will be of little value unless there is sufficient domestic or foreign demand for Congo DR's cassava, now or in the future (the same land may be, of course, suitable also for other crops).

Declaring a piece of land as suitable for producing a certain crop, implicitly assumes that people find it worthwhile to exploit it for this purpose. In other words, land must not only possess minimum biophysical attributes in relation to the requirements of the crops for which there is, or will be, demand, but it also must be in a socio-economic environment in which people consider it an economic asset. For example, in low income countries, people will exploit land even if the yields or, more precisely, the

---

[13]Adapted from Alexandratos and Bruinsma [13].

returns for their work, are low compared to those prevailing elsewhere, but they are still remunerative relative to the urgency to secure their access to food. This means that the price of food is high relative to their income and that the opportunities of earning higher returns by doing something else are limited as well. Thus, what qualifies as land with an acceptable production potential in a poor country, may not be so in a high-income one. An exception would be if poor quality of land is compensated by a larger area per person with access to mechanisation[14] where returns for work in farming must generate income not far below what can be earned in other work. Obviously, the socio-economic context within which a piece of land exists and assumes a given value or utility, changes with time: what qualifies today as land suitable for farming may not be so tomorrow.

It is no easy task to account fully for all these factors in arriving at conclusions concerning how much land with crop production potential there is. For example, if food became scarce and its real price rose, more land would be worth exploiting and hence be classified as agricultural, than would otherwise be the case. Therefore, depending on how one intends to use such information, one may want to use different criteria and hence generate alternative estimates.

---

[14] Relatively low-yield rainfed but internationally competitive agriculture (wheat yields of 2.0-2.5 tons/ha compared with double that in Western Europe) is practised in such high income countries as the USA, Canada and Australia. But this is in large and fully mechanized farms permitting the exploitation of extensive areas that generate sufficient income per holding even if earnings per ha are low.

Chapter 18

Food Security Prospects in the Maghreb

S. Benjelloun
Institut Agronomique et Vétérinaire Hassan II
BP 6202 Rabat-Instituts, 10101 Rabat, Morocco

Food security in the Maghreb countries appears to have been improving in the past thirty years, at least in terms of availability. Although there is little consistent data available for certain countries, calorie availability in Morocco has increased from just over 1700 Kcals in 1964/1966 to almost 3000 Kcal in 1996/1998. However, food distribution is not even, and economic access among the poorer groups does not ensure consumption of an adequate, varied diet, especially by the vulnerable groups. This paper will examine the current trends and discuss key factors, which determine food security in the Maghreb countries and the limitations of the information currently available that would permit a comprehensive projection of food security to 2050.In spite of the large state investment in the agricultural sector, and the importance of the latter in the GDP (Algeria: 10%; Morocco: 12.6%; Tunisia: 12.2%), staple foods are still largely imported. In the three countries, agricultural production fluctuates greatly each year, as most of it is rainfed. According to FAOSTAT, for the period 1997/99, imports covered large portions of the total supply of cereals (Algeria: 68%; Morocco: 42%; Tunisia: 59%). The figures are even higher for sugar (Morocco: 54%; Algeria and Tunisia: 100%) and vegetable oils (Algeria: 80%; Morocco: 64%; Tunisia: 79%). The situation is not better for non-staple foods. Domestic production of milk hardly covers the effective demand and is far from meeting the nutritionally recommended level of intake in all three countries. In Morocco, the Ministry of Agriculture has set a strategy for production through 2020. It aims to increase the level of self-sufficiency for most agricultural products: soft wheat (70%), durum wheat (80%), barley (90%), corn (30%), sugar (76-100%), vegetable oils (50-60%) and legumes (70%). The strategy also sets objectives for agricultural exports (vegetables, fruits and olive oil), as this sector contributes to the agricultural balance of trade. This strategy of self-sufficiency appears ambitious in the face of the natural and economic constraints on the horizon. Moreover, the concept of food self-sufficiency has long been superceded by that of food security, which requires, in addition to availability, stability and access to food. In this framework, a country's food security is not assessed solely in terms of supply and demand (national level), but also in terms of household access (income, regions) and individual access within the household (male/female, adult/child). Using this analysis, the indicators for Morocco are troublesome. In terms of household food security and population nutritional status, if the situation is better than in most other African countries, it is still far from adequate. While stunted growth is still prevalent among children (Algeria: 18%; Morocco: 23%; Tunisia: 8%), all three countries are now going through an important nutrition transition. Indeed, increasing proportions of the population (more than one third in Morocco and Tunisia) are overweight or obese. These nutrition conditions are likely to put even more strain on public sector resources.

A. Marquina (ed.), Environmental Challenges in the Mediterranean 2000-2050, 301–317.
© 2004 *Kluwer Academic Publishers. Printed in the Netherlands.*

1. Food and nutrition

The last World Food Summit (WFS 2001) revealed that the objective of its previous edition (1996) is still far from being reached [8]. Actually, there is a large gap between the target of the 1996 WFS (400 malnourished people by 2015) and the possible outcome if the current trend continues (around 700 malnourished people).

By regions, the summit showed that in the Near East and North Africa region, the number of undernourished people increased from 25 millions in 1990-92 to 32 million in 1997-99. During the years 1996-2000, agricultural production growth in this region has fluctuated between positive and negative growth. This reflects mainly the oscillation in the rainfall, the latter being the main factor affecting agricultural production [17].

In addition, the attempts by governments in the region to achieve food self-sufficiency have created perverse incentives to agricultural mismanagement, resulting in resource depletion. For example, producer subsidies have encouraged mechanized cereal cultivation on marginal lands, degrading the environment and increasing vulnerability to droughts. At the same time, as these policies were not targeted, they tended to favor larger farmers [14].

The present paper attempts to analyze the situation of food security in the Maghreb countries, using the concept of food and nutrition security.

2. Conceptual Issues

2.1. FOOD SELF-SUFFICIENCY

In the seventies, food security issues were examined at a country level and were analyzed essentially in terms of food self-sufficiency. The latter represents the ability of a country to produce domestically all the food it needs. It is appraised through the use of food balance sheets which show the country's food gap. This is the difference between aggregate consumption needs and actual food supplies, through either production or imports. This approach enables the determination of food deficit countries. However, within a country, it does not permit the determination of food deficit regions nor groups of population which may suffer from food deficiency.

By the eighties, it became clear that food availability at a national level did not reflect necessarily the availability in all sub-regions and in all households. In fact, countries such as Brazil or India which export cereals still had localized areas of food deficits.

2.2. FOOD AND NUTRITION SECURITY

The weakness of the concept of national food security has helped move the analysis to the household level. This analysis evolved later to be called food security. It reflects the ability of a country to procure sufficient food for all its population. Hence, this concept is broader than that of food self-sufficiency. FAO defines it as "physical and economic access to adequate food for all household members, without undue risk of losing such access" [8]. In this definition, the term "adequate" carries two components: the nutritional quality of the food and the health and sanitation environment for the body to absorb the nutrients contained in the food. Hence, the new concept includes, in fact, both food and nutrition security.

The concept of food security encompasses three dimensions: availability, access and stability, corresponding each to a different level: national, household and individual. Table 1 displays these dimensions and their levels of measurement.

Table 1. Food security: dimensions and measurement

Aspect	Level	Measurement	Factors
Availability	National	Food demand and supply	Population, Resources
Access	Household	Effective demand	Income, region,
		Household food consumption surveys	urban/rural
Stability	Individual	Nutritional status of the population: nutritional surveys	Gender, age

Food availability is measured at a national level using food balance sheets. It is mainly affected by population growth and the resources (production and importation). In the food security concept, available supplies of food are only a prerequisite for household food security. That is, households must also have access to food.

Access to food can be measured through household food consumption surveys. The factors which affect it are region, income level and milieu of residence. Obviously, food security at the household level does not imply that all members in the household are food secure. Equally, a food insecure household may contain food secure members.

Stability, which is the durability in time of food security, reflects on reducing risks and uncertainty of access of individuals to food. One proposed indicator to measure the stability of food security in a country is the nutritional status of the population. Indeed, the latter being the end result of food consumption, the simplicity of its assessment makes it a good indicator of the variability over time of food security.

3. Food and Nutrition Security in the Maghreb

3.1 FIRST DIMENSION OF FOOD SECURITY: AVAILABILITY

3.1.1. Food Availability
Figure 1 shows the evolution of cereal availability per person between 1962 and 2000 in a trend of three-year moving averages. Availability has been increasing in all three Maghreb countries. The highest level has been observed in Morocco (from 196 kg/person/year in 1962/64 to 246 kg/person/year in 1998/2000). However, a large part of this availability is supplied through imports. The latter, in fact, fluctuates from year to year due to oscillations in rainfall. Algeria appears to be the most dependent on imports (between 26 and 110%[1]) followed by Tunisia (between 29 and 109%) and Morocco (between 12 and 58%).

---

1 The proportion may be higher than 100% because part of the imports may be oriented toward uses other than human consumption.

Figure 1. Cereal availability

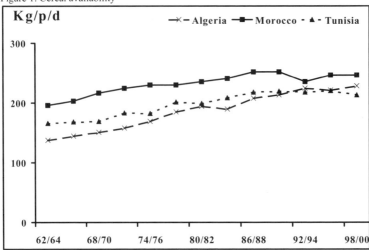

Source: FAO, 2002 [13]

Concerning sugar availability, it is quite high in all three countries. In 1998/2000, it was 24.3 kg/person/year in Algeria, 34.9 kg/person/year in Morocco and 30.2 kg/person/year in Tunisia (Figure 2). These high levels reflect the food habits in the Maghreb where sweet tea is the national drink. Because of this strong food habit, all three countries depend heavily on imports to satisfy the demand. In 1998/2000, Morocco imported 52% of total availability, Tunisia 106% and Algeria 123%.

Figure 2. Sugar availability

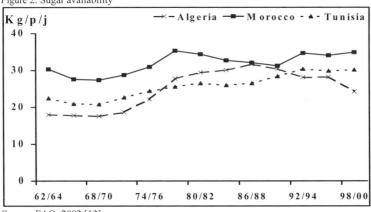

Source: FAO, 2002 [13]

For oils, availability has also been steadily increasing (Figure 3) but this was possible only with heavy imports (Morocco: 69-96%, Algeria 42-115% and Tunisia: 52-106%).

Figure 3. Oil availability

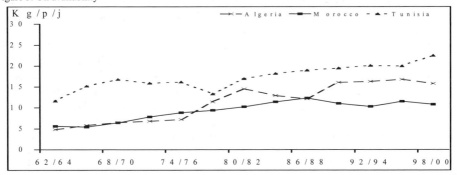

Source: FAO, 2002 [13]

Another product, usually not considered among staple foods, but nutritionally important is milk. Its availability in the three countries steadily increased between 1962 and 2000 (Figure 4). Morocco had the lowest availabilities (28 kg/person/year in 1962/64 and 32 kg/person/year in 1998/2000) while Algeria had the highest level (42 kg/person/year in 1962/64 and 105 kg/person/year in 1998/2000). Tunisia made significant progress, increasing its level of availability from 39 kg/person/year in 1962/64 to 94 kg/person/year in 1998/2000. In fact, this progress is reflected in the decrease of the proportion of available milk which is imported (from 20% in 1962/64 to 9% in 1998/2000). Algeria depends heavily on imports (between 30 and 68% of total availability) while Morocco imported between 24 and 12% of its consumption. Since milk is not considered a staple food, countries import it only when solent demand can not be covered through national production. That is, the nutritionally recommended level of intake is not taken into account.

Figure 4: Milk availability

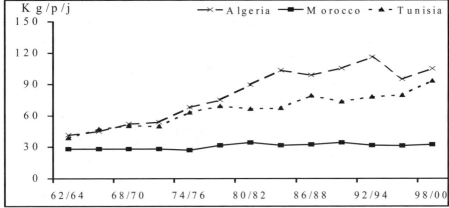

Source: FAO, 2002 [13]

3.1.2. Wheat Production and Consumption Projections

According to projections made by Schmitt[2] [19], the level of per capita wheat consumption will slightly decrease due to the already present high level in all three countries. These high levels are maintained due to the food habits where cereals continue to increase their weight in the diet even when the household income increases.

What is more important to notice in Figure 5 is the widening gap between consumption and production, the latter continuing to be lower. This is more the case for Algeria whose production would cover a mere 17% of the projected consumption in 2025. The figures would be higher for Morocco (45%) and Tunisia (35%).

Figure 5. Wheat production and consumption projections (1995-2025)

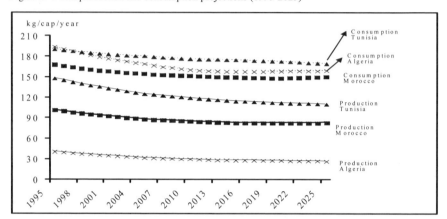

3.1.3. Calories and nutrients availability

In terms of calories, all three countries have increased their availabilities between 1962/64 and 1998/2000 (Algeria: 1701 – 2963 Kcal/person/day; Morocco: 2279 – 2971 Kcal/person/day; Tunisia: 2276 – 3367 Kcal/person/day) [13].

Similarly, per person availabilities of proteins have steadily increased in all three countries. In Algeria, protein availability increased from its level of 45 grams/person/day in 1962/64 to 82 grams/person/day in 1998/2000. Morocco increased its availability from 61 grams/person/day in 1962/64 to 82 grams/person/day in 1998/2000 while Tunisia's figures are 58 and 92 grams/person/day respectively.

Finally, the general increase in availability also refers to fats. The latter, expressed in grams/person/day, increased between 1962/64 and 1998/2000 with the following figures: Algeria: 31 – 69; Morocco: 35 – 59; Tunisia: 58 – 102.

3.1.4. Population and land projections

The main factors affecting the ability of a country to increase the share of domestic production in food availability are its population and the available resources. Constraints

---

[2] Schmitt projection model used the following assumptions: 1) positive growth of per capita GDP, 2) Consumption and production prices will continue increasing in constant terms for the next ten years (USDA hypothesis), 3) Population growth as in the middle United Nations hypothesis, 4) the model does not take into account the potential change in food consumption (westernisation of diets).

facing Moroccan agricultural production include the shrinking of arable land, the weak farm structure, the lack of financing and poor professional organisation of farmers [3].

In terms of population growth, it is still relatively high in the region in spite of the substantial efforts achieved by the three countries in family planning. Table 2 shows that the North Africa region has a growth rate of 1.8% per year for the period 2000-2005 [24]. The figure is the same for Morocco and Algeria while it is significantly lower for Tunisia (1.1%).

Table 2. Population projections

Country/ Region	Total population 2001 (million)	Total population 2050 (million)	Average growth rate 2000-05	% Urban	Average growth rate Urban 2000-05
World	6134.1	9322.3	1.2	47	2.0
North Africa	177.4	303.6	1.8	51	2.9
Morocco	30.4	50.4	1.8	56	2.8
Algeria	30.8	51.2	1.8	60	3.2
Tunisia	9.6	14.1	1.1	66	2.3

Source: UNFPA, 2002 [24]

In the face of population growth, resources are steadily worsening. Per capita arable land in Tunisia has decreased from 0.72 hectares in 1961 to 0.31 in 2000. The figures are lower for Algeria (from 0.59 to 0.29) and Morocco (from 0.55 to 0.25). According to population projections for 1995-2025, these figures are expected to attain very low levels, jeopardizing the potential for agricultural production and hence decreasing even more the prospects for food self-sufficiency (Figure 6). In fact, the per capita arable land will be, in 2025, 0.16, 0.22 and 0.21 hectares respectively in Algeria, Morocco and Tunisia. In addition, most of the arable land in the region is of poor agricultural quality, rainfed and semi-arid.

Figure 6. Projection of per capita arable land (1995-2025)

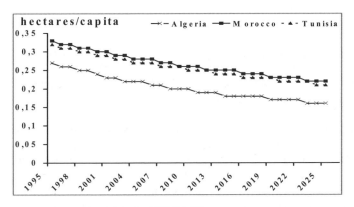

Source: FAO, 2002 and Schmitt, 2002 [13,19]

3.2. SECOND DIMENSION OF FOOD SECURITY: ECONOMIC AND SOCIAL FACTORS

Access to food includes the physical access and the economic access.

3.2.1. Physical Access

It is usually reflected in the spatial distribution of food, among different regions of the country, and particularly among rural and urban areas. Because of difficult access to remote areas, several food products, especially the perishable ones do not reach these regions. Hence, food products such as fish and sea products tend to be more consumed in the coastal areas than in the inner regions. For example, in Morocco, fish consumption varies from an average of 1.2 kg/person/year in the South to an average of 9.77 kg/person/year in the Eastern region [10]. Similarly, the figure is almost three times lower in rural areas (3.5 kg/person/year, on average) than in urban areas (9.9 kg/person/year, on average.

The difference in access to food between rural and urban areas can be partially explained by the impact of agricultural policies. These policies increased the share of agricultural production oriented to the market encouraging rural producers to sell their products to the food industry, including those which have been traditionally subsistence products [6]. The result is that, more often than not, the processed products do not find their way back to the rural markets, the reason being the lack of infrastructure (roads, electricity, refrigerators, etc.). This has been termed "urban bias" of national food policies. In fact, numerous studies have shown that while these policies boosted the level of food consumption among urban people, their rural counterparts have seen their diet worsen. Particularly, dairy development projects have lead to the improvement of the consumption of milk and dairy products in urban areas while they reduced the intake, especially of dairy products, in rural areas. Attracted by monetary income, rural producers prefer to sell most or all their milk production to the dairy industry [5]. In Morocco, the intake of dairy products decreased in the rural areas between 1970 and 1984 (from 27 to 20 kg/person/year) while it increased among the urban population (from 31 to 44 kg/person/year). It was during this period that most milk collecting units were installed in the rural areas [10].

3.2.2.Economic Access

Economic access to food reflects the purchasing power of the different socioeconomic classes. Poverty rate in a country is an indicator of the size of the population of households which may be unable to purchase the food. To assess this dimension of food security for the Maghreb countries, the human development approach will be used.

Human Development Index (HDI) is a measure introduced by the United Nations Development Program to complement the previously used indicator of Gross Domestic Product. HDI is based on three distinct components: life expectancy, education and per capita income. It is intended to represent better the level of human development in a country, and not only the economic component. It also enables the classification of countries, and hence it is an indicator of how a country is doing compared to others in the world.

Concerning the Maghreb countries, if the HDI has steadily improved from 1975 to 2000 as shown in Figure 7, it is however still low. Morocco was ranked 123 in 2000 among 173 countries [23]. The other two countries are slightly better ranked (Algeria 106 and Tunisia 97).

Figure 7. Human development Index Trend (1975-2000)

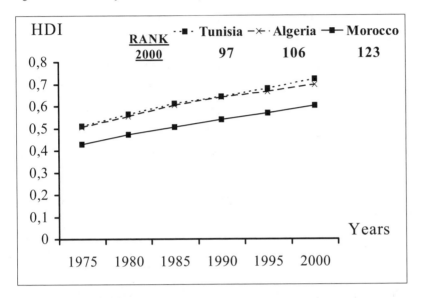

Source: UNDP, 2002 [23]

Disparities are also reflected in the proportion of the population living below the poverty line. In Morocco, this proportion has even increased from 13% in 1990 to 19% in 1998[3]. The figure is higher in rural areas (18% in 1990 and 27% in 1998) than in urban areas (8% in 1990 and 12% in 1998) [11].

To the disparity between urban and rural areas one should add the disparity between expenditure classes. According to the 1998 Moroccan national survey, the richest 10% of the population spend 12 times more than the poorest 10%. This figure is higher in urban areas (10 times) than in rural areas (7 times). Table 3 shows the level of food expenditures by food groups and expenditure classes, as reported by the 1998/99 national survey. The magnitude of disparity in expenditure among classes is reflected in the fact that total food expenditure of the highest quintile (6549DH/pers/year) is almost five times that of the lowest quintile (1439DH/pers/year). Also, meat expenditure is 7.5 times higher in the highest expenditure quintile than in the lowest. Those of fish are 10 times.

---

[3] In Morocco, the poverty line has been estimated for the National 1998/99 Survey at 10.75 Dirhams per person per day for urban areas and 8.32 Dirhams per person per day for rural areas (1 Dirham ≅ US$0.09) (Direction de la Statistique, 2000).

Table 3. Food expenditure by food groups and expenditure classes in Morocco
Unit: Dirhams/person/year

	Expenditure quintile (Per capita total annual expenditure)[1]					
	1	2	3	4	5	All
Cereals & cereal products	444.4	577.1	675.4	766.8	989.7	690.5
Dairy products & eggs	58.2	111.2	182.4	265.9	573.8	237.2
Fats	140.7	191.0	244.1	289.1	418.1	256.5
Meats	257.7	466.7	673.5	1029.9	1935.9	872.5
Fish	23.0	42.1	69.2	89.9	233.6	91.5
Vegetables	165.2	236.5	313.1	387.9	575.7	335.6
Legumes	55.5	83.1	106.0	128.6	192.0	112.9
Fruits	42.4	75.4	117.2	187.6	427.3	169.9
Sugar	89.1	109.7	122.1	137.1	156.6	122.8
Other products	162.9	265.2	385.3	556.1	1045.8	482.6
Total	1439.0	2158.0	2888.3	3839.0	6548.5	3372.0

[1] Lowest economic class: 1; Highest economic class: 5.
Source: Direction de la Statistique, 2000 [10]

3.3. THIRD DIMENSION OF FOOD SECURITY: NUTRITIONAL ASPECTS

As discussed above, the nutritional status of the population is used to reflect on the third dimension of food security, namely its stability. In addition, as one of the indicators of health status, it contributes to the general well-being of the population. Two groups of nutritional problems are considered: nutritional deficiencies and overweight and obesity.

3.3.1. Nutritional Deficiencies

According to the World Health Organisation projections, the proportion of stunted[4] preschool children in North Africa would decrease from its level of 33% in 1980 to 17% in 2005 (Figure 8). These figures are lower than those for the other developing countries [9]. The decrease reflects the policies implemented in the region toward the improvement of health and nutrition of children. However, the figure of 17% is considered to be high, if compared, for example, to the prevalence in South America (12%).

In Morocco, the evolution of the proportion of stunted preschool children has hardly changed between 1987 and 1997. In particular, its level in the rural areas is rather high (29% in 1997). Again, this reflects the disparity between rural and urban areas, the latter having a figure of 15%, that is half the rural figure [1].

---

[4] Stunting is assessed by the proportion of children whose height is too low for their age, that is their z-score for height-for-age is lower than –2 standard deviations.

By contrast, Tunisia has reduced both the percentage of stunted children and the disparity between rural and urban areas. In fact, the figure in rural areas went from 33% in 1994 to only 11% in 1997 and that in urban areas went from 15% to 6%. Hence, the disparity between rural and urban areas decreased from 18 points (33-15) to 5 points (11-6) [25].

Figure 8. Trend and projection of stunting among preschool children (1980-2005)

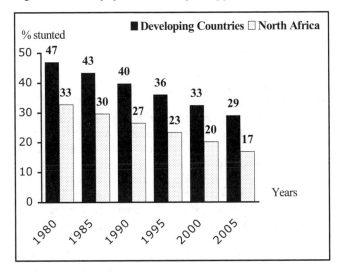

Source: de Onis Blöswer 2002 [9].

In Algeria, according to the reported figures, while the proportion of stunted children is still high (18.3% in 1995 and 18.9% in 2000), the disparity urban/rural is almost null [25].

Another indicator of nutritional status is the prevalence of micronutrient deficiencies. These are known as "hidden hunger". Unlike "visible famine", which represents a general lack of food, these deficiencies reflect unbalanced diets. For example, iron deficiency anemia may be caused by a diet containing too few animal products combined with too many fiber-containing components. Favoring factors include multiple pregnancies in women and parasitic diseases in children. In the Maghreb, anemia affects more than two-fifth of pregnant women (Morocco: 45%, Algeria: 42% and Tunisia: 41%) [15] [21]. A more recent survey in Morocco showed that anemia is still prevalent in spite of programs implemented by the Ministry of Health: women of procreating age: 33%, pregnant women: 37%, children 6-59 months: 32% and men: 18% [20].

Goiter which reflects iodine deficiency is also prevalent, especially in Morocco with a national figure of 22% in 1994. The prevalence is lower in Algeria (9%) and Tunisia (4%) [21].

Another micronutrient deficiency is that of vitamin A. In Morocco, the 1996 survey showed that 41% of children aged 6 months to 6 years suffer from subclinical deficiency [16].

3.3.2. Nutrition Transition
The three countries are going through nutritional transition, that is the transition from the problems of malnutrition to those related to obesity and cardiovascular diseases. This transition has occurred simultaneously with the transition in other processes, namely demographic, epidemiological and dietary transitions. These will be briefly discussed below.

3.3.3. Demographic Transition
The demographic transition, that is the shift from a pattern of high fertility and high mortality to one of low fertility and low mortality (typical of modern industrialized countries), began in the Maghreb countries in the first half of the seventies. In Morocco, total fertility rate reached 3.1 children per woman during the period 1995-97 [1]. The figure projected by UNFPA for the period 2000-05 is slightly lower (3.03) and is even lower for Algeria (2.79) and Tunisia (2.10) [24].

Life expectancy at birth in Morocco reached 70 years in 1999 from 47 years in 1962 [18]. The figures reported for 2000 are similar for both Algeria (69 years) and Tunisia (70 years) [22]. This is the consequence of both the decline in infant and juvenile mortalities and the improvement in adult longevity.

Table 4 shows the proportions of the over 65 population for the years 2000, 2010 and 2020, as projected by UNDP [22]. It shows that these proportions will steadily increase. The degenerative diseases being more prevalent among the elderly, it is expected that, in the future, the Maghreb countries will experience a higher prevalence of these diseases.

A second aspect of demographic transition is reflected by urbanisation which is expanding rapidly. Because of the characteristics of their diet and their sedentary lifestyle, urban people have a higher prevalence of overweight and nutrition-related non communicable diseases.

From its level of 29% in 1960, the urban population reached 53% in 1997 in Morocco and is expected to represent 65% in 2014 [23]. The other two Maghreb countries have higher proportions of urban population (Algeria: 60% and Tunisia: 66%). These figures will increase rapidly as shown by the average growth population in urban areas (Morocco: 2.8%, Algeria: 3.2% and Tunisia: 2.3%) [22].

Table 4. Over 65 population proportion projections

	2000	2010	2020
Algeria	3.93	4.51	5.07
Morocco	4.69	5.26	5.70
Tunisia	5.39	6.33	6.67

Source: UNDP, 2002 [23]

3.3.4. Epidemiological Transition.

Overweight and obesity are known to be in themselves risk factors for cardiovascular diseases. Hence, developing countries which are undergoing the nutritional transition are also going through an epidemiological one. The latter is the shift from a predominance of infectious diseases to the rise in degenerative diseases, namely, diabetes mellitus, cardiovascular diseases and some types of cancer.

In Morocco, a survey conducted by the Ministry of Health in 2000 on risk factors of cardiovascular diseases indicates that hypertension affects around one third of the over 20 population, and is higher among women than among men. Diabetes affects equally 6.6% of men and women while it is higher in urban areas. Hypercholesterolemia concerns 29% of the population and is more frequent among women and in urban areas.

In Tunisia, diabetes mellitus affects 8.8% of men and 7.9%,of women [25]. No data could be found for Algeria.

3.3.5. Dietary transition

Table 5 shows the evolution of food expenditures by food groups through four national surveys in Morocco.

Table 5. Evolution of food expenditures between 1970/71 and 1998/99 (in reel prices, Base 1989) Unit: Dirhams/person/year

	Urban				Rural				National			
	1970	1984	1990	1998	1970	1984	1990	1998	1970	1984	1990	1998
Cereals & cereal products	405	507	506	499	418	536	689	482	415	523	512	491
Vegetables & Fruits	475	376	440	445	231	210	368	288	316	281	404	373
Fats	145	190	233	195	126	172	307	201	132	180	270	198
Sugar & sweets	179	130	154	123	194	138	166	128	189	134	160	126
Animal products	850	757	849	916	416	383	724	498	557	546	787	723
Total	2489	2560	2723	2464	1701	1800	2790	1743	1963	2129	2756	2133

Source: Direction de la Statistique, 1972-2000 [10].

In this table, it is interesting to note that while in urban areas, animal product expenditure has steadily increased throughout the period, they rather fluctuated in rural areas. At the same time, cereal expenditure has varied very little in urban areas while it increased up to 1990 in rural areas before dropping in 1998. This reflects the dietary behavior of Moroccan households whose cereal intake does not automatically decrease as they increase their animal product intake. This is further confirmed by the examination of annual per capita food group expenditures by expenditure classes (see Table 3 above). It shows that higher quintiles have higher expenditures of all food groups, including cereals. While the lowest expenditure quintile spends an average of 444DH/pers/year on cereals, the highest quintile spends an average of 990DH/pers/year, that is more than double the former [11].

Hence, even when households are better off economically, their cereal consumption does not drop as a consequence while their consumption of animal products (mainly meat, chicken and fish) increases. These products are, in fact, prepared in tagine (a sauce-containing dish)

and tend to induce more bread consumption. Moreover, these dishes are prepared in oil, the usual cooking medium for tagine. The sauce is usually eaten with bread. It is therefore expected that with the improvement in economic status, people tend to consume more animal products but also a continuing high quantity of cereals, and a higher quantity of oil, resulting in overall higher calorie consumption.

In terms of actual intake, only 1970/71 and 1984/85 Moroccan national surveys provide data. Table 6 and Figure 9 show the shift in the levels of intake of different food groups and of caloric structure of the diet between the two surveys. Mean daily caloric intake increased from 2466 in 1970 to 2606 kcal/person in 1984. Fats which represented only 18% of total caloric intake in 1970 provided 22% in 1984 [10]. In spite of being outdated, these figures indicate the trend of caloric intake and its increasing proportion of fats, a main dietary factor in the onset of obesity among the population.

Table 6. Evolution of intake by food groups in Morocco Unit: kg/person/year

	National surveys	
	1970/71	1984/85
Cereals & cereal products	216.4	210.4
Vegetables	83.7	107.39
Legumes	5.0	5.8
Fruits	46.1	31.8
Fats	13.1	14.6
Sugar	29.7	27.0
Dairy products	28.3	30.3
Eggs	1.3	2.9
Meats	15.5	10.4
Poultry	2.3	5.6
Fish	3.6	6.2
All animal products	51.0	55.4

Source: Direction de la Statistique, 1992 [10]

Figure 9: Evolution of caloric intake and its structure in Morocco

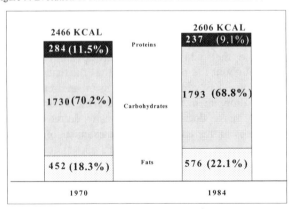

Source: Direction de la Statistique, 1992 [10]

3.3.6. Anthropometric Transition

The prevalence of obesity among the under-five children is quite high in the Maghreb countries. Algeria has the highest figure (9.2%), followed by Morocco (6.8%) and Tunisia (3.5%). These prevalences are all higher than the overall prevalence in developing countries [9].

In Morocco, more recent data show that while undernourishment persists (23% stunting and 10% underweight), overweight is expanding (9.2%). This prevalence is slightly higher among boys (9.5%) than among girls (8.8%) and among urban children (10.1%) than among rural children (8.6%) [1].

But overweight is more spread among the adult population as shown in Figure 10 for Morocco and Tunisia. Unfortunately, no data could be found for the Algerian adult population.

Figure 10: Prevalence of overweight and obesity in Morocco and Tunisia

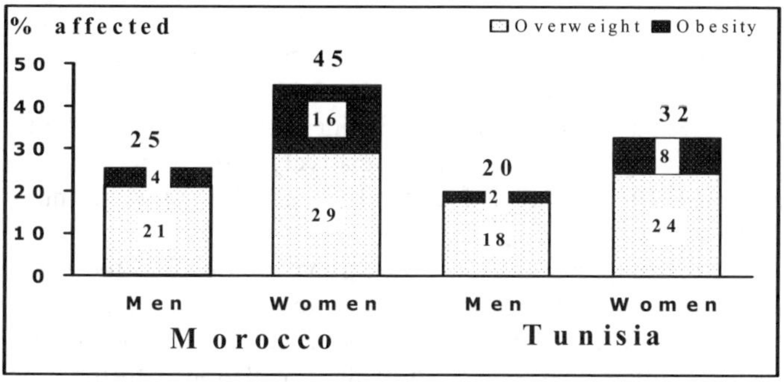

Source: Benjelloun, 2002 (Morocco) [4] WHO, 2002 (Tunisia) [25]

For Morocco, the comparison of data from 1984 and 1998 national anthropometric surveys shows the rapid increase in the prevalence of overweight and obesity, particularly among urban female population. In all categories, prevalence of underweight decreased and that of overweight increased. For both periods, underweight is more frequent among men (5.7%) than among women (4.8%) and in rural areas (6.5%) than in urban (4.4%). Similarly, the highest prevalence of overweight and obesity is found among urban women (50.2%) and the lowest among rural men (20.8%) [4].

The higher prevalence of overweight and obesity among women may be partially explained by cultural factors. In fact, in Arab culture, fatness has long been considered a sign of beauty. In Tunisia, a study by Beltaifa et al. [2] showed that obesity among Tunisian women was related to sociocultural determinants.

The difference in anthropometric status between urban and rural population may be explained by the difference in their diet quality and in their lifestyle. In terms of diet, even if total energy intake is higher in rural areas (2746Kcal/pers/day compared to 2423 in urban areas), it has a lower contribution of fats (19% compared to 28% in urban areas) and of animal products (4.6% compared to 9.7% in urban areas) [4]. As with lifestyle, rural people have higher calorie expenditure in relation to their agricultural occupation and their means of transportation. Because of lack of roads and of transportation means, they generally walk to fields, to the market and inside villages. By contrast, urban people ride motorcycles, cars or buses.

4. Conclusion

The Maghreb countries have achieved national food security in terms of national availability, mainly through imports. The prospects for the next twenty-five years indicate that the dependence on imports will persist, due to the growth in population, the shrinking agricultural resources and in particular the heavy weighting of cereals in the population diet.

In terms of food and nutrition security, the countries in the Maghreb are marked by strong regional and economic disparities, particularly between the urban and the rural areas, the latter being at a disadvantage. Rural populations and low income groups have limited access to food due to physical as well as economic barriers. This results in pockets of under-nutrition, affecting mainly preschool children, and confined mostly to the rural areas and among the poor income groups.

Micronutrient deficiencies, also called hidden famine, are still prevalent in the Maghreb. Goiter and iron deficiency anemia affect a relatively significant share of the population. They are caused, among other factors, by unbalanced diet, multiple pregnancies and parasitic diseases.

Like other emerging developing countries, the Maghreb countries face today a new set of nutrition and health problems, namely obesity and non-communicable diseases. These are the consequence of the demographic, epidemiological and dietary transitions. Overweight and obesity affect more than one third of the population with the highest prevalence reported for urban women.

Hence, while they have not yet eradicated the problems of malnutrition, the Maghreb countries are already affected by those of obesity and degenerative diseases. This situation calls for adapted policies to face both problems.

References

1. Azelmat, M., and Abdelmoneim, A. (1999) *Enquête nationale sur la santé de la mère et de l'enfant (ENSME) 1997.* Ministère de la Santé, direction de la planification et des ressources financières, service des études et de l'information sanitaire et PAPCHILD, Rabat.
2. Beltaifa, L., Traissac, P., Lefèvre, P., Ben Romdane, H., Gaigi, S., and Delpeuch, F. (2001) Nutrition transition in Tunisia and its implication on obesity in women., Poster presented at The International Congress of Nutrition, August 27-31, Vienna.
3. Benjelloun, S. (1998) Expansion démographique et sécurité alimentaire, in: CERED (ed), *Population et développement au Maroc*, Centre d'Etudes et de Recherches Démographiques, Rabat, 217-232.
4. Benjelloun, S. (2002) Nutrition transition in Morocco, *Public Health Nutrition 5(1A)*, 135-140.
5. Benjelloun, S, Rogers, B.L. and Berrada, M. (1998) Income and consumption effects of milk commercialization in the Lukkos area of Morocco, *Ecology of Food and Nutrition 37*, 269-296.
6. Benjelloun, S. Rogers, B.L. and Berrada, M. (2002) Impacts économiques, alimentaires et nutritionnels des projets de développement agricole: le cas du projet d'irrigation du Loukkos au Maroc, *Cahiers Agricultures 11*, 45-50.
7. CERED (1997) *Situation et perspectives démographiques du Maroc.* Centre d'Etudes et de Recherches Démographiques, Rabat.

8. De Haen (2001) The state of food and agriculture. FAO Conference, 2-13 November 2001. Slides from the FAO website (www.fao.org).

9. De Onis, M., and Blössner, M. (2000) Prevalence and trends of overweight among preschool children in developing countries, *A J Clin Nutr 72*, 1032-1039.

10. Direction de la Statistique (1992) Consommation et dépenses des ménages 1984/85. 1, 5, 6 and 7. Statistics Office, Rabat.

11. Direction de la Statistique (2000) Enquête nationale sur les niveaux de vie des ménages 1998/99. Premiers résultats. Statistics Office, Rabat.

12. FAO (1997) Implications of economic policy for food security. A training manual, *Training materials for agricultural planning 40*. FAO, Rome.

13. FAO (2002) Data extracted from Food Balance Sheets in the statistical database in the FAO website (www.fao.org).

14. Hazell, P. , Oram, P., and Nabil, C. (2001) Managing drought in the low rainfall areas of the Middle East and North Africa. *EPTD discussion paper 78*. Environment and Production Technology Division, IFPRI, Washington D.C.

15. Ministère de la Santé (1995) Enquête nationale sur la carence en fer et en iode, INAS, Rabat.

16. Ministère de la Santé (1999) Enquête régionale sur la carence en vitamine A, Direction de la population, Rabat.

17. Nordblom, T., and Shomo, F. (1995) Food and feed prospects in *West Asia/North Africa, Social Science Paper 2*, ICARDA, Aleppo.

18. Nouijai, and Attané, (1998) Bilan de l'expérience marocaine, in: CERED (ed), *Population et développement au Maroc*, Centre d'Etudes et de Recherches Démographiques, Rabat, 15-25.

19. Schmitt, F. (2002) Equilibre alimentaire en Méditerranée : les enjeux pour le XXIe siècle. Thèse de Doctorat, Université de Montpellier I, Faculté des Sciences Economiques, Montpellier.

20. SEIS (2001) Enquête nationale sur la carence en fer, l'utilisation du sel iodé et la supplementation par la vitamine A. service des études et de l'information sanitaire, Ministère de la Santé, Rabat.

21. The World Bank (1999) Towards a virtuous circle: a nutrition review of the Middle Easet and North Africa, *Working Paper Series 17*, Human Development Group and MNESED, the World Bank, Washington D.C.

22. UNDP (2002a) *Arab Human Development Report 2002*, UNDP website (www.undp.org).

23. UNDP (2002b) *Human Development Report 2002*, UNDP website (www.undp.org).

24. UNFPA (2002) *The State of the World Population 2002*, UNFPA website (www.unfpa.org).

25. WHO (2002) Data extracted from the statistical database of the WHO website (www.who.org)

Chapter 19

EU Agriculture and Common Agricultural Policy: Prospects for the 21st Century and Implications for Mediterranean Countries[1]

J.M. Garcia-Álvarez-Coque
Universidad Politécnica de Valencia
Camino de Vera, s/n. 46022 Valencia

There are overwhelming reasons for Mediterranean countries to strengthen their links. Centuries of common history, cultural roots and human exchange should provide a base for avoiding any sort of *apartheid* between the different shores of the Mediterranean. This is, indeed, the official goal of the Barcelona Process, launched in 1995[2] with the aim of creating a Euro-Mediterranean partnership, including a Free Trade Area between the EU and 12 Mediterranean partners by 2010. Dissatisfaction about how agricultural trade has been managed in the Barcelona process has become a constant during the negotiations and the reviews of the trade arrangements between the EU and the Southern Mediterranean Countries (SMCs). Such dissatisfaction has appeared on both sides of the Mediterranean basin as reflected by: (i) the claim for a larger EU market access by Southern Mediterranean exporters and (ii) EU producers fears of increased competition from a loss of community preference. The Barcelona process has called for a progressive liberalisation of agricultural trade based on traditional flows. However, the Euro-Mediterranean Agreements (EMA) do not consider full liberalisation of agricultural trade. As a result of this 'controlled' approach, agriculture has been given a low profile during the definition and putting into practice of the EMAs. It has been usual for agricultural negotiations to be left as the final step of the bilateral trade talks between the EU and the Mediterranean partners. The treatment of agriculture under the Euro-Mediterranean Free Trade Area (EMFTA) has been largely *ad hoc* and commercial concessions have varied according to the sensitiveness of the product on EU markets and to the export competitiveness of each particular partner. Top-level meetings, e.g. Ministerial Conferences, have not undertaken discussions in depth about the pros and cons of a common approach for agricultural trade and rural development in the Mediterranean basin. In the following pages we will consider the future prospects for EuroMediterranean agricultural trade relations, by assessing how a reform of the EU's Common Agricultural Policy (CAP) might contribute to overcome the existing political constraints on the full inclusion of agricultural trade in the Barcelona process. Next, we shall proceed to review the current status of agricultural trade between the EU and the SMCs. Then, we will refer to the current treatment of agriculture in the EMAs. We will later study the policy options for the CAP in the coming years to finally assess how CAP

---

[1] This chapter draws from the author's contribution to the project conducted at CIHEAM leading to the annual reports on "Development and agro-food policies in the Mediterranean region" [1]. Comments by an anonymous reviewer contributed to a significant improvement of the preliminary draft.
[2] The Barcelona declaration was signed by the 15 EU Member States and 12 Mediterranean countries (Algeria, Cyprus, Egypt, Israel, Jordan, Lebanon, Malta, Morocco, Syria, Tunisia, Turkey and the Palestinian Authority).

A. Marquina (ed.), Environmental Challenges in the Mediterranean 2000-2050, 319–328.
© *2004 Kluwer Academic Publishers. Printed in the Netherlands.*

reform could affect the future of Euro-Mediterranean integration. We will use the term "South" in a political way rather than in a geographical way. Indeed, the North-South asymmetries in the Mediterranean region are outstanding: According to the World Bank (World Development Indicators), the gross national income *per capita* ratios, measured at PPP[3] rates, is 7 between France and Morocco, 6 between Italy and Egypt and 4 between Spain and Algeria. On average, the per capita income in the EU is about ten times that of the SMCs.

1.Euro-Mediterranean agricultural trade

Asymmetries are not just a matter of income differences. They also have to do with varying political systems, social structures, demographic trends, and trade. Thus, the EU is a more important trading partner for the SMCs than these countries are for the EU[4]. This is clear when agricultural trade is considered. While, in 2000, the EU was the destination for 47,2 per cent of SMCs' agricultural exports and the source of almost 37,7 per cent of SMCs agricultural imports, only 10,2 per cent of the total extra-EU agricultural exports went to SMCs and only 7,1 per cent of the total extra-EU agricultural imports originated in SMCs.

From the political point of view, regional integration has the advantage of lowering the costs of negotiating international trade agreements, compared to multilateral negotiations. The EU is, by far, the SMCs' main trading partner. This gives political significance to a trade negotiation based on the bilateral exchanges with the EU. However, that is not the case for the EU. Trade asymmetry provides part of the explanation as to why Mediterranean partners are not at the top of the EU agenda as far as trade negotiations are concerned.

SMCs are highly dependent on their trade with the EU, although trade dependence varies throughout the countries in the region. Trade intensity indexes[5], in Figure 1, reflect the varying nature of interests that the different Mediterranean countries could show with respect to their integration strategy with the EU. On the one hand, high intensity indexes might reflect an interest in maintaining strong bilateral trade relations with the EU. On the other, low intensity indexes might suggest that the association with the EU could become an opportunity to diversify exports towards high value markets. Some SMCs, such as Morocco, Algeria, Tunisia and Israel present relatively high trade intensities. The Maghreb area appears to be, compared to the Middle East and Turkey, more dependent on EU agricultural markets.

Trade dependence is not only a consequence of the strong historical and trading links between the SMCs and the EU, but also due to the lack of South-South integration [10]. While intra-regional integration has received a recent push under the Agadir process and other regional initiatives, intra-regional trade among SMCs remains quite

---

[3] Purchasing Power Parity.
[4] Trade asymmetry in the Mediterranean region is considered, from the view point of overall trade, by Handoussa and Reiffers [7].
[5] Trade intensity, or degree of specialization of a given country's exports on the EU market, is measured by Iij = (Xij/Mj)/(Xi/Mw), being Xij the exports from country i to the EU; Xi the total country i's exports; Mj total EU imports and Mw total world trade. If Iij >1, country i's trade intensity with the EU is higher than what could be theoretically expected if country i accounted for a similar share of EU imports (Xij/Mj) as country i's share in world imports (Xi/Mw). The intensity measure can be expected, other things being the same, to have higher values for the countries and products which enjoy concessions in the EU.

marginal. Although wider EU market access has been claimed by the SMCs for a long time, breaking the strong EU-Mediterranean bilateral link (through the development of a regional market) remains a challenge for SMCs.

Figure 1. Trade intensity of agricultural exports to the EU (2000)

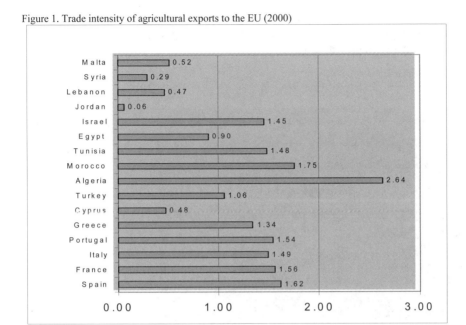

Sources: Comext for Euro-Mediterranean trade, MEDAGRI for total trade; author's calculations.

The bilateral agricultural trade balance between the EU and SMCs remains clearly favorable to the EU. In the period 1998-2000, the average value of EU agricultural exports to SMCs was 1.3 billion Euros larger than the average value of EU agricultural imports from SMCs. While the full opening of the EU market still represents an issue of the Barcelona process, the SMCs currently supply a larger market for EU agricultural exports than the EU for SMCs'. The overall positive balance for the EU is significant for cereals, dairy products, sugar, and meats and it is worth mentioning that this balance has significantly increased for the two first products between 1996-1998 and 1998-2000 (figure 2).

Trade balance is only positive for SMCs for fresh fruits, fresh vegetables and preparations of fruit and vegetables (see Figure 2). Again, the situation is not the same for all the SMCs. The region could be subdivided into two subgroups, at least regarding their bilateral trade patterns with the EU. The first group represented by the major exporters in the region, Turkey, Israel, Morocco and Tunisia, shows a positive and improving bilateral trade balance with the EU, while the second group, formed by Egypt, Algeria, Libya, Lebanon, Syria and Malta, shows a negative trade balance.

Agricultural trade in the EU is, to a great extent, of an intra-regional nature. In 2000, over 70 per cent of the total (intra-EU + extra-EU) imports originated in EU countries. Moreover, a significant part of EU imports is supplied by Southern European countries. Thus, in 1998-2000, Southern European countries[6] accounted for 12.8 percent of the EU total agricultural imports while SMCs only accounted for 2.2 per cent (percentages calculated on average triennial values). In fact, between the periods 1996-1998 and 1998-2000, EU agricultural imports from Southern Europe have shown to be more dynamic than EU agricultural imports from SMCs. Consequently, the performance of SMCs in the EU agricultural market appears to be relatively weak compared to that of Southern European countries.

Figure 2. EU trade balance with SMCs

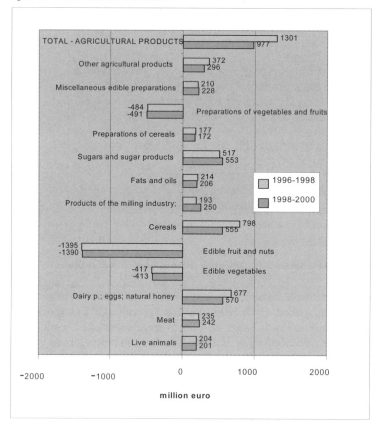

It is true that some of the existing constraints for SMCs' exports are supply-related; and that it is unlikely that fully open market access for SMCs will significantly boost their exports to the EU, at least in the short term. Quality specifications and high marketing costs (including logistics, post-harvest operations, transport, etc.) still hinder their export competitiveness. Only a few countries, not necessarily those most endowed with favorable climatic conditions and abundant labor, are able to export the quality products demanded by high-income consumers. However, market access becomes a

---

[6] Spain, Greece, Portugal and Italy.

necessary condition for improving export performance. Otherwise, efforts to improve export competitiveness in the South would not be worthwhile.

2.Agriculture and the Euro-Mediterranean Association

Why is agriculture relevant in the context of the Barcelona Process? One of the current problems of the Euro-Mediterranean Associations rests on the failure to make progress with the liberalisation of bilateral agricultural trade. The EMFTA will surely involve reciprocal concessions in industrial trade. However, the EU has not offered significant new concessions to the SMCs in terms of market access for their agricultural exports, and these continue to be limited to "traditional flows". Therefore, we see here a sort of asymmetric reciprocity. It is true that agriculture is not the only pending matter of the Barcelona process. Other sensitive issues include fiscal losses; rules of origin, migration and the remaining restrictions on the trade of textiles. Solutions are being found for all these issues, yet not always the best. But most of these solutions are worked out at a regional level. In a sense, the Barcelona process should play the role of producing global public goods, through regional institutions.

In contrast, agriculture has frequently received an *ad hoc* treatment in Euro-Mediterranean negotiations and rural development is far from being considered as a global public good in the region. This is in spite of the official messages (Agenda 2000 and CAP Mid-Term Review) that place rural development at the top of the EU priorities in the CAP reform process. EU trade preferences are negotiated on a bilateral basis and tend to be more generous for products and countries where the EU does not anticipate a real danger of competition. As far as agriculture is concerned, trade liberalisation in the Mediterranean regions appears as the result of a continuous negotiating process, where the first countries to sign an Association Agreement with the EU are also the first ones to benefit from an improvement in the agricultural provisions under further revisions. And this results in significant distortions between the Mediterranean partners. Euro-Mediterranean trade rules are an expression of a form of nepotism fashioned to political interests.

In fact, it is still early to have an *ex post* assessment of the Barcelona process. This is because there have been significant delays in the negotiation of the Association Agreements. Agreements are still waiting ratification for Egypt, Algeria, and Lebanon. And Syria is still negotiating. Tunisia is perhaps the partner where some conclusions on actual implementation can be drawn, but only on a preliminary basis. Therefore, discussion on advantages, costs and choices related to the implementation of the agreements is still needed.

Four main results can be mentioned from recent studies focused on agricultural trade in the region.

The first is that agricultural trade in the Mediterranean region is less free than what many people believe, especially among EU farming organisations. Studies in the second half of the nineties [7,9] were useful in explaining why the border measures applied by the EU on horticultural trade have a lot in common with the "pre-Uruguay Round World". The management of border measures usually involves "red tape" that reduces transparency, and this effect is difficult to grasp by nominal rates of protection and similar indicators. With the remaining constraints (entry prices, exchange of letters, tariff-rate quotas, anti-dumping actions), tariff-preferences produce only small static benefits. For some horticultural products, seasonal market variations supply some "windows" in the EU markets and they may give an impression of openness for horticultural imports. However, to give an example, in December when Moroccan tomato exports challenged the EU market, EU producers still accounted for more than the 80 percent of the EU (intra + extra) trade in this product.

A second result is that agricultural trade liberalisation will not produce substantial losses for European agriculture, compared to the potential benefits for the SMCs. Under the assumption that quantitative restrictions do not constrain exports, calculations performed by the FEMISE network for a group of 4 Mediterranean exporters (Egypt, Morocco, Tunisia and Turkey) suggest that free market access for these countries would increase their exports of sensitive products (fruit and vegetables, flowers and olive oil) to the EU by an equivalent amount of 11% of total intra-EU trade of these products [8]. This change does not seem to be dramatic for the EU as a whole over a gradual transition, although the local impact on Southern European exporters would require further examination.

A third result is that any unilateral liberalisation of agricultural imports by SMCs without reciprocal concessions from the EU would prevent SMCs from obtaining net benefits from the Euro-Mediterranean partnership. This has been illustrated by quantitative research based on Computable General Equilibrium (CGE) modelling, sponsored by OECD and the World Bank [2]. Differences in land productivity between the EU and SMCs suggest that a drastic agricultural liberalisation in the SMCs would significantly increase rural poverty. Considering that the Association Agreements, in their current shape, largely exclude agriculture from the liberalisation provisions, most of the burden on SMCs will rest on manufacturing. However, quantitative simulations fail to take sufficiently into account the short-term adjustment costs and social impact of trade creation. Some further assessment has to be done on the potential distribution of the modest welfare gains.

Most of the simulations referring to the Association Agreements show a fall in the relative price of manufacturing against agricultural commodities. This would suggest a distributive impact from urban to rural households in the South, but the models are not informative of the relative distribution of these gains among rural income groups. A fourth result is that Foreign Direct Investment in the region is constrained by a "hub and spoke" system where the EU acts as the hub and the partners as the spokes. The attractiveness of the SMCs for FDI has declined during the nineties, diverting FDI to the EU candidates. The region's share in global FDI directed to developing countries dropped from 6 percent in 1990 to 3.7% in 2000. "South-South" integration may be quoted here as an effective way of breaking this vertical system.

3.Prospects for the CAP reform and Euro-Mediterranean integration.

Apparently, the game of Mediterranean integration has few "win-win" solutions. Not including agriculture in the EMFTA would mean for SMCs that the uncertain benefits would be a promise arising from long-term economic reforms. On the other hand, Southern European farmers fear Mediterranean competition and find it difficult to understand why they receive less support than Northern European farmers do.

Reforming the CAP might help to change the picture. But, is this possible? This depends on: (i) The prospects for agricultural markets; (ii) The political viability of reform proposals.

As regards the *market outlook*, any projection is hampered by the same uncertainties created by agricultural policies in industrial economies. One example is given by the new US farm bill which, through higher loan rates and target prices for US cereals, creates a downward pressure on the level of world prices. Within the EU, the implementation of the Agenda 2000 has contributed to correct some of the historical imbalances of the CAP. According to the DG-Agri projections, wheat prices received by EU producers declined by 29 %, in nominal terms, over the 1992-2000 period as the result of the MacSharry reform and before the implementation of Agenda 2000. After allowing for annual variations, beef prices declined by 14 % during the same period, while pork prices (which benefited from the drop in feed costs due to the cereal reform) also declined by 16 %. The EU is currently able to export both wheat and barley without export refunds, and is expected to continue to do so in the near future. In any of the possible future scenarios, the EU would continue to be a significant net exporter of cereals, beef and dairy products.

However, prospects for EU agriculture reflect the need for further reforms, as recently argued in the Mid-Term Review (MTR) proposals submitted by the European Commission in July 2002 [3] leading to a final settlement by the Council in July 2003[7].

With regard to the *political viability* of the reform process, the CAP experience does not suggest that the changes will speed up in the next few years. Thus, it is unlikely that dramatic CAP changes will take place over the Agenda 2000 time span (until 2006).

The outlook for the coming CAP is that reforms will be inevitable but gradual. The reform process will be *inevitable* due to the two following reasons. First, the standard protectionist CAP no longer supplies what the European society demands. There is increasing social pressure for transforming the CAP into a policy targeted at environmental concerns and rural development.

Secondly, the dynamics of globalisation will oblige the CAP to move to a pattern of agricultural support that makes the CAP consistent with trade liberalisation. International developments include the EU Enlargement, the WTO negotiations, the EBA initiative[8], and the bilateral agreements, including the EMAs. Market opening foreseen in the free trade agreements will have important implications for European markets, the CAP regulations and the budget [5]. For example, sugar and rice regimes will have to be reformed by 2008 when the EBA agreement is fully implemented.

The reform will be *gradual* because the European society will not accept any scenario including the "radical" elimination of the CAP. Moreover, it seems unquestionable that a CAP reform will have distributive effects, among groups of farmers, territories and the EU Member States. This makes CAP reform even more difficult. Some groups of farmers, in particular the larger ones, are still reluctant about

---

[7] See http://europa.eu.int/comm/agriculture/mtr/index_en.htm.
[8] See details about the "Everything But Arms" in http://europa.eu.int/comm/trade/pdf.eba_ias.pdf.

"*modulation*", that is to say, to the transfer of direct aids to the CAP's "*second pillar*" (rural development)[9].

However, the CAP will probably be subjected to substantial reforms in the next decades, and the MTR will probably not be the last. CAP reforms will contribute to enhance the role that the EU already plays in granting access to developing countries. In particular, for SMCs, it will help to give significance to the Barcelona Process, as the EU will have to liberalize its agricultural sector. Thus, the CAP of future years will face the challenge of making support to EU rural areas consistent with rural development in developing countries. Consequently, the question now is not if there will be an opening up of EU agricultural markets in the coming years, but *how* and *when* this will happen.

Although EU market access for olive oil and horticultural products remains restricted, Southern European farmers blame the CAP for granting much more direct support to Northern European producers. Horticultural products, which account for 16 per cent of final agricultural production, receive only 3.5 per cent of the total CAP expenditure. Cereals, by contrast, make up 12 per cent of the final agricultural production and receive almost 40 per cent of the total budget. This overall imbalance has been a constant of the CAP, which has not been corrected by the Agenda 2000 package and by the MTR.

The CAP should aim at breaking the 'North-South' conflict of interests, inside the EU. An acceleration of the CAP reform after 2013 (period for implementing the MTR package) will enhance rural development in a compatible way with freer trade. Southern European farmers might be in favour of CAP reform if the reform had the indirect effect of reallocating agricultural support on a new basis. A CAP model based on rural development could devise support targeted to the contribution of agriculture to employment, added value, quality, environment and other non-product specific variables. This could eventually lead to re-balancing the support between the North and the South of the EU, although this development would not be immediate. In fact, the MTR [3] has included the stabilisation (and further decoupling) of CAP direct payments depending on farmers' historical receipts. However, as CAP support becomes more transparent, this will result in European society understanding the need for a new approach to agricultural policies in all the EU territory, in order to concentrate the funds in the less favored areas, which are precisely those where the risk of erosion and depopulation is higher.

To summarize, further adjustments will have to be made to the CAP. Firstly, for the sake of equity and coherence. Secondly, because the CAP financial framework, even after the MTR, remains uncertain given the international trade liberalisation under the Doha Round and the enlargement of the EU. Moreover, further adjustments in the CAP towards rural development policies may favor Mediterranean farmers if the new support measures are centred less on production references and more on territorial and human factors like employment on farms, and the contribution of agriculture to preserving the environment and the rural landscape.

---

[9] The Mid Term Review deal reached by the EU in June 2003 established a modulation rate of 5 per cent on direct payments received by farms with more than € 5 000 direct payments a year. According to the Commission this will result in additional rural development funds of € 1.2 billion a year.

4.Desirable developments.

Four developments might contribute to approximate opposing regarding about an eventual end to the agricultural exception in the Euro-Mediterranean partnership [4].

The first is that EU enlargement will probably expand the domestic EU market for horticultural products. Between 1997 and 2000, EU horticultural exports to Central and Eastern Europe have increased at an average annual rate of 10 percent for fresh fruit and 7 percent for fresh vegetables. Economic growth in Central and Eastern Europe would boost this trend after the enlargement. With these prospects, an enlarged EU market might reduce the objections of those who today are opposed to wider market access of the SMCs.

A second desirable development would be a substantial CAP reform towards rural development. The present set-up of Euro-Mediterranean relations is, in fact, a direct outcome of the current CAP. Further reforms could make rural development compatible with trade liberalisation. This development would possibly require compensatory policies in specific areas along both Mediterranean shores. The idea refers to the interest of a "Mediterranean agricultural pact" that would compensate potential losers of the FTA in rural areas in the EU countries as well as in the South.

A third development would be the increase in intra-industrial flows and diversification of trade flows. Evidence from Portugal and Spain suggests that European integration has brought about a rapid increase of intra-industrial trade in agricultural products. Moreover, intra-Arab integration could help to increase intra-industrial flows in non-traditional goods such as fruit and vegetables and processed agricultural products.

The fourth development would be the liberalisation of services, which could benefit both sides of the Mediterranean. The Association Agreements do not address explicit commitments on this matter, although it is implicit that economic reform and WTO negotiations will improve the regulatory environment to encourage FDI and competition. Trade liberalisation may have positive effects on the whole agro-food system, including logistics and transport in those areas like Almeria and Valencia, most affected by competition.

5.Concluding remarks

Agricultural trade continues to receive specific treatment in Mediterranean integration. Full consideration of agriculture in the Barcelona process is correlated with the CAP reform. However, there are two areas of research where further efforts are needed.

First, quantitative results are usually too general and of little use to assess local impacts, which are also sector-specific. Little research has been done in clarifying links between trade liberalisation and poverty alleviation in the South. One complication comes from the lower integration of the SMCs in the world economy, in comparison with other developing regions. Not many lessons from real case studies can be extracted from past experiences of trade reform in these countries.

Second, research is relatively poor with regard to institutional design, governance and policies dealing with globalisation. Benefits from the Euro-Mediterranean partnership do not seem to come automatically. Standard literature points out the convenience of domestic policies anchoring liberalisation as well as creating favorable incentives for investment. However, it is striking that much of the responsibility is put onto the SMCs and not on the need for a substantial CAP reform beyond the Agenda 2000 and the MTR.

References

1. CIHEAM (2001) Développement et Politiques Agro-alimentaires dans la Région Méditerranéenne, *CIHEAM Rapport Annuel*, Paris.
2. Dessus, S., Devlin, J. and Safadi, R. (eds.) (2001) *Towards Arab and Euro-Med Regional Integration*, Development Centre of the Organisation for Economic Co-Operation and Development Economic Research Forum for the Arab Countries, Iran And Turkey, The World Bank, Development Centre Seminars, Paris.
3. EU Commission (2002) Mid term review of the Common Agricultural Policy. Com(2002) 394 final.
4. Garcia-Alvarez-Coque, J.M. (2002) Agricultural trade and the Barcelona Process. Is full liberalisation possible?, *European Review of Agricultural Economics 29, 3, 399 – 422.*
5. Garcia-Azcarate, T., Mastrostefano, M. (2002) Mediterranean Integration and the Future of the CAP, X European Congress of Agricultural economists. Plenary session on Mediterranean integration, Zaragoza, August 30.
6. Grethe, H. and Tangermann, S. (2000) EU trade preferences for agricultural exports from the Mediterranean Basin: Evolution and Outlook. Paper prepared for the Seminar on "Mediterranean Agriculture within the Context of European Expansion" organised by EUROSTAT and the Spanish Ministry of Agriculture, Valencia, 8–10 November 2000.
7. Handoussa, H, and Reiffers, J-L. (2002) The FEMISE Report on the Evolution of the Structure of Trade and Investment Between the European Union and its Mediterranean Partners, The Euro-Mediterranean Forum of Economic Institutes, FEMISE Network, http://www.femise.org.
8. Lorca, A. (2000) L'impact de la libéralisation commerciale Euro-Méditerranéenne dans les échanges agricoles, *FEMISE Research Programme. Universidad Autónoma de Madrid.* http://www.femise.org/PDF/A_Corrons_09_00.pdf (retrieved December 2001).
9. Swinbank, R. and Ritson, C. (1995) The impact of the GATT Agreement on EU fruit and vegetable policy, *Food Policy, 20, 339-357.*
10. Zarrouk, J. E. (2001) Linkages Between Euro-Mediterranean and Arab Free Trade Agreements, in Dessus, S. Devlin, J. and Safadi, R. (eds.) (2001) *Towards Arab and Euro-Med Regional Integration*, Development Centre of the Organisation for Economic Co-Operation and Development Economic Research Forum for the Arab Countries, Iran and Turkey, The World Bank, Development Centre Seminars, Paris, pp 225-254.

Section VII

Urbanisation

Chapter 20

Urban change in Cairo

Hassan Abdelhamid
Ain Shams University, Cairo, Egypt

Cairo constitutes the urban center of Egypt. It has experienced numerous transformations, which have shaped its social and spatial structure in the course of its long history. A complex urban structure exists, which bears the imprint of different phases of development. Tracing the major developments in Egyptian urbanisation over the last 20 years, one can observe two distinct trends. On the one hand, there has been a stabilisation and diffusion of urbanisation, on the other a stabilisation of rural-urban migration. It is a double movement of deconcentration at both metropolitan and national levels. In fact, Cairo has ceased to attract a large proportion of the migratory population. Greater Cairo now constitutes 17% of the total population, the same proportion as in 1966. This is a phenomenon of out-migration from Cairo and covers two aspects; the informal cities for poor people (informal agglomeration surrounding metropolitan areas- *ashwaïyyat*) and the private cities for rich people (such as *al-Rehab, New Cairo, Mena Garden City, Dream Land, Utopia, Beverly Hills*...etc) This trend points to the transition of Cairo from the European model of a compact city to the American pattern of vast diffused spatial development. During the last 20 years, the new population map of Greater Cairo has shown the impact due to the relation between urbanisation and the government's ideology. From the analysis of the Egyptian urban context, this paper examines the factors of the emergence of Cairo's new urbanity and attempts to answer the following three questions: What are the links between informal agglomerations and the emergence of the private cities? What are the links between the existence of this abnormal phenomenon (informal cities and private cities) and the Egyptian concept of urbanisation? What are the prospects for the future urbanisation of Greater Cairo in the coming years ?

1. The demographic changes during the last 20 years in Egypt

The demographic indicators show a remarkable decline in population growth, from 3% in the mid-eighties to 2% by the end of the nineties (Ministry of Information, 2000). Moreover, the demographic indicators demonstrate the following facts:

- The Population

The total population over 19 years increased from 43.914.000 in 1981/82 to 65.156.000 in 1999/2000, with an increase of 21.242.000 at an annual growth rate of 2.22%. The total number of families in 1999 were 13.300.000 against 9.700.000 in 1986. Average life expectancy rose from 58.1 years in 1981/82 to 66.6 years in 1998/999 for males and from 60.6 years in 1981/82 to 70.5 years in 1998/99 for females.

- Birthrate, mortality and natural increase

The rate of birth declined from 38 per thousand according to the 1986 census to 26.4 per thousand in 1999/2000. As a result of the effective ongoing health programmes

A. Marquina (ed.), Environmental Challenges in the Mediterranean 2000-2050, 331–344.
© 2004 *Kluwer Academic Publishers. Printed in the Netherlands.*

and media awareness campaigns, the mortality rate dropped to 6.4 per thousand while natural growth fell to 20 per thousand in 1999/2000.

- Workforce

Egypt's workforce increased from 11.092.000 in 1981/82 to 18.818.000 in 1999/2000 with an annual increase of 2.98 % . The workforce – population ratio during that period rose from 26.4 % to 29.8 %. This indicates the society's ability to benefit from the population increase, turning it into a catalyst for development. The employment workforce ratio showed 92.6 % in 1999/2000 against 90.8 % in 1991/92.

- Housing

Investments allocated to the housing sector are estimated to be about 66.6 billion (E.P) used in building 2.823.400 housing units, including 684.600 units carried out by the private sector, accounting for 95 % of the total units built, broken down as follows:

  - 1.766.000 urban economy-level housing units,
  - 634.800 urban medium-level housing units,
  - 230.600 urban upper-medium-level housing units,
  - 192.000 rural housing units in other rural areas built between 1997/98 and 1999/2000.

- Reconstruction areas

These include Sinaï, the Red Sea, the northern coast, New Valley and Greater Cairo pivots. These areas were provided with infrastructure facilities including potable water, sanitary drainage, electricity, transport and communication, in addition to other public services.

- Achievements in construction areas:
  - 3.100 Km of water lines and network were extended, 36 water treatment units and stations and 35 water desalination units were built and 532 wells were drilled.
  - 505.9 Km of sanitary drainage network were extended and 3 treatments units and stations built.
  - 1,120 Km of electric cables and network and wireless lines were extended, 39 electric power generators plants and transformers erected.
  - 3,731.4 Km of main and subsidiary roads were paved and five fishing harbours in addition to five ferry boats were built.
  - 26,200 housing units were also built; 44 buildings and construction training centres were established, providing courses for 251.600 trainees.

- New Cities and Urban Communities

These communities are intended to develop about 600.000 feddans of desert areas in order to establish population-attracting areas, accommodating about seven million, thus promoting industrial development. Number of cities established over the period 1981/2000 reached 19 distributed as follows:

  - 5 cities within Greater Cairo district
  - 5 cities around the ring area
  - 5 cities in Lower Egypt, east and west Delta
  - 4 cities in Upper Egypt,
  - And initial phases for nine other cities have been completed.

2.Demographic profile and demographic shift of urbanisation

Table 1.Demographic Profile.

	Population (000)			Annual population growth rate		Crude birth rate	Crude death rate	Contraceptive Privative rate	% Net lifetime internal migration	Demographic dependency ratio
	1960	1986	1996	60/86	86/96	1998	1998	1998	1996	1998
Cairo	3349	6069	6801	2.3	1.1	23.1	6.8		3.2	52.6
Alexandria	1516	2927	3339	2.5	1.3	23.7	6.7		6.4	54.5
Port-saïd	245	401	472	1.9	1.6	22.3	6.1		10.2	53.2
Suez	204	328	418	1.8	2.5	25.4	5.5		17.4	63.2
Urban govs	**5314**	**9725**	**11030**	**2.3**	**1.3**	**23.3**	**6.7**	**62.1**		**53.6**
Damietta	388	740	914	2.5	2.1	28.3	5.7		-1.4	62.5
Dakahlia	2015	3484	4224	2.1	1.9	25.7	6.2		-5.6	66.0
Sharkia	1820	3414	4281	2.4	2.3	27.6	6.1		-4.2	72.2
Kalyoubia	988	2516	3301	3.6	2.8	25.0	5.3		11.9	67.6
Kafr-el-sheikh	973	1809	2224	2.4	2.1	23.3	5.9		-1.6	69.4
Gharbia	1715	2885	3406	2.0	1.7	24.2	6.1		-4.2	64.9
Menoufia	1348	2221	2760	1.9	2.2	26.2	6.4		-10.4	70.3
Behera	1686	3249	3994	2.5	2.1	25.9	5.6		-0.8	70.3
Ismailia	284	545	715	2.5	2.8	30.3	6.1		13.5	66.2
Lower-Egypt	**11217**	**20863**	**25819**	**2.4**	**2.2**	**25.8**	**5.9**	**59.2**		**68.3**
urban	**2432**	**5750**	**7117**	**3.3**	**2.2**	--		**62.2**		
Rural	**8785**	**15113**	**18702**	**2.1**	**2.2**	--		**58.1**		
Giza	1336	3726	4784	4.0	2.5	26.6	6.2		16.6	66.5
Benisuef	860	1449	1859	2.0	2.5	33.5	7.8		-4.3	88.9
Fayoum	839	1551	1990	2.4	2.5	31.8	6.3		-5.3	88.7
Menia	1560	2645	3310	2.0	2.3	34.7	7.8		-3.5	86.5
Assyout	1330	2216	2802	2.0	2.4	34.5	7.7		-8.1	86.9
Suhag	1579	2447	3123	1.7	2.5	33.8	7.4		-9.3	86.1
Quena	1352	2259	2803	2.0	2.2	32.4	8.2		-7.1	84.6
Aswan	385	809	974	2.9	1.9	22.6	6.8		-0.9	72.8
Upper-Egypt	**9240**	**17102**	**21646**	**2.4**	**2.4**	**31.6**	**7.2**	**36.5**		**81.2**
Urban	**1905**	**5415**	**6659**	**4.1**	**2.1**	--		**50.8**		
Rural	**7335**	**11687**	**14987**	**1.8**	**2.5**	--		**29.9**		
Red sea	25	90	157	4.9	5.7	25.7	3.4		31.6	53.5
New- valley	34	113	142	4.7	2.3	27.3	4.7		4.6	68.9
Matrouh										
North- Sinai	104	161	212	1.7	2.8	36.7	4.1		13.8	74.6
South- Sinai	50	171	252	5.4	4.0	31.2	4.6		10.7	75.5
	-	29	55	--	6.6	24.7	4.1		34.4	54.4
Frontiers gov	**213**	**564**	**818**	**3.8**	**3.6**	**30.4**	**4.2**			**68.0**
Urban	**213**	**326**	**480**	**1.6**	**3.9**	--				

Rural	--	238	338	--	3.6	--			
Egypt	**25984**	**48254**	**59313**	**2.4**	**2.1**	**27.5**	**6.5**	**51.8**	**69.7**
Urban	**9864**	**21215**	**25286**	**3.0**	**1.8**			**59.3**	
Rural	**16120**	**27039**	**34027**	**2.0**	**2.3**			**45.6**	

Sources: Arab Republic of Egypt (2000) *Egypt Human Development Report 1998/99,* Institute of National Planning, Cairo.

Table 2. Urbanisation

	Urban population (as % of total)				Urban population annual growth rate (%)			Population of largest city (as % of total urban)			Households/ electricity %
	1960	**1976**	**1986**	**1996**	**60/76**	**76/86**	**86/96**	**1960**	**1976**	**1986**	**1999**
Cairo	100	100	100	100	2.6	1.8	1.1	100	100	100	99.9
Alexandria											
Port-saïd	100	100	100	100	2.7	2.4	1.3	100	100	100	99.4
Suez	100	100	100	100	0.4	4.3	1.6	100	100	100	99.8
	100	100	100	100	-0.3	5.4	2.5	100	100	100	99.5
Urban govs	**100**	**100**	**100**	**100**	**2.4**	**2.2**	**1.3**	**63.0**	**64.6**	**62.4**	**99.7**
Damietta	24.9	24.8	25.2	27.4	27.4	2.7	3.0	74.2	65.5	47.8	100
Dakahlia											
Sharkia	18.1	24.0	26.2	27.8	27.8	3.3	2.6	41.5	39.5	34.6	99.3
Kalyoubia	16.2	20.0	21.1	22.5	22.5	3.1	3.0	42.1	38.2	34.0	99.5
Kafr-el-Sheikh	25.4	40.8	43.8	40.6	40.6	4.9	2.0	40.2	57.5	64.7	100
Gharbia	17.0	20.7	22.8	22.9	22.9	3.5	2.2	23.9	26.6	25.0	100
Menoufia											
Behera	28.2	33.3	32.7	31.1	31.1	2.1	1.2	38.0	38.2	38.2	99.7
Ismailia	13.6	19.7	20.1	19.9	19.9	2.9	2.1	29.9	30.5	29.7	99.4
	18.2	24.1	23.4	22.8	22.8	2.5	1.8	41.2	28.7	25.5	98.9
	100.0	49.2	48.8	50.3	50.3	-3.0	3.1	79.0	83.3	80	100
Lower-Egypt urban Rural	**21.7**	**26.4**	**27.6**	**27.6**	**3.8**	**3.2**	**2.2**	**8.0**	**9.4**	**12.4**	**99.9**
Giza	32.4	57.0	57.5	54.1	54.1	4.5	1.9	57.8	89.3	88.8	99.4
Benisuef	21.4	24.9	25.1	23.5	23.5	2.8	1.9	42.9	42.7	41.9	98.3
Fayoum	19.3	24.1	23.2	22.5	22.5	2.7	2.2	63.1	60.6	59.2	100
Menia	17.2	21	20.8	19.4	19.4	2.5	1.6	35.2	34.0	32.6	96.7
Assyout	21.8	27.7	27.9	27.3	27.3	2.8	2.2	44.0	45.4	44.2	98.5
Suhag	18.1	21.3	22.0	21.7	21.7	2.7	2.4	21.7	25.1	24.8	98.4
Quena	13.7	22.9	23.4	24.4	24.4	3.0	2.6	31.1	23.9	23.9	99.5
Aswan	25.4	37.9	39.6	42.6	42.6	3.2	2.6	49.4	61.7	59.8	99.1
Upper-Egypt Urban Rural	**20.6**	**30.5**	**31.7**	**30.8**	**30.8**	**3.4**	**2.1**	**13.1**	**31.8**	**34.5**	**98.9**

Red sea	100	87.4	85.5	74.4	74.4	4.7	4.4	25.1	25.9	30.8	99.3
New-valley	100	40.8	44.5	48.3	48.3	3.8	3.1	36.4	76.6	76.4	97.8
Matrouh	100	46.0	55.5	55.5	55.5	4.7	3.7	29.6	52.4	52.4	100
North-Sinai	100	100	62.6	59.1	59.1	28.2	3.5	58.9	64.0	64.0	100
South-Sinai			39.5	50	50.0		9.1		38.6	38.6	96.8
Frontiers gov Urban Rural	**100**	**55.0**	**57.8**	**58.7**	**58.7**	**7.9**	**4.0**	**13.8**	**19.1**	**20.8**	**99.3**
Egypt Urban Rural	**38.0**	**43.8**	**44.0**	**42.6**	**42.6**	**2.8**	**1.8**	**34.4**	**31.6**	**28.6**	**99.5**

Sources: Arab Republic of Egypt (2000) *Egypt Human Development Report 1998/99,* Institute of National Planning, Cairo.

Tracing the major developments in Egyptian urbanisation over the last 20 years, one can observe two distinct trends.

First, there has been a stabilisation and diffusion of urbanisation, and on the other hand, a stabilisation of rural-urban migration; in other words, Egypt is currently experiencing a double movement of deconcentration at both metropolitan and national levels.

Between 1960 and the year 2000 Egypt's population rose from 25.984.000 to 65.156.000 an increase equal to 39.172.000 (one million per year).

Interestingly, this high growth rate is associated with an end to urban polarisation. Contrary to the prevailing idea of a continuous rural-urban influx, the urbanisation process in Egypt has been both stabilized and diffused. The urban proportion of population declined from 44 % in 1986 to 42.6 % by 1996 [5].

This stabilisation is associated largely with the urbanisation of big villages and the rapid growth of small towns. This will be discussed below. The migration to these villages and small towns serves as an important stabilizing factor.

3.Demographic profile of Cairo

Table 3. Population of Greater Cairo Region, 1960-2000.

Governorate	1960	1986	1996	2000
Total Population (000s)				
Cairo	3.349	6.069	6.801	
Kalyoubia	988	2.516	3.301	
Giza	1.336	3.726	4.784	
GCR Total	5.673	12.311	14.886	
Urban Population (000s)				

Cairo	3.349	6.069	6.801
Kalyoubia	251	1.102	1.341
Giza	433	2.142	2.590
GCR Total	4.033	9.313	10.732

United Nations Defined Urban Agglomeration	**1960**	**1985**	**1995**	**2000**
Cairo/Shubra El Khemia	3811	8321	9707	10399

Sources: Institute of National Planning (1998), Arab Republic of Egypt (2000) *Egypt Human Development Report 1998/99,* Institute of National Planning, Cairo, Arab Republic of Egypt (1999) Statistical Year Book 1992-1998, Central Agency for Public Mobilisation and Statistics (CAPMAS), Cairo, United Nations (2002), World Urbanisation Prospects: The 2001 Revision.

Table 4. Primacy Indicators (Greater Cairo Region)

	Egypt	Greater Cairo Region	GCR as % of Egypt
Total Population (1996)	59.312.914	14.886.335	25.1
Urban Population (1996)	25.286.335	10.731.614	42.4
GDP (LE Billion) (1996/97)*	255.4	85.0	33.3
Air Traffic (1996)			
Aircraft Movements	174.796	90.469	51.8
Air Passengers (000s)	13.492	7.900	58.6
Licensed Vehicles (Dec 1998)			
Private Cars	1225.479	758.057	61.9
Taxis	291.600	118.298	40.6
Buses (Public and Private)	33.029	14.356	43.5
Telephones (1995/96)*	2.959.706	1.331.543	45.0
Hospital Beds (1997)*	1.245.571	484.456	38.9
Industrial Labour Force (1996)*	3.619.591	1.403.625	38.8
University Students (1996/97)	1.034.539	501.711	48.5

Sources[1]: Institute of National Planning (1998), Arab Republic of Egypt (2000) *Egypt Human Development Report 1998/99,* Institute of National Planning, Cairo and Arab Republic of Egypt (1999) Statistical Year Book 1992-1998, Central Agency for Public Mobilisation and Statistics (CAPMAS), Cairo.

[1] Estimates based on published per population and per household ratios applied to 1996 Census population and household totals.

Table 5. Social and Economic Indicators, Greater Cairo Region[2]

Governorates of Greater Cairo Region

Index	Cairo	Kalyoubia	Giza	Egypt Total
Life expectancy at birth (years) - 1996	66.5	68.1	66.4	66.7
Adult literacy rate (%15+) - 1996	72.9	59.6	61.5	55.5
GDP per capita (LE) - 1996/97	7004	3299	5540	4306
Human development index – 1996	0.792	0.597	0.701	0.631
Households with access to piped water (%) - 1996	97.2	72.6	87.9	82.6
Households with access to sanitation (%) - 1996	92.0	41.9	59.7	45.1
Households with electricity (%) - 1996	98.6	95.6	98.3	95.1
Combined basic and secondary enrolment (%) - 1996/97	91.3	78.8	81.0	80.9
Infant mortality rate (per 1,000 live births)* - 1996	28.8	27	32.3	34.0
Female adult literacy rate (%15+) – 1996	65.6	46.6	50.5	43.5
Female life expectancy at birth (years) 1996	68.1	69.8	66.9	67.9
Maternal mortality rate (per 1,000 live births) – 1992	200	103	221	107
Under five mortality rate (per 1,000 live births)* - 1996	34.5	35.5	41.7	44.3
Physicians (per 10,000 people) - 1997	10.3	3.9	8.7	7.1
Hospital beds (per 10,000 people) – 1997	46	23	20	21
Telephones (per 1,000 households) - 1995/96	664	77	163	233
Cars (per 1,000 people) – 1993	73	11	36	21
Women in labour force (% 15+)	19.7	15.0	12.4	15.3
Percentage of labour force in: Agriculture - 1996	1.1	17.6	13.3	28.3
Percentage of labour force in: Industry - 1996	32.4	30.5	30.5	20.5
Percentage of labour force in: Services - 1996	66.5	51.9	56.2	51.2
Professional and technical staff (as % of labour force 15+) – 1996	31.6	20.5	24.4	21.2
Unemployment rate: Total (%) - 1996	7.3	7.3	5.3	8.9
Unemployment rate: Females (%) - 1996	11.2	15.9	7.9	20.3
Unemployment rate: Adults aged 15-29 (%) – 1996	19.0	15.9	12.9	20.1

Income share: Lowest 40% - 1995/96	20.6	25.2	22.1	21.9
Ratio of highest 20% to lowest 20%) - 1995/96	4.9	3.3	4.4	4.4
Poor persons (% of total) - 1995/96	10.8	28.3	12.0	22.9
Ultra poor persons (% of total) - 1995/96	2.9	8.8	2.6	7.4
Population (% urban) - 1996	100.0	40.6	54.1	42.6
Annual urban population growth rate (%) - 1986-96	1.1	2.0	1.9	1.8
Population (000s) - 1996	6.801	3.301	4.784	59.313
Annual population growth rate (%) - 1986-96	1.1	2.8	2.5	
Crude birth rate (per 1,000 population) - 1996	24.6	24.8	27.3	28.2
Crude death rate (per 1,000 population) - 1996	7.3	5.4	6.0	6.5
Net lifetime internal migration (as % of total population) - 1996	3.2	11.9	16.6	
Demographic dependency ratio (%) - 1996	52.6	67.6	66.5	69.7
Population density (per km2)	31.993	3.358	57	60

Sources: Institute of National Planning (1998), Arab Republic of Egypt (2000) *Egypt Human Development Report 1998/99,* Institute of National Planning, Cairo and Arab Republic of Egypt (1999) Statistical Year Book 1992-1998, Central Agency for Public Mobilisation and Statistics (CAPMAS), Cairo.

From these profiles, we can see that there is a stabilisation of rural migration to Greater Cairo, and the metropolitan areas are engaged in a structural movement of centrifugal redistribution from the core areas of Cairo to the peripheries. As the core area of Cairo loses population, new agglomerations rise around it. Over the past ten years, Cairo's central districts (for example West Bank, Dokki and Guiza), have progressively lost a large proportion of their inhabitants.

Starting from the 1960s, six *qism* or districts, had lost population by 1966. This number had increased to 17 by 1976, 18 by 1986, and had reached 22 by 1996. On the whole, central Cairo lost some 580, 000 inhabitants between 1986 and 1996 [5].

In short, since the 1970s, Egypt has been subject to the deconcentration of population at both metropolitan and national levels. Urbanisation has started to slow down throughout the country and the rural exodus appears to belong to the distant past. Already, in 1986 some 80 % of migrants recorded in the cities came not from the countryside but from other urban centers. Generally, the share of inter-provincial migrants, that is, people born outside a given province, decreased from 11 % in 1960 to 7.5 % in 1986. Thus, the permanent population movement paved the way for an increasingly circular migration pattern [5].

Greater Cairo is a typical mega-city [14], with 15 million inhabitants living in three municipalities (Cairo, Guiza and Kalyoubia). Within this relatively small but agglomerated area, various groups with immense income disparities live next to each other. The agglomeration is still growing but at the same pace as the Egyptian population as a whole.

However, Greater Cairo faces one of the highest population densities in the world (32.000 people/km2). In some neighbourhoods, more than 100,000 people live in one square kilometre-often in buildings no more than five or six storeys high. Located in an arid landscape, there is thus precious little space available which could function as green space, and there are almost no private gardens suitable for cultivation [14].

Although the conservation of agricultural land has long been a priority of Egyptian development policy, much of the critically needed arable land in Greater Cairo is being lost to urban development; half of which consists of illegal construction and the remainder is planned as new development in the desert. Although a housing crisis has been forecast by the international news media, it is estimated that Greater Cairo may have a surplus of some one million housing units.

Cairo has a serious air pollution situation due to motor vehicles (licensed vehicles in December 1998 were 758.057 in the Greater Cairo Region, 61.9 % of the total licensed vehicles all over the country) and industry. The city government frequently monitors air pollution to measure concentration of noxious gases. In addition, it utilises all forms of media to promote environmental conservation within the city, including celebrations like the annual Nile Inundation to encourage environmental awareness.

Thus, Greater Cairo has ceased to attract a large proportion of the migratory population. Greater Cairo now constitutes 17 % of the total population [1]. The prime reason for this is the apparent "saturation" of Greater Cairo to accommodate the low-income groups. The current urban conditions have caused many inhabitants to seek residence outside major urban centres. While Greater Cairo still provides opportunities for employment, the high price of land, population density and the shortage of affordable accommodation associated with the partially free-market cost of housing, force many newcomers as well as long term residents to think again about staying in the city [5].

In addition, the unaffordable prices for newly built formal housing exclude the low-income groups from the housing market. Thus, there remains no other option for young people, particularly those intending to start a family, but to seek housing on the informal market [4]. Hence, they venture out to join the "outsiders" who live in the large "*ashwaiyat*", those informal agglomerations surrounding metropolitan areas, some of which already accommodate groups of indigenous populations such as villagers or tribal people [9].

Many people living in these communities still depend on job opportunities offered in the metropolitan areas, to which they commute daily. However, their residential communities are more than just functional dormitories. Rather, they are the focus for the family, a network of friends and recreation as well as job location. In addition, these informal agglomerations perform a significant function in the national economy [5]. They accommodate cheaply paid labour, subsidised by low-cost housing and provide basic necessities such as land and low rents and food, in particular agricultural products. They offer people the possibility of maintaining strong family networks, security and protection. In 1996, on average, there were 1,600 people per square kilometre, and local units had an average of 4,500 inhabitants.

These informal agglomerations are perceived as "abnormal" places where, in modern conventional wisdom, the non-modern and thus non-urban people, who are the villagers, the traditionalists, the nonconformists and the unintegrated, live. It is indeed puzzling that over 20 % of all Egypt's and half of Greater Cairo's population, who reside in these areas, are seen as outsiders, living in abnormal conditions; most of those places were built without permits, the streets have no formal names, men wear the traditional "*galabia*", women sit and socialize in front of their homes and adults are largely active in the informal economy and considered as non-modern people.

But the question is what is a modern or a normal city where modern people or normal people live? We will find out the answer later on.

On the other hand, throughout the city, there has been a surge in private sector building, the pace of which is likely to outrun overall governmental control. Perhaps, the epitome of the reorientation towards the power of capitalism is the creation of cities like "*al rehab*", the first private city in the country.

Similar projects are planned by other private developers on the outskirts of Cairo and are facilitated by the creation of the Cairo ring road. Exclusive residential areas with names such as "*Golf city*", "*New Cairo*", "*Mena Garden city*" and "*Dream Land*", have been newly opened or are under construction. For 250.000 US$, a homeowner can purchase a modest villa in *Golf city* and live among 500 acres of artificial lakes and golf links [13].

This trend points to the transition of Cairo from the European model of a compact city such as London, to the American pattern of vast diffused spatial developement such as Los Angeles where identity, history, memory and symbolism are lost to the diversified sub-centres of the vast urban plain [5].

Many factors have led to the emergence of this new urbanity. Cairo's super-rich people are escaping from the high density, traffic congestion, air and noise pollution, and spatial contraints which are transforming even the upmarket districts. Casual observation would reveal how quickly the old, spacious villas in the *zamalk* and *maadi* suburbs of Cairo are being demolished and turned into high-rise buildings [7]. *Zamalek* and *maadi* are no longer the status symbols but, rather, these new private cities with their new financial sources (from lucrative private business), more efficient means of private transport and communication, and the new ring road around Cairo, all of which have enabled rich people to pursue this historical exodus; with this exodus, they can keep their distance from having to see the severe poverty and protect themselves from the political spectacle of *ashwaiyyat*. In addition to all these factors, the ineffective planning and the urban management have also had a part in the problem of Cairo's urbanisation.

4.The Egyptian Concept of Urbanisation:

4.1.A COMPLEX URBAN STRUCTURE

The large and complex urban structure existing nowadays in Cairo, bears the imprint of different phases of development. Actually, Cairo's urban, political and economic history has been divided into four periods: Islamic, Imperialist, Arab Socialist and the current one, labelled Transitory because of the ideological transition transforming the economic system from socialism to capitalism [12]. Successive and divergent political economic regimes have left their imprint on Cairo's landscape in greatly dissimilar patterns.

Through the analysis of Cairo's one thousand years of history [10] four major periods have been identified, as mentioned before. Each of these periods contains a dominant ideology stamped on the landscape and can still be located within modern Cairo.

In addition to this general spatial facets, distinct dominant features such as architectural style and specific examples of construction are identified as symbolic representations of the primary characteristics of each political economic regime.

The landscape represents a way whereby certain classes of people have recorded their lives and their world through their imagined relationship with nature and by which they have underlined and communicated their own social role and that of others with respect to external nature. The urban landscape can be viewed as a social product reflecting social changes, values and perceptions. In reading the urban landscape, it is important to recognize that the best way is likely to belong to those who hold power and therefore dictate the dominant urban ideology [12].

The transition from an Islamic political economy to one dominated by European imperialism started with the French expedition to Egypt in 1798. At this time, European values were quickly appropriated by the Egyptian elite who began incorporating them into their daily life.

By the middle of the nineteenth century, they were already establishing a new European section in Cairo, leading to a division of the city into two sections, the traditional one and the modern one. A new "European Cairo" had been created by 1882 when formal British rule replaced Ottoman control.

By creating a European Cairo, the actions of Mohamed Ali's family show an almost total rejection of traditional Islamic architectural values. Ismaïl's vision of modernisation was to turn Egypt into a piece of Europe, with Cairo as its capital.

In 1952, we had an independent Egypt with another political system, dominant Arab socialism. Under this new regime, urban policy experienced a fundamental shift towards large scale urban projects, designed to cope with the rapidly expanding Cairo and industrialisation, and there was a massive effort to "redesign the population map of Egypt".

Nasserism witnessed a marked reduction of European influence on the landscape of Cairo. The government became deeply involved in the large scale development of housing. There was a critical need for affordable habitations for the lower economic classes, many of whom had recently migrated to Cairo and were living in the large slum and squatter areas on the urban periphery. The government created a public company, and built thousands of apartments in the poor areas of Cairo, such as *zeinhom, helwan, imbaba* and *shubra el khima*. These monoliths came to dominate the landscape in many areas, including those of the new socialist era as well as the former Islamic or imperialist eras, where they tended to crowd out the urban morphology of previous periods. These buildings differed considerably from traditional housing in Egypt. Presently, they are in a pitiful state of degradation and dilapidation in terms of both physical structure and services.

After Nasser, Sadat began the process of reconnecting Egypt with the world economy by his policy, called "*infitah*", or the reopening of Egypt to the world economy. But, if Paris and Rome were favourite models for Ismaïl, Los Angeles and Houston were the favourites for Sadat who let loose private developers and speculators.

New luxury high-rise buildings mushroomed all over the city, replacing private villas through massive slum clearance.

From Sadat up to the present time Egypt has lived in a state of transition switching between socialism and capitalism. The most important change is the rapid and agressive integration of Egypt into the world economy [12].

4.2.OFFICIAL DEFINITION OF URBAN AND ITS RESULTS.

From this evolution of Cairo's landscape, the identification of dominant ideologies and their representation on the landscape is particularly important.

The concept of urbanisation depends on the dominant ideology. Actually, what is the concept of urbanisation? What is urban life?

The official definition of urban in Egypt is arbitrary and restrictive. It is an administrative definition. This definition in mere administrative terms (as *markaz*) conceals an important trend of urbanisation. Under this definition, urban has an administrative role. According to such a definition, there are only about 200 cities in Egypt, a clear underestimation. Indeed, if Egypt adopted the Indian definition of urban (communities with more than 5.000 inhabitants), around 80 % of Egyptians would be urbanites [5].

The official Egyptian definition may be functional for administrative purposes but it hides an alternative process of urbanisation, namely that which concerns mostly small towns and the struggling urban villages with a population of 10.000 or more. Perhaps, this pattern of unrecognized urbanism in Egypt is expressed by *Jamal Hamden* that urbanism "*al-umran*" begins in the village [8]. Actually, we do have urban villages or agro-towns which have begun to acquire urban characteristics.

This idea points to a shift from a universal state managed and planned urbanisation to a more private and spontaneous one. This "post-metropolisation" should be seen as a new trend in and a challenge to Egyptian political economy at the end of the twentieth century. The post-metropolisation signifies a diffusion of urbanisation over a vast area (through the growth and interlinking of agro-towns, urban villages and new industrial towns) outside, but close to, Greater Cairo, with the latter retaining its dominance.

It is the contradiction between massive urban diffusion on the one hand and economic and social polarisation on the other that characterises current Egyptian "post-metropolisation" [5]. The unplanned urbanisation in Egypt highlights not only a concentration of population but also the needs, concerns and possible urban-type conflicts which would directly involve the state. It is not therefore surprising that the state refuses to recognise these agglomerations as urban, since doing so would oblige them to make expensive urban provisions such as sewerage, paved roads and running water.

Indeed, the strict official definition of what constitutes an urban unit and the invention of the concept of "*ashwaiyyat*" as a political category tends to produce new spatial divisions which exclude many citizens from urban participation.

As we described before, *ashwaiyat* are perceived as abnormal, but more than 20 % of Egypt's total population and half of Greater Cairo's population live there. We have to admit that this kind of abnormality is not true because the fact is that the populations of informal settlements like *ashwaiyat* are involved in the complex urban economy and divsion of labour, and constitute a significant component of the diversified whole, which is the city. In the old sociological tradition, what defines urban is primarily the organic ensemble in a particular space, with a variety of lifestyles and economic activities, and those of the *ashwaiyyat* are significant components.

This tendency to see "informality" and the "*ashwaii* way of life" as producing outsiders has been highlighted by a shift in emphasis from public to private spatial development. This is exemplified by the new highly exclusive townships exhibiting global styles of urban planning.

This duality of peripheral informalisation on the one hand and planned exclusive suburbanisation on the other is a stark manifestation of the urban polarisation and social cleavage present in Egyptian society today. Indeed, Egyptian urbanism is characterised by closure or the surrounding-wall paradigm; it is not a shared space, rather, it produces outsiders [5].

5. Conclusions (Uncertain Future)

An article on mega-cities remarks that, four decades ago, cities such as Mexico City and Cairo were relatively attractive places to live in, with little traffic along their spacious, cleanly-swept boulevards [6]. Now that their populations have quadrupled and their quality of life has greatly degenerated (they are considered as the first and the second most polluted capitals in the world), their immigration rates have declined, although people continue to flock to smaller cities. This trend suggests that living conditions in Greater Cairo could eventually become intolerable. Given the various scenarios of decentralisation in response to a combination of market forces and governmental action, it becomes questionable whether future population growth projections of Cairo will in fact materialize. Most developing countries perceive the spatial distribution of their population and the resulting city patterns as unacceptable, and many governments have attempted to change such patterns by indirect national policies or explicit spatial development strategies.

If such trends continue, the problems of deconcentration at both national and regional levels may eventually give way to the reverse process and problems of deconcentration and dispersion. This is clearly contingent on active governmental efforts as well as the market processes that tend to spread development outside Greater Cairo.

The fact remains that no one can be certain exactly what Greater Cairo is like now, or how rapidly it is expanding or going to expand in the future. However, Greater Cairo will probably continue to grow rapidly in the foreseeable future. In the process, such a growth expands the city's influence and functions over a much wider region, including other cities and rural settlements and a frequently uncontrolled and unplanned periphery. This requires a redefinition of the nature and structure of "future Greater Cairo" as a mega-city [6].

The system of Greater Cairo now is an overload system. Greater Cairo's estimated 15 million people are currently being served by an infrastructure planned for 4 million at the most. In Greater Cairo, toxic fumes make it difficult to breathe the air along certain sections of the Autostrada because of burning rubbish. Greater Cairo's housing needs for a 20-year period were estimated at 3.6 million units. Unacceptable densities and unmanageable patterns of growth often result in overcrowded housing, insanitary conditions, extremely long and cumbersome journeys to work, loss of agricultural land and of open space, and haphazard peripheral development, among others.

The actual phenomenon in Greater Cairo is the "out-migration". This out-migration takes two forms: informal cities for the poor people and private cities for rich people. This phenomenon in Greater Cairo shows that Egyptian urbanism is characterized by closure; it is not a shared space anymore.

References

1. Arab Republic of Egypt (2000) *Egypt Human Development Report 1998/99,* Institute of National Planning, Cairo.
2. Arab Republic of Egypt (2001) *Year Book 2000*, Ministry of Information, Information State Service, Cairo.
3. Arab Republic of Egypt (1999) *Statistical Year Book 1992-1998*, Central Agency for Public Mobilisation and Statistics (CAPMAS), Cairo.
4. Assawi, A. (1996) *Al aswaiyyat wa namazeg el-tanmiyya* (The Informal and Development Patterns), Center for the Study of Developing Countries, Cairo University, Cairo.
5. Bayat, A., and Denis E. (2000) Who is afraid of aswaiiyat? Urban change and politics in Egypt, *Environment and Urbanisation, 12, 2, 2000*, 185-199.
6. El-Shakhs, S. (1992) The future of mega-cities: Planning implications for a more sustainable development, Hamm, B. et al., eds., *Sustainable Development and the Future of Cities*, Universitat Trier Press, Trier.
7. Farag, F. (1998) The demolition Crew, in *Al-Ahram Weekly, 6, 11*, 15.
8. Hamdan, J. (1980-1984) *Shakhsiyat misr (Egypt's Character)*, Alam al-Kutub, Cairo, four volumes.
9. Nicholas, H., and Westergaard, K. (1998) *Directions of Change in Rural Egypt*, American University in Cairo Press, Cairo.
10. Rodenbeck, J. (ed.) (2000) Cairo, APA Publications, London.
11. Stewart, D.J. (1996) Cities in the desert: the Egyptian New Town Program, *Annals of the Association of American Geographers, 86, 3*, 459-480.
12. Stewart, D.J., (1999) Changing Cairo: The political economy of urban form, *International Journal of Urban and Regional Research, 23, 1*, 128-146.
13. Stewart, D.J. (2001) Middle East Urban Studies, Identities and Meaning (Progress Report), *Urban Geography, 22, 2,* 175-181.
14. Sutton, K., and Fahmi, W. (2001) Cairo's Urban Growth and Strategic Master Plans in the Light of Egypt's 1996 Population Census Results, *Cities, 18, 3*, 135-149.
15. United Nations, Population Division (2002) World Urbanisation Prospects: The 2001 Revisions, New York.

Chapter 21

Implications of Urbanisation for Turkey. The Case of Istanbul

R. Keles
Eastern Mediterranean University
Gazimagusa - TRNC, Mersin 10

Istambul is the major city of a rapidly urbanising country, Turkey. Physical, environmental, economic and socio-cultural problems facing the city are described and analysed in this chapter, with special emphasis on the local and national impact of the policies to be implemented. An attempt is also made to see how future developments in the world in general and in the region in particular will affect the quality of the environment and urban management. It seems almost certain that unless the present rate of population growth is curbed in the coming decades, the hopes for sustainable development may just remain on paper.

1. Turkey's Urban Profile

Turkey is one of the most rapidly urbanising countries in the Mediterranean region with more than half (65 percent in 2000) of its population living in urban centers. The annual rate of urban growth, over the last forty years, has been around 6 percent. In absolute figures, the urban population increased from 6.9 million in 1960 to nearly 44 million in 2000. It is estimated that it will reach 57.3 million by the year 2005 [18], increasing the share of urban population to 80 percent.

The socio-economic characteristics of Turkey have been greatly influenced by the demographic trends facing the country. The average rate of increase in the population was 2.49 percent during the period 1960 to 1990, but it has declined slightly during the last decade [25]. It decreased to 1.83 percent by the year 2000.

Living and working conditions in the countryside together with the inheritance legislation, which enhanced the fragmentation of rural landholdings, caused a rapid migration to cities following World War II. Rural to urban migration is chiefly directed towards major urban centers, and especially to the five largest cities, including Istanbul, the major city [4]. Both rural to urban migration and natural growth played an important role in Turkish urbanisation. However, the generally accepted hypothesis that declining birth rates make the migration component of the urban population growth the most significant determining factor does not hold true in the case of Turkey where the birth rates are still quite high even in the metropolitan centers.

A. Marquina (ed.), Environmental Challenges in the Mediterranean 2000-2050, 345–359.
© 2004 *Kluwer Academic Publishers. Printed in the Netherlands.*

The relative share in the urban population of the cities with a population of 100.000 or more increased from 45.3 percent in 1960 to 70.5 percent in 2000, and it is estimated that it will be around 80 percent by the year 2010. During the 1960-2000 period, more than 30 million people began to live in cities, and it can be safely assumed that 20 million more people will be living in urban areas in ten years time. Out of these 20 million people, 16 million will be settled in cities of 100.000 and more inhabitants [13].

There is a significant increase in the number of cities in the size-category of 100.000 or more inhabitants. The number was 56 in 2000: this figure may reach 80 by the year 2010. This tendency can be interpreted as a process operating at the expense of the prevailing supremacy of Istanbul on the urban scene. Istanbul is still ahead of other cities not only in the number of industrial establishments, but also in all other social, cultural and economic institutions.

Ankara, the State Capital, preserved its secondary position despite its spectacularly rapid growth rate. Although the population of all the larger cities are far below the level where they should be normally according to the so called "rank-size rule", the distribution of secondary regional centers throughout the national territory shows, at the same time, a remarkable tendency towards a more even distribution. In other words, the primacy of Istanbul in the hierarchy of settlements has been weakening since the early 1960's. In fact, a change in the primacy of Istanbul between 1950 and 2000 is noticeable in Table 2 below. This change may be explained by the relative efficiency of the redistributive policies implemented during the First (1963-1967) and the Second (1968-1972) Five Year Development Plan periods. These considerably influenced the number and relative importance of other large cities and have given the spatial pattern of urbanisation a relatively balanced appearance. This does not mean, of course, that the unbalanced pattern of urbanisation has been entirely changed. What it means instead is that some regional centers like Eskisehir, Mersin, Bursa, Adana, Gaziantep, Konya and Diyarbakir are gradually becoming counterbalancing magnets in Anatolia, attracting the population flow that is otherwise expected to concentrate in Istanbul and Ankara.

Table 1. Stability in the Ranking of the Five Biggest Cities (1950-2000)

Cities	1950	1985	1997	2000
Istanbul	1	1	1	1
Ankara	2	2	2	2
Izmir	3	3	3	3
Adana	4	4	4	4
Bursa	5	5	5	5

Source: State Institute of Statistics, Population Censuses

This trend is in line with the observation of S. El-Shaks emphasizing that the form of the primacy curve seems to follow a consistent pattern in which the peak of primacy is reached during the stages of socio-economic transition. His reference to the decentralisation and the spread effects in the development process, namely the increasing importance of the periphery, also holds true in Turkey's case, with the minor difference that this has not happened spontaneously, but has come about as a result of deliberate state action.

Table 2. Change in the Primacy of Istanbul

Share of the Major City (%)	1950	1985	2000
Ankara and Izmir (second and third largest) as (%) of Istanbul	53.8	68.1	61.7
Ankara, Izmir and Adana (second, third and the fourth largest) as (%) of Istanbul	65.8	82.2	74.6
Ankara, Izmir, Bursa and Adana (second, third, fourth and the fifth largest) as (%) of Istanbul	76.4	93.4	88.2

Source: Rusen Keles, *Urban Poverty in the Third World,* IDE, Tokyo, 1988, p.56; and the State Institute of Statistics, Population Censuses

The growth of urbanisation is at the same time a reflection of the pace of urbanisation in the country and in the differential demographic growth of regions. Among the seven geographic regions of Turkey, the Marmara region, which has Istanbul as its center, represents the highest ratio of urban population which was 78.1 percent in 2000. Reciprocal effects of the urbanisation of the region and of Istanbul are not negligible. In the priority list, the Marmara region is followed by the Aegean (57.8 percent), Central Anatolia (64.8 percent), and the Mediterranean (56.7 percent). The least urbanized regions are the Southeast (61.6 percent), the East (49.2 percent), and the Black Sea (42.6 percent) regions.

Even after the relative slowing down of its rate of population increase, Istanbul is still one of the main urban centers, which attracts migrants from all parts of the country. It is interesting to see that only 37.3 percent of its 1997 population was born in Istanbul. The rest (62.7 percent) is composed of immigrants. Out of 5.5 million inhabitants in 1985, over a million (18.5 percent) originated in the following Black Sea provinces: Kastamonu, Trabzon, Çorum, Ordu, Sinop, Rize, Smsun, Tokat, and Gümüshane. Nearly 700.000 (12.2 percent of the total migrants) came from the following eastern provinces: Sivas, Samsun, Kars, Malatya, Erzurum, and Erzincan. The fact that more than 60 percent of the inhabitants of Istanbul were born in other provinces seems surprising in view of the observation that, for the whole country, the population born in places other than their place of residence was only 20.9 percent for that year.

Table 3. The Population of Istanbul (000)

Years	Greater Istanbul (A)	Turkey (B)	(A/B)
1927	705	13.648	5.2
1950	1.102	20.947	4.8
1960	1.506	27.755	5.4
1970	2.203	35.605	6.2
1980	2.909	44.707	6.5
1990	6.229	56.473	11.0
2000	8.803	67.804	13.0

Source: State Institute of Statistics, Population Censuses. The impressive increase in the figure of 1990 is due to the consolidation that was realized in 1984

2. Ankara versus Istanbul

The proclamation of the new Turkish Republic in 1923 can be regarded as a turning point with regard to the location of the State Capital. Istanbul was distrusted by the rulers of the new Turkey. It was considered as a symbol of the old regime and of colonial exploitation. Therefore, Atatürk, the founder of the Republic, and his associates wanted to find a new State Capital, whose inhabitants were dedicated to nationalism and the modernisation goals of the Republic, rather than continue with Istanbul with its "cynicism and corruption".

As a result, Istanbul gradually lost its socio-economic and political importance. An indication of this trend was the decrease in the city's population from 1 million at the beginning of the 20th century to 704.825 in 1827 and to 758.488 in 1935. On the contrary, Ankara was a small Anatolian town and a trade center in 1923 when it was designated as the new State Capital. It was also regarded as strategically much safer than Istanbul. Besides, the latter was regarded by the founders of the Republic as much too cosmopolitan, representing the interests of the imperialists rather than those of the Anatolian people that supported the Turkish liberation movement from when it began in the late 1910's.

Following the transfer of the capital to Ankara, Istanbul remained as the nation's industrial and business center, its greatest port, the focus of the transportation network and the center of cultural and intellectual life. It attracted most of the private industrial and commercial investments during the initial decades of the Turkish Republic. Its population continued to increase rapidly during the 1960's and 1970's. Many of the agencies responsible for economic development and public investments remained in Istanbul, and a large amount of imported goods required for the development of new industries flowed inevitably through Istanbul. This necessitated, in turn, public investment in the city's port facilities which further reinforced its position as the principal channel for foreign trade.

3. Istanbul: A Short History of a Ruralising Metropolis

Istanbul served as the capital of the East Roman Empire from 395 A.D. It is the only major city in the world to lie on two continents and the 8 kilometer long Golden Horn cuts through the heart of the European side. Across the Bosphorus lies the Asian section of the city between the Black Sea to the north and the Sea of Marmara to the south [1]. This one-time capital of three empires (Roman, Byzantine and Ottoman) stretches out to embrace a diversity of population and natural and man-made environments [17].

Under the highly centralized Ottoman system, Istanbul grew steadily. By the end of the 18th century, its population exceeded 870.000. Concern over the rapid increase in the number of its inhabitants led to imperial policies to check the flow from the provinces of migrants. European experts were also invited to work out master plans to guide the city's development. At the same time, foreign capital was channeled into Istanbul as the city's economic dominance over its vast hinterland increased. European businessmen and their local agents were heavily concentrated in the metropolis. The city became a European outpost, the place where the imperialist powers had a chance to control their growing share of the Ottoman economy. This was the typical pattern of urban development, called *dependent urbanisation (*urbanisation dépendente) by Manuel Castells, taking place in almost all Third World countries which were more or less integrated with the world capitalist system [28].

4. The Contemporary City and its Economy

Modern Istanbul spreads far beyond the ancient city's seven hills, covering about 6.600 square kilometers on both sides of the Bosphorus. About 67.8 percent of Istanbul's population live on the European side. But, in recent years, the Asian portion of Greater Istanbul has been growing more rapidly. The basic settlement pattern has been linear, with development spreading from the old city north for 40 kilometers on the European shore of the Sea of Marmara, and east for 100 kilometers on the Asiatic sea coast.

Although the historical business center was located on the European side in the historic city and adjacent areas to the north, Istanbul has become, during the last four decades, a multi-centered metropolis with several business districts located in various parts of its territory, containing institutions of finance, imports and exports, corporations, retail, wholesale, etc. In other words, Istanbul is a mosaic of commercial districts, industrial plants, strip development, residential zones, transportation facilities, military bases and forests that ring the waterways. Industrial development is more intensive in the extreme west and east of the city's linear development pattern: Sefaköy, Avcilar, Çekmece in the west, and Kartal, Pendik, Gebze in the east. Heavy industry is particularly concentrated in the eastern industrial estates which have suitable transportation connection with the rest of Turkey [8].

While greater Istanbul accommodates about 13.6 percent of the country's population, the GNP it creates provides an important input into the country's economy. The share of the active population working in industrial occupations in Istanbul increased from 23 percent in 1955 to 41.4 percent in 1985, while corresponding figures for the whole country during the same period were 6 percent and 14.4 percent, respectively. The economy of Istanbul represents 30.3 percent of the whole industry and 47 percent of all the services. The share of Istanbul in per capita GNP is two and a half times that of Turkey [19]. The share of Istanbul in total investments in Turkey dropped from 27 percent in 1965 to 19 percent during the 1970's with a view to fostering development in the less developed eastern regions.

The economic sectors in which the share of Istanbul is disproportionately high are paper and paper products (39.5 percent of Turkey's figure), chemicals, petroleum, coal and plastic products (38.7 percent of the total) and metallurgy (35.7 percent). On average, the share of the city in manufacturing industries is 26.7 percent. In these sectors, the number of workshops represented by Greater Istanbul is 50 percent of the total. In fact, Istanbul represented 64.3 percent of foreign trade in 1986. It is also interesting to note that a considerable part (38.3 percent) of the tax revenue in the country is collected in Istanbul. For some kinds of taxes, this percentage is still over 50 percent. In heavy industry, Istanbul represents 43 percent of all the factories in Turkey. As far as the number of workers is concerned, this percentage is 30 percent. It becomes 27 percent in terms of the total economic output. For light industry, on the other hand, the percentages indicating the share of Istanbul in Turkey's total are 24.4 for the number of plants and 30.5 for the number of workers [22].

The share of the active population who work as salaried workers is much higher in Istanbul compared with Turkey's figure in general. Similarly, the ratio of the self-employed is three times smaller than the national average. Another indicator of the change in the economic structure of the city is the distribution of the active population among economic sectors. As expected, the share of the non-agricultural sectors is incomparably high in the major city. In the distribution of the active population among professions and status categories, several occupations seem to occupy the most important places. These are employees, small businessmen (the self-employed), the retired, and civil servants, in that order.

The main central business district (CBD) is situated on the west side of the Bosphorus, hence creating daily journeys between both sides. Water, together with other modes of transport, plays an important part in mass transportation. In the CBD, which has nation-wide economic importance, there are traditional activities which do not need to be located there. Therefore, the Planning Authority of Istanbul proposed, in the early 1980's, the relocation of such functions to sub-centers to be established along the 100 kilometer length of the metropolitan area. A great many administrative agencies of the many commercial and industrial establishments of the country are active in Istanbul. The CBD, at first located in the historical core, has lately been expanded towards the north. The public sector agencies are also situated in the CBD or in the near periphery.

Despite the tendency to decentralize, encouraged and guided by the planning authorities, the sub-centers are still not sufficiently developed in Istanbul. There does not exist an efficient distribution of activities among these newly structured sub-centers and the main CBD. In practice, all development has been guided by market forces. As a result of the white collar and skilled force being largely concentrated in the CBD, in areas where economic feasibility and prestige value coincide, there emerged an increase in housing demand. As the public sector has not been able to cope with and regulate the demand in time, neighborhoods have emerged with no planning.

Industrial establishments are unevenly distributed through the city. Their locations were not chosen as a result of the optimisation of economic, social and physical location factors, taking into consideration their likely impact upon the environment. The excessive concentration of industrial establishments in the central zones became a matter of concern from the point of view of national security and strategic considerations during the 1960's and, consequently, on the suggestion of the National Security Council, the city council adopted a new industrial settlement plan which aimed at the decentralisation of new industrial establishments to areas away from the center. Although the industries already settled in the central zones are trying to migrate out of the city, this does not result in a considerable decrease in the number of industrial factories within the city. The reason is that the factories are still viable [27].

5.Housing and Squatting

Meeting the housing needs of low-income families in Istanbul has been a major task of the governments since the late 1940's when the rural to urban migration began to accelerate. The number of gecekondus, unauthorized dwellings, increased from 10.000 in 1948 to more than a million in 2000.The Governor of Istanbul stated in the early 1980's that the average number of gecekondus built annually was around 22.000 [9]. Roughly 60 percent of the city's population live at present in them.

According to the estimates of the Planning Authority of Greater Istanbul, the population of the city is expected to reach 12 million by 2010 [23]. This is rather a conservative figure. Because if the annual additions to the present population, estimated to be about 1 million, are taken into account, the city's population may have already surpassed the estimate referred to above [7]. When the fact that the number of informal subdivisions have increased from 200.000 to over 3 million since 1975 is taken into account, one could safely assume that the population of the city may double in the first two decades of the 21st century [7]. As to the distribution of the additional population of the city, it is assumed that one third of the increase will take place on the east side and two thirds in the west.

Greater Istanbul has about 1.8 million dwellings with an average household size of 4.5 persons. This leads, together with the yearly increase in population of 250.000 people, to a yearly housing need of about 60.000 dwellings. When added to the number of gecekondus built prior to 1970, one can assume that there exists about 1 million gecekondus in the city, housing nearly 60 percent of the city's population.

Gecekondus are often mentioned and defended as a solution to the acute housing shortage in the metropolis. There is no doubt that they help to fill the gap in the supply of social housing. Gecekondu construction is an outcome of the creative intelligence of self-help efforts of the homeless urban poor. It also reinforces the self-reliance of the community residents. Of course, it ensures the use of the labor force at a considerably low cost, but it is doubtful as to what extent it helps to reduce unemployment. This is a structural illness, as in other Third World countries.

However, the gecekondus are built on lands unsuitable for human habitation with inadequate or no infrastructure and services. Therefore, the need for partial or full renovation or the relocation of their residents requires repetitive investments and makes this method prohibitively costly [10].

In Istanbul, the gecekondu building process is no longer a kind of self-help activity, carried out by the homeless. Rather, it has become partly or totally commercialized during the last three decades, and has lost its social significance entirely. Presently, the urban poor who can barely pay the rent of a gecekondu are no more than tenants of the professional gecekondu builder [13].

6. Governing Greater Istanbul

Greater Istanbul is one of the 81 provinces of Turkey. A Governor, appointed by the Council of Ministers, is in charge of establishing law and order. On the other hand, Istanbul is the largest local government unit, a municipality, in the country. Until 1984, it consisted of the city of Istanbul and about thirty smaller municipalities surrounding it. Following the attempt by the government to incorporate peripheral settlements administratively within a single metropolitan administration, the Constitution of 1982 permitted setting up metropolitan municipalities in the largest urban centers. Using this power given by the Constitution, the government that came to power in late 1983 managed to pass legislation through Parliament to create a metropolitan municipality in Istanbul (1984), the same as with two other major cities, namely Ankara and Izmir, based on a two-tier model of administration.

Istanbul originally had a municipal Organisation founded in the middle of the 1850's under Ottoman rule, and modified later in the Republican period in the 1930's, but the structure could not meet the needs of more efficient local services and better participation in the decision-making of today's Istanbul. At present, certain major functions, such as master planning, large-scale undertakings of water supply, transportation, sewerage, central heating, parks, car parks and terminals and the like, are carried out by the metropolitan (upper-level) municipality of Istanbul. All the local services, other than the ones mentioned above and those that are described in the municipal legislation as the duties of municipalities, are left to 32 district (sub-provincial) municipalities and 42 smaller non-district municipalities within the metropolitan area [11]. The increase in the number of the smaller municipalities which are not districts in the administrative sense, grew parallel to the increase in the number of squatter settlements which mushroomed without facing strong preventive measures during the 1960's and 1970's. For example, the following non-district municipalities, were set up during the ten years between 1966 and 1977: (1966): Avcilar, Güngören, Yakacik, (1967): Sefaköy, Alibeyköy, (1969): Hadmköy, Celaliye, Soganlik, (1970): Esenler, (1971): Kemerburgaz, Selimpaça, Yenibosna, Dolayoba, (1972): Çinarcik, (1975): Yesilbag, (1976): Kocasinan, Halkali, (1977): Yahyalar [25].

The municipalities at both upper and lower levels elect their mayors and city councils. All units in the metropolitan area enjoy special revenue sources provided by special legislation. The new system of local administration has achieved its first goal to a certain extent, which consisted in providing better and more efficient public services. However, the goal of participation is still far from being realized. The main reason is that the optimal size of the district municipalities in population and territory is not adequately defined. In almost all district municipalities, the size of the population and the territory exceed the optimum levels that are suitable for active public participation. In fact, it is obvious that there can be no participation, in the real sense of the word, in municipalities such as Bakırköy which had a population closer to two million in 1990. Others like Kadiköy, Fatih, Sisli and Kartal each had a population of between half a million and one million.

The size of district municipalities and the complexity and anonymity of social interactions that characterize these sub-centers are not suitable to producing a participatory atmosphere. This seems to justify the observation of some social scientists that called such a state of affairs "the decline of community and "the decline of citizenship" [2].

There are also important differences in the sizes of the population and territory of these municipalities. Since the number of inhabitants determine the amount of revenue that each district municipality receives from the central government budget, these discrepancies give rise to enormous injustices, in the absence of compensatory mechanisms.

The Council of Greater Istanbul Municipality is made up of the mayor of the metropolis, who is elected by the inhabitants of all of Istanbul for a period of five years, mayors of the district municipalities and one fifth of the members of each of the district councils, who are also elected. The metropolitan council is constituted for five years. The members of the council keep their positions in the lower level councils while serving on the metropolitan council. Istanbul has a metropolitan council which consists of nearly 150 members.

There is no doubt that the quality of the environment in the metropolitan area will be affected in years to come by the extent to which the central government and the local authority cooperate efficiently in carrying out of the public services directly or indirectly related to the environment. It seems that attempts to reform the local government system will enable the sub-national government to respond better in the future to the increasing concern of the inhabitants as to the quality of the environment. In accordance with the goals set by the Rio and Habitat II Conferences, the government is currently trying to put into effect a local government reform which is, to a large extent, in line with the principle of subsidiarity.

7. Urbanisation and Environment

Istanbul was one of the cleanest cities in Turkey, not polluted by industry or urbanisation until recently. However, it has changed during the last two and a half decades. The municipality has had to set up a special department of environmental protection to deal with water and air pollution. It has dealt with the removal of more than 4000 unhealthy industrial establishments along the coast. According to the findings of several applied researches, the most important source of noise in Istanbul is the traffic. Measurements of 1986 show that the average SO_2 mgm per cubic meter was 193.9 for the European side, and 164.4 for the Asian side. For the whole metropolitan area, the value was found to be around 92 mgm per cubic meter. On selected sites on the historical peninsula, the noise level was 73.9 db for Sisl and, 74.5 db for Kadiköy. As far as solid waste discharged is concerned, the major centers of economic activity like Eminönü, Beyoglu, Sisli and Bakirköy are at the top, with per capita figures well above Istanbul's average figure.

The green spaces of past centuries were rapidly turned into apartment blocks in Istanbul in the second half of the 20th century. Per capita green space, which should not be less than 7 square meters according to the existing planning standards, is only 2.2 square meters. When allowance is made for public squares in the city and cemeteries, the amount of actual usable green space becomes much smaller. Only 33.9 percent of the available green space is actually usable by the inhabitants of the city. There is a definite adverse impact of building activities upon the natural environment.

Squatting itself can be regarded as an environmental problem in the metropolitan area where most of the adverse consequences of unplanned and disorderly urbanisation are reflected. One million out of the two million gecekondus existing in the country are situated in Istanbul. Regardless of their economic and social characteristics, at least three features of squatter settlements in Istanbul affect the environment directly: The first is the visual pollution created by unplanned and unlawful construction; the second is the inadequacy of urban infrastructure; and the third is the deterioration of hygienic conditions in these quarters. In fact, according to the findings of applied research carried out by the State Planning Organisation [20], the percentages of the households in these unauthorized settlements deprived of basic urban public services are quite high, as shown below:

a)Those that do not receive their drinking water from the city's regular water network (9.0 percent)

b)Those whose liquid wastes are not discharged into the city's sewerage system (16.5 percent)

c)Those that discharge their solid wastes into streets or to the sea, but not to the garbage baskets (12.3 percent)

d)Those whose solid wastes are not collected regularly by the city authorities (28.0 percent)

A major environmental issue related to the water supply system is that nearly half of the 7000 kilometer long pipe system is either unhealthy or in need of repair or complete renewal. A considerably large part of the system is broken and suffers leaks. Similarly, almost one third of the population and one fifth of the area of Istanbul is deprived of a modern sewerage system. Accelerated urbanisation during the 1950's and 1960's filled the *Golden Horn* shores with haphazardly located factories, manufacturers, work premises and warehouses. Chimneys started to dirty the air, and water polluted the sea. The quantity of daily liquid waste discharged into the Golden Horn by the industries in the region is approximately 200.000 tons. Of this, 67 percent is chemicals, 27 percent washing water, 4 percent cooling water, and the rest other kinds of wastes. More than half of the industrial premises were found to discharge solid wastes.

It was observed, in the early 1970's, that the ratio of sulphur dioxide in the air, pollution during the days where the ratio exceeded the limit harmful to human heath, reached 62 percent at Silahtar, 67 percent at Eyüp and 58 percent at Hasköy. Little left of the *Golden Horn* of past centuries, much praised by local and foreign poets and authors like Baron Durand, Pierre Loti, and Charles Diehl. *The Golden Horn* has been reduced to a simple *horn,* no longer possessing the adjective *golden.*

Within one or two decades, Turkey's membership of the European Union could materialize. Starting from the end of 2004, Turkey will probably begin negotiations for full membership. This requires full adaptation to the rules, standards and regulations concerning running water, services, various kinds of pollution , and infrastructures like those prevailing in the member states of the European Union. One can expect, as a result, a sharp improvement in all these and related standards, and in the quality of life in general.

8. Impact of Globalisation upon the Environment

Since the early 1970's, the world has witnessed a rapid Globalisation with at the same time a decline in the power of the national state. It is capital that largely determines the success and failure of individual economic sectors as well as the totality of national economies. Since capital recognizes no national boundaries, the traditional concept of national sovereignty has lost much of its meaning. Globalisation requires that all obstacles to the free flow of international capital be removed even at the expense of environmental values. As a result, sustainability is to a considerable degree adversely affected by Globalisation. On one hand, it puts all sorts of sophisticated technological opportunities at the service of mankind, but on the other hand, it exerts, directly or indirectly, deteriorating effects upon natural and historical values, because capital has a single, global, *sui generis* logic which everyone is forced to obey. This observation is true not only for transnational capital, but also for national and local capital.

Turkey had to open its doors to the influence of Globalisation as early as the beginning of the 1980's. Since the strategic points of connections are the cities and their economic, financial, socio-cultural and political institutions, Istanbul, as the only city which possessed most of the characteristics of a *world city*, assumed an increasing role in promoting the interests of international capital. For example, according to the Mayor of Istanbul in the middle of the 1980's, "Istanbul is a city 2.500 years old and has more than 3000 historical buildings. If nearly one hundred of them prevent the construction of a major transport artery serving international and national trade, they must be demolished. We do not share protectionist views which prevent development". [5] This policy began to dominate all aspects of the city's development from the early 1980's: five-star hotels, luxury residential zones, university campuses and the like have been created in the city's green areas but inadequate attention has been paid to the preservation of natural assets and resources. The sites where the Yapi Kredi Plaza, Sabanci Center, Konrad Hotel, Hyatt Regency, Swiss Hotel and the like have been built are a few of the examples of this kind of development.

The services necessary for the Globalisation of capital have been concentrated in the high-rise buildings in the trade centers that were built with no regard to the zoning regulation of the city. It has been suggested that Istanbul is in a position to undertake the activities required by Globalisation, such as banking, insurance, legal and accounting services, consultancy, communications, publicity, marketing and engineering services for neighboring countries like Georgia, Azerbaijan, Ukraine or Romania [14, 15, 16]. The same logic emphasizes that unless Istanbul turns its attention to the demands of the new world system, it will face the risk of becoming a marginalized city with a severely curtailed potential for further evolution [14, 15, 16].

On the other hand, Turkey's international economic policies and its likely membership of the European Union have already adversely affected development of the cities and urban environments. In order to ease conditions favorable to the free flow of capital, important amendments have been made in the Constitution and the corresponding legislation. For example, the MAI (Multi Party Agreement) that was accepted by Turkey requires, among other things, that "the provisions in the previously signed environmental agreements that are not favorable to free trade be removed". Similarly a new Law on Privatisation (No: 4046) in 1999 authorized the newly established Privatisation Authority to work on master plans for the communities where the privatized economic enterprises were located. The condition was that the central or local authorities could not modify the plans for five Years. Furthermore, Articles 125 and 155 of the Constitution have been amended with a view to permitting international arbitration to resolve legal conflicts that might arise from administrative contracts of concessions. In other words, this amendment puts an end to the role played so far by the Council of State for the protection of the environment. Parliament, in its turn, passed the necessary legislation accordingly without much delay (Law No: 4501, October 2000). These examples indicate that the dynamics affecting urban sustainability will no longer be controlled by domestic institutions but will rather be shaped internationally as a result of Globalisation. There is no doubt that the impact of these changes is badly affecting the historical identity of Istanbul and will continue to be detrimental to the environment, as in the last two decades. We believe that it is not possible to agree with the views of the proponents of Globalisation. They prefer integration within international capitalism through a network of global cities which implies sacrificing the cultural, historical and natural identity features of such cities as Istanbul in order to contribute to the successes of world capitalism.

9. Conclusion

Urbanisation in Turkey is still proceeding at a considerable speed. Push and pull factors behind rural to urban migration movements are still strong enough to distort the rural-urban balance in favor of the latter. Unless an effective system of regional planning exists, inequalities between geographic regions, particularly between the east and the west, will not be reduced naturally by time. Inequality in the distribution of population in economic activities and investment is the very reason which causes a concentration of pollution in the highly urbanized Istanbul and surrounding settlements. Although the degree of urbanisation in the Marmara region in the west of the country is as high as 78.1 percent, this figure is less than 50 percent in Eastern Anatolia and the Black Sea regions. Such a striking imbalance, that is also reflected in all economic and social indicators of development, requires immediate handling through planning at national and regional levels. This would also ease the task of Istanbul in the long run.

Since the pace of industrial development lags behind the speed of urbanisation, rural migrants in major urban centers face an inescapable problem of unemployment or underemployment. Even in Istanbul, which is supposed to be the most industrialized city in the country, the rates of unemployment and underemployment are much higher than the corresponding figures for Turkey. As a result, the share of the informal sector in the economy of the city grows and the number of inhabitants living in the illegally formed squatter settlements increases. The informal sector and informal settlements become integrated with time, giving the false impression that real integration is taking place in both space and society. In fact, it is happening without spatial and socio-cultural integration. Unemployment, poverty and alienation are the very sources of the deterioration of environmental conditions in general. It seems that it will take at least a decade to overcome such conditions before joining the European Union as a full member.

When mentioning the informal sector one refers to unplanned and disorderly development from the standpoint of City Planning and of urban public services. Plans to upgrade the informal settlements formed through informal initiatives will never lead to a future urban development in the metropolis. These initiatives usually remain on paper and the main course of urban development is determined by the pressures of informal forces.

Changing the internal dynamics of the society during the 1980's and 1990's, as well as the effects of Globalisation, formalized the informal sector, and allowed it to acquire a permanent feature [12]. It seems reasonable to assume that the permanent effects of

Globalisation on cities and the environment are more negative than positive in nature. The main characteristics of urbanisation in general are reflected in the general socio-economic, cultural and spatial patterns almost equally. All the features characterizing Istanbul indicate that it is a premature metropolis. The growth of Turkey during the last century has made Istanbul a great city of quantities with insufficient qualitative additions. Istanbul is proud of having spectacularly attractive historical assets and natural beauty, but the city is largely lacking the managerial capacity of providing the basic urban services and infrastructure for its residents. It underwent a phenomenal urbanisation in the demographic sense, during the second half of the 20th century, but too fast to enable it to cope adequately.

As an industrialized city, which still possesses pre-industrial traces in its historical nuclei, Istanbul has been decentralising in an unplanned manner, with the great majority of its residents concentrated in its surrounding peripheries. The core grows relatively slowly, but the periphery receives the lion's share of the increase of both population and industry. Between 1950 and 1990, the center grew by 250%, while the size of the periphery's population increased by 4500%. When the features of the new migrants are taken into consideration, one can admit that Istanbul is *a great ruralizing metropolis,* not only in terms of the origins of the great majority of its inhabitants, but also in terms of the peculiarities of its daily life, human relations and economy.

Patterns of informal economic mechanisms, as exemplified in *gecekondus, dolmuses* (shared taxis), and small-scale economic enterprises, prevent Istanbul from maturing into an industrialized world city, serving better the needs of international capital. Old and new, east and west co-exist in Istanbul within the framework of multiculturalism. As a result, its social fabric is composed of many different, and most of the time, contrasting social elements. As such, it basically reflects the prevailing characteristics of the socio-economic and cultural structure of Turkey as a whole.

As far as the implications of urbanisation are concerned, one can see certain points of optimism as well as several points of pessimism. One of the most important factors that allows considerable optimism is that population growth may be stabilized by the year 2020, as estimated by the yearly development reports of the World Bank. It has been estimated that the total population of Turkey will reach 90 or 100 million by 2020 and the annual rate of growth would be reduced to less than one percent. This would consequently affect the amount of surplus population in rural areas and would slow down rural to urban migration considerably. This means that more people may stay in the villages without increasing environmental and other negative externalities in the biggest metropolitan centers like Istanbul.

Another point of optimism is the impact of the Southeast Anatolian Regional Development Project (known as the GAP) that has already begun to check the population flow from the East to the West, and to reduce regional inequalities between geographic regions. No less important than this is the likely membership of the European Union which may produce positive consequences in the improvement of environmental protection standards. Furthermore, the role of international and supra-national Organisations, like the UN, UNEP, UNESCO, UNCHS, is immense in fostering cooperation. Next is the increasing emphasis on the principle of subsidiarity, in other words on the need for public services to be carried out by the authorities closest to the people. Increased emphasis on this principle in European countries may encourage members of local communities to take part in efforts to protect the environment more efficiently. Reliance upon people and increasing public awareness may produce beneficial results for the orderly development of cities and towns.

It is hoped that the decreasing political and military tensions in the Mediterranean region will help reduce the enormous defense expenditure that could otherwise be used for development purposes.

But on the other hand, a considerable part of the geography of the region is vulnerable to natural disasters. Therefore, preparedness of the societies in the region must be increased in order to avoid or mitigate the damages of such natural events. Similarly, doubts about the role of the state in the society, and an increasing emphasis on Privatisation and deregulation as a result of world-wide Globalisation, seem to create and strengthen an anti-planning attitude towards urban management. Consequently, public land in Turkey and elsewhere is being privatized systematically. This may create negative conditions for the orderly growth of the metropolis. It is assumed that market solutions are the best tools for the effective implementation of urban development schemes. It seems that in the absence of the intervention of public authorities, the natural and historical values of greater Istanbul would suffer most from haphazard development. Sustainable development, which requires harmonisation of economic growth and environmental protection, must be based on the planned guidance of urbanisation. The historical and natural values of Istanbul cannot be sacrificed to an uncontrolled pace of urbanisation.

References

1. Aru, K. A. (1971) Istanbul, in *Rural-Urban Migrants and Metropolitan Development,* Laquian, A. Istanbul Technical University, Istambul
2. Bookchin, M. (1997) *The Rise of Urbanization and the Decline of Citizenship,* Sierra Club Books, San Francisco.
3. Castells, M. (1975) *La Question Urbaine,* Maspero, Paris.
4. Danielson, M. N. and Keles, R (1985) *The Politics of Rapid Urbanization: Government and Growth in Modern Turkey,* Holmes and Meier, New York.
5. Ekinci, O. (1994) I*stanbul'u Sarsan On Yil: 1983-1993 (*The Ten Years that Have Shaken Istanbul: 1983-1993), Anahtar Yayinlari, Istanbul.
6. Eussner, A. (1989) *Türkiye'nin Avrupa Toplululu u Tam Üyelilinin Bölgesel Politika Açisindan Etkileri (*Effects of Turkey's Full Membership of the EC upon Regional Development Policy), Friedrich-Ebert Foundation, Istanbul.
7. Görgülü, Z. (1993) *Hisseli Bölüntü ile Oluçan Alanlarda Yasallaçtirmanin Kentsel Mekana Etkileri (*The Effects of Legalization of the Process of Informal Subdivision on the urban Space), Yildiz Technical University, Faculty of Architecture, Istanbul.
8. Güvenç M. (1999) Istanbul'u Haritalamak: 1999 SayImIndan Istanbul Manzaralari (Mapping Istanbul: Profiles of the City reflected in the 1990 Population Census), *Istanbul 36 .*
9. Istanbul Valiligi *(*1983) Planlama ve Koordinasyon Müdürlügü, *1. Istanbul Sempozyumu,* The Governerate of Istanbul, Directorate of Planning and Coordination, *The First Istanbul Symposium.*
10. Keles, R. (1988) *Urban Poverty in the Third World: Theoretical Approaches and Policy Options,* Institute of Developing Economies, Tokyo.
11. Keles, R (2000) Urbanisation and Urban Policy in Balland, D. (ed.), *Hommes et Terres d'Islam ,*Melanges Offerts a Xavier de Planhol, Tome II, Institut Français de Recherche en Iran, Teheran, 127-133.
12. Keles, R. (2001) Urbanisation, City and the Informal Sector, State Institute of Statistics, *Informal Sector (I),* Tuncer Bulutay, Ankara, 11-17.
13. Keles, R. (2002) *Kentlesme Politikasi,* (Urbanisation Policy), Imge, Ankara

14. Keyder, Ç., and Öncü A. (1994) Istanbul at the Crossroads, *Biannual Istanbul, Selections, 93, 2*, 1, Turkish Historical Foundation, Istanbul, 38-44.

15. Keyder, Ç. (1996) (ed.) I*stanbul: Küresel ile Yerel Arasında (*Istanbul: Between the Global and the Local), Metis, Istanbul.

16. Keyder, Ç. (1996) Marketing Istanbul, *Biannual Istanbul, Selection Turkish Historical Foundation* Istanbul, V. 4, 2, 87-91.

17. Kuban, D. (1996) The Growth of a City: From Byzantium to Istanbul, *Biannual Istanbul, Selections 1996, Spring,,* V. 4, 2, 73-76.

18. National Report of Turkey to the Habitat II Conference (1996) Prime Ministry, Istanbul.

19. OECD *(1988) Regional Problems and Policies of Turkey,* OCDE, Paris.

20. State Planning Organisation (1993) *Gecekondu Arastirmasi* (Survey on Squatter Settlements), Ankara.

21. Sönmez, M. (1996) A Statistical Survey: Istanbul in the 1990's, *Biannual Istanbul, Selections 1996* Spring, Istanbul, V 4, 2, 104-108.

22. Sönmez, M. (1996) Istanbul and the Effects of Globalisation, *Biannual Istanbul, Selections 1994 and 1995,* Winter, Istanbul, V. 3, 1, 83-85.

23. Tekeli, I. (1994) *The Development of the Istanbul Metropolitan Area: Urban Administration and Planning,* IULA-EMME, Istanbul.

24. The Increasingly Growing Giant: The Geographic Anatomy of Istanbul (1996) *Biannual Istanbul, Selections 1996,* Winter, Istanbul.

25. Tuncer, B. (1977) *Turkey's Population and Economy in the Future*, The Development Foundation of Turkey, Ankara.

26. Turgut, S. (2000) *Metropoliten Alanlarda Planlama-Kent Yönetimi Iliskileri ve Istanbul Metropoliten Alaninda Planlama Yönetimine iliskin Bir Model Denemesi (*The Relationships between Planning and Urban Administration and a Modeling Experiment concerning Planning Administration in the Istanbul Metropolitan Area), Yildiz Technical University, Istanbul.

27. Tümertekin, E. (1972) *Analysis of the Location of Industry in Istanbul,* Istanbul University, Geographical Institute, Istanbul.

28. Zevelyov, I. (1989) *Urbanization and Development in Asia,* Progress Publishers, Moscow.

Section VIII

Pollution

Chapter 22

The Pollution of the Mediterranean: Present State and Prospects

F. S. Civili
Senior Environmental Affairs Officer
MED POL Programme Coordinator, UNEP/MAP. Athens

The paper reviews the main anthropogenic pressures and the present status of the marine and coastal pollution of the Mediterranean basin and also describes the policy responses made and those expected to be made in the future by the riparian countries. The overall regional policy/legal framework considered in the paper is the 1995 Barcelona Convention on the protection and sustainable development of the Mediterranean region and in particular its Protocol on land-based sources of pollution and its implementation instruments. The Mediterranean countries have formally agreed to implement a pollution reduction programme that should gradually bring about the elimination of polluting releases into the sea by the year 2025. The paper reviews the actual state of implementation of the programme and describes the first achievements and results and those expected in the medium- and long-term.

1. Land Based Pressures

1.1. CONCENTRATION ON THE COAST AND IN BIG CITIES

The resident population of the Mediterranean coastal states has almost doubled in the last 40 years, reaching 450 million in 1997 and is expected to reach approximately 600 million in 2050 and possibly 700 million by the end of the 21st century. At present, one third of the Mediterranean population, around 145 million people, is concentrated on the narrow coast. At the same time, the rural areas of the Mediterranean are being increasingly abandoned in favour of the large urban agglomerations on the coast. This is particularly evident in the southern and eastern Mediterranean .

The Mediterranean urban population, 40 % of which is currently concentrated on the coast, is expected to double by the year 2025. The pressure resulting from this population load is not equally shared between the northern and the southern Mediterranean countries but has shown a shift in the last 50 years from the North which accounted for two thirds of the Mediterranean population in 1950 to the South which at present accounts for 50% of the Mediterranean population.

A. Marquina (ed.), Environmental Challenges in the Mediterranean 2000-2050, 363–376.
© 2004 *Kluwer Academic Publishers. Printed in the Netherlands.*

The pressure exerted on the Mediterranean coast from the heavy resident population load, is dramatically amplified by the intensive seasonal increase in the population from tourism, which in some countries represents up to 90% of the total population. At least 50% of the tourists arriving in the Mediterranean are concentrated on the coast. The pressure of mass tourism, which is heaviest on the north-western coast of the Mediterranean, translates into further stress on the already burdened natural resources, further land-use conflicts and further deterioration of historic sites, fragile natural habitats and coastal and marine ecosystems. The pressure is likely to increase in the future, with an estimated doubling of tourism related development in the Mediterranean, escalating from 135 million arrivals in 1990 to a projected 350 million in 2025.

The dense human settlements established along the Mediterranean coast produce large amounts of municipal wastewater, which is usually, though not always, conveyed to municipal sewer systems. These systems may, or may not (as is the case for a large number of small and medium size communities in the Mediterranean), be connected to wastewater treatment facilities so that wastewater is discharged into the sea either untreated or after various degrees of treatment. It enters the sea through outfalls, or by seepage resulting from leaks or other faults in the sewerage system. At major tourist resorts with intense, short-term population pressure during the peak summer season, the sewage treatment plants are frequently quite unable to cope with the additional loads, discharging water that is still highly polluted into the sea. Improperly discharged sewage carrying increased loads of nutrients, such as nitrogen and phosphorus, a heavy load of micro-organisms and a variety of chemical wastes, impacts heavily on marine and freshwater ecosystems, on human health and on fishing and other economic and recreational activities.

Figure 1. Number of Mediterranean cities served by wastewater treatment plants

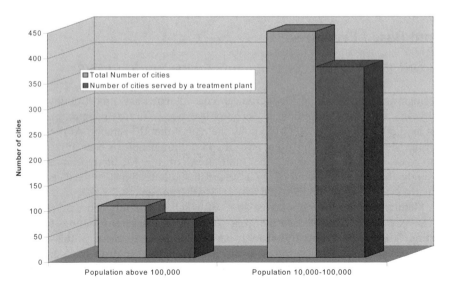

Source: UNEP/WHO, 2000 [5]

Wastewater treatment plants serve around 55% of 545 coastal cities with more than 10,000 inhabitants, in 19 Mediterranean countries reviewed by WHO/EURO as part of the MED POL Programme [5]. Figure 1. shows the cities that are served by wastewater treatment plants according to the resident population density. Figure 2. illustrates the treatment facilities situation of the Mediterranean coastal population by country. Around 30% of the population of the coastal cities surveyed is served only by a sewerage network.

Figure 2. The status of treatment facilities in Mediterranean countries

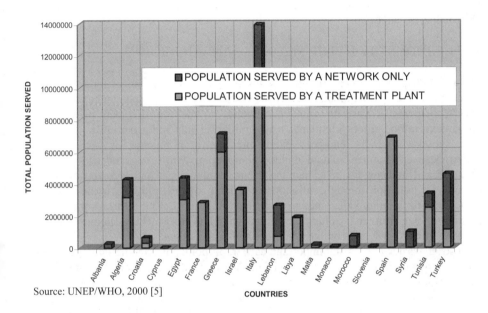

Source: UNEP/WHO, 2000 [5]

1.2. AGRICULTURE: THE LARGEST NON-POINT CONTRIBUTOR OF POLLUTANTS TO THE MEDITERRANEAN

In most Mediterranean countries all agricultural practices and land use, including cultivation, irrigation, dairy farming, pastures and animal feedlots, are considered as non-point sources of pollution. Agriculture may be only one of a number of non-point sources of pollution through run-off water, sediment transport and leaching, carrying phosphorus, nitrogen, pesticides, metals, pathogens, salts and trace elements, but has become the largest non-point contributor of pollutants to the Mediterranean. These pollutants gradually find their way, through groundwater, wetlands and rivers to the sea in the form of sediment and chemical loads.

With the intensification of agriculture the use of pesticides has greatly increased in the Mediterranean in the last twenty years, threatening the quality of ground and surface waters, human health and ecosystems. This is most evident in the northwestern region, especially in Spain, Italy and France, closely followed by Turkey in the northeast. Aerial transportation makes a considerable contribution of pesticides to the marine environment, particularly organochlorinated compounds. Agricultural run-off through rivers (notably the river Rhone in France, the Ebro in Spain, the Po in Italy, the rivers Axios, Loudias and Aliakmon in Greece and the Nile in Egypt) is the most important point-source of pesticides to the Mediterranean.

By 2025 countries south and east of the basin are expected to show a five-fold increase in their agro-food activities. These countries will therefore be the most vulnerable to increased pollution and environmental pressures from the development of the agro-food sector. Land-based pollution from the increasing use of large quantities of fertilizers is the most obvious possible outcome in the countries to the south and the east of the Mediterranean basin.

1.3. THE INDUSTRY IN THE MEDITERRANEAN REGION

More than 200 petrochemical and energy installations, basic chemical industries and chlorine plants are located along the narrow Mediterranean coast and catchment basins of rivers, including at least 40 major oil refineries, in addition to cement plants, steel mills, tanneries, food processing plants, textile mills and pulp and paper mills.

The activity of these industries exerts pressure in a number of ways on the Mediterranean environment. These include competition with other land uses such as agriculture and housing for an expanding urban population, a high demand on water resources, air pollution through emissions of sulphur oxides, nitrogen oxides, carbon dioxide and carbon monoxide, land, air and water pollution from industrial solid wastes consisting of slag from coal mining, coal processing and steel making, sludge from the processing of ores, dust and combustion ashes and mine tailings.

Industrial wastewater is an important carrier of pollutants, including oils, heavy metals, detergents, solvents and organic chemicals as well as heated cooling water to the Mediterranean aquatic environment. Industrial wastewater is either discharged directly to the sea or through municipal sewerage systems, outfalls, uncontrolled disposal sites and rivers. There is documented evidence of contamination of the Po, Ebro and Rhone, large rivers draining into the Mediterranean, with polychlorinated biphenyls (PCBs), polycyclic aromatic hydrocarbons (PAHs) and solvents [1].

Of the substances that are produced by industry or released as a result of its activity, the most harmful to human health, marine ecosystems and biodiversity are the toxic, persistent and bioaccumulative pollutants known as TPBs. These include the heavy metals mercury, cadmium and lead, some organometallic compounds (compounds where one metal atom is bound to at least one carbon atom, tributyltin being the most widely known) and numerous organic compounds, called Persistent Organic Pollutants (POPs). POPs have been grouped together as a result of their important toxic effects, including effects on the function of the endocrine system, their propensity for long-range transport and deposition and their typical low water solubility and high accumulation in fatty tissue. All these properties combined translate into serious potential adverse effects on the environment, wildlife and human health at locations near and far from their source. Looking at the export specialisation of the Mediterranean countries, which is a fairly accurate reflection of their industrial activity, provides an insight into the most important sources of pressure currently exerted on the environment from industry, on a country basis. There is still a considerable gap in industrial development between the northern and the southeastern countries of the basin. A different picture is likely to emerge in the course of the 21st century well into its shift south and eastward. These likely shifts in industrial production imply a potential increase in industry related environmental pressures in the southern and eastern part of the Mediterranean basin in the new millennium.

1.4. RIVERS: AN IMPORTANT SOURCE OF ALL TYPES OF LAND-BASED POLLUTANTS TO THE MEDITERRANEAN

The Mediterranean river flow regime greatly influences the timing and extent of the input of significant pollution loads, which are carried to the Mediterranean Sea by around eighty identified large and small rivers [4].

The heavy metal load of Mediterranean rivers is generally lower than in most western European rivers. This may be the result of the dilution of urban and industrial loads by the high levels of suspended solids in Mediterranean rivers in combination with a highly erosive environment. Natural variations can sometimes account for even a doubling of heavy metal concentrations in comparison to reference values, but above this level, pollution is the most likely explanation. This is the case for Pb concentrations in many rivers (Rhone, Tevere, Herault, Brenta, Martil), as well as for Zn (Adige, Herault, Martil, Po, Tevere), Hg (Po, Rhone), Cu (Ebro, Herault, Orb, Rhone, Tevere) and As (Orb, Herault). Hundreds of reservoirs from the damming of rivers of the Mediterranean basin are probably retaining much of the sediment-bound metals originating from human activities [4].

The load of organic micropollutants carried by Mediterranean rivers has not been thoroughly assessed even though rivers are an important source of input of these chemicals into the Mediterranean [4]. Industrial chemicals such as polychlorinated biphenyls (PCBs), polycyclic aromatic hydrocarbons (PAHs) and solvents are known to contaminate large rivers draining into the Mediterranean such as the Po, Ebro and Rhone [1]. Rivers carrying agricultural run-off (notably the river Rhone in France, the Ebro in Spain, the Po in Italy, the rivers Axios, Loudias and Aliakmon in Greece and the Nile in Egypt) are the most important point source of pesticides discharged into the Mediterranean. A number of small Mediterranean rivers running close to land used for intensive agriculture have been found to contain elevated levels of pesticides (> 1mg/l). Residues of the new generation pesticides atrazine, simazine, alochlor, molinate and metolachlor have frequently been detected in important rivers draining into the Mediterranean, though only up-to 3% of the quantities of these pesticides applied to cultivated land are exported by rivers [4].

The levels of nutrients in Mediterranean rivers are about four times lower than those of western European rivers. Nevertheless, nitrate levels in Mediterranean rivers are on the increase and the trend in ammonia levels is variable, depending on the degree of sewage collection and treatment.

The level of bacterial contamination of rivers is not well documented in the Mediterranean. It ranges from negligible in rivers running through sparsely populated land to quite severe in some southern rivers. With a few exceptions of some major rivers in Greece and Italy, the situation in EU Mediterranean countries has improved in the last two decades owing to improved sewage treatment and disposal [4].

River discharges can carry important amounts of the radionuclides caesium-137 and plutonium-239, 240 to the adjacent continental shelves, especially in areas where the deposition from the Chernobyl accident was significant on the hinterland (northern Adriatic and Liguro-provencal areas). Riverine geochemical processes may delay the output to the sea of the effluents from the Mediterranean nuclear facilities that are located along rivers. In any case, river input, along with the input from the nuclear industry and exchanges through the straits amount to no more than 10% of the total ^{137}Cs and ^{239,240}Pu delivery to the Mediterranean Sea from fallout. The constructions of dams and the diversion of river flow for irrigation have severely reduced the river inputs into the Mediterranean over the last 40 years. The highest retention of river water (probably more than 90%) is the Nile. The current estimated reduction in river water discharge for the Mediterranean basin as a whole is between 30 and 40 per cent.

2. The Response of the Mediterranean Countries: A Common Strategy to Address Land Based Pollution

As early as the 1970's it became obvious to the countries surrounding the Mediterranean Sea that although human activities at sea led to marine pollution, its origin was mainly to be found in land based activities. As a result, the countries devoted particular attention to the preparation of an appropriate legal instrument to cover this aspect of marine pollution. Shortly after adopting the Mediterranean Action Plan (MAP) and the Barcelona Convention for the Protection of the Mediterranean Sea Against Pollution, the Contracting Parties adopted and signed the Protocol for the Protection of the Mediterranean Sea against Land Based Sources (LBS Protocol).

The LBS Protocol entered into force in June 1983 and a calendar of activities for its implementation was set by the countries for the period 1985 to 1995 through the MED POL Programme (the pollution assessment and control programme of the Mediterranean Action Plan).

The 1992 Earth Summit in Rio signalled a change in the pace of events that eventually consolidated the shift in the direction of the MED POL Programme towards action for the prevention and control of pollution from land based activities. Shortly after the Rio summit the Mediterranean States, wanting to give effect at the Mediterranean level to the Agenda 21 resolutions, approved an Agenda MED 21 promoting the integration of environmental concerns in environmental policies in the Mediterranean. This was closely followed by the revision of the Barcelona Convention in 1995, to give legal status to the commitments made at Rio. In the same year, 108 countries and the European Commission adopted the Washington Declaration, a commitment to protect and preserve the marine environment from the impacts of land based activities through *inter alia* giving priority to the implementation of the Global Programme of Action for the Protection of the Marine Environment from Land-Based Activities (GPA). Subsequent events in the Mediterranean in the field of land based pollution control show that the Mediterranean countries are putting into practice the goals of the GPA on a regional level.

In 1996, the Contracting Parties to the Barcelona Convention signed a revision of the Protocol for the Protection of the Mediterranean Sea Against Pollution from Land-Based Sources (LBS Protocol).

Under the revised LBS Protocol, the Mediterranean States agreed to take measures to prevent and control the degradation of the Mediterranean Sea area caused by land based sources and activities originating in their territories, including discharges from rivers, outfalls and coastal establishments. Substances that are toxic, persistent and liable to bio accumulate are placed first in the list of priority substances to be phased out.

The amended protocol, under the new title of "Protocol for the Protection of the Mediterranean Sea Against Pollution from Land-Based Sources and Activities" covers not only the Mediterranean Sea itself, but also the entire watershed area within the territories of the riparian states draining into the Mediterranean Sea, the waters on the landward side of territorial boundaries as well as communicating brackish waters, marshes, coastal lagoons and ground water.

The signature of the amended LBS Protocol, is a milestone in the history of the Mediterranean Action Plan, as it sets the legal framework for a progression from land based pollution assessment to taking strong action on pollution control.

2.1. THE NEW APPROACH TO COMBAT LAND-BASED POLLUTION: THE STRATEGIC ACTION PROGRAMME (SAP)

One of the major breakthroughs in the Mediterranean countries' efforts to combat land-based pollution, which was prompted by the signature of the revised LBS Protocol, is the formulation and adoption by the Contracting Parties of a Strategic Action Programme (SAP) to Address Pollution from Land-based Activities [2].

The SAP is an action-oriented MED POL initiative identifying priority target categories of substances and activities to be eliminated or controlled by the Mediterranean countries through a timetabled schedule for the implementation of specific control measures and interventions. The SAP, adopted by the Contracting Parties in 1997, is the basis for the implementation of the Land Based Sources Protocol by the Mediterranean countries in the next 25 years. In addition, the SAP represents the regional adaptation of the principles and aims of the Global Programme of Action (GPA) to address pollution from land-based activities, adopted in Washington in 1995.

The key land based activities addressed in the SAP are linked to the urban environment, (particularly municipal wastewater treatment and disposal, urban solid waste disposal and activities contributing to air pollution from mobile sources) and to industrial activities, targeting those responsible for the release of persistent toxic and bioaccumulative substances into the marine environment, giving special attention to persistent organic pollutants (POPs). Also addressed are the release of harmful concentrations of nutrients into the marine environment (originating from municipal sewage, industrial waste water, agriculture and atmospheric emissions), the storage, transportation and disposal of radioactive and hazardous wastes and activities that contribute to the destruction of the coastline and coastal habitats.

For each of the above issues, the SAP indicates the main environmental targets (quantitative and qualitative) and describes the activities that need to be implemented at the national and the regional levels in order to gradually achieve them. In addition to interventions such as the use of more advanced industrial processes or the construction of sewage treatment plants by a set date, the SAP also stresses the need for the use of Best Available Technologies (BAT) and Best Environmental Practices (BEP) and the full compliance with existing Agreements and Conventions. Considering the large technical, scientific and institutional gaps still existing in the region as well as the large differences existing between countries, the SAP includes a programme of capacity building aiming, among others, at increasing the capacity of the countries to apply cleaner production technologies, BAT and BEP.

The SAP has a built-in scope for the review of detailed operational timetables at two to three year intervals, and a detailed work-plan and time-schedule for the first biennium has been prepared by the Secretariat and approved by the countries. The activities being implemented during the biennium 2002-2003 are creating the infrastructure for the implementation of the SAP and are essential to equip all the countries with the necessary tools (regional guidelines, strategies, plans and programmes for sharing technical information and advice, priority capacity building and preparatory public participation activities) that will allow them to eventually fulfil their long-term priority objectives under the SAP. The adoption of the SAP and the actual initiation of its implementation even before the entry into force of the amended LBS Protocol, is a clear indication of the determination of the countries to take concrete action to combat land-based pollution and at the same time contribute to maintaining and restoring marine biodiversity, safeguarding human health and promoting the sustainable use of marine living resources.

Shortly after its adoption, the SAP was recognised by the Council of the Global Environment Facility (GEF) as an important programme dealing directly with some of the major concerns relating to international waters. As a result of this recognition the GEF Council approved in 1998 a four-year Mediterranean GEF Project, started in January 2001, entailing a contribution of six million US$ for the realisation of a number of important groundwork activities of the Strategic Action Programme that are essential for the Programme's long-term success. One of the main expected outputs of the first years of implementation of the SAP is the formulation and adoption of National Action Plans specifically designed to tackle land-based pollution. The National Action Plans (NAPs) are intended to follow on from the adoption in each country of all the targets and activities of the nationally relevant components identified in the SAP. Funds from the Mediterranean GEF Project will be used to support national inter-ministerial committees in the development and implementation of NAPs in countries. The concrete implementation of country specific National Action Plans to combat pollution from land-based activities is the operational long-term output of the SAP.

One of the most important components of the SAP, which is expected to lead directly to a reduction in polluting inputs to the Mediterranean Sea from land-based activities, is the package of actions which accompany the evaluation of the impacts of pollution hot spots (Figure 3.) and the environmental audit of pollution-sensitive areas (i.e. those areas of natural or socio-economic value at risk of becoming future pollution hot spots) in the Mediterranean, taking into consideration their regional and trans-boundary significance.

Pre-investment studies are being conducted in the most important hot spots and detailed environmental assessment reports will be soon ready for the most important pollution sensitive areas from a regional perspective.

Figure 3. The distribution of the identified Mediterranean pollution hot spot

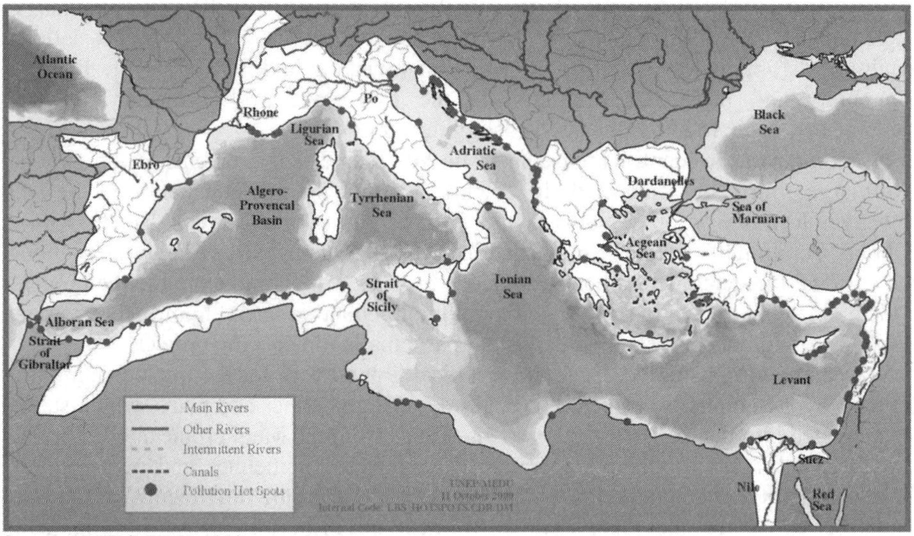

Source: UNEP/MEDU, 1999

This package of actions is expected to lead to investments by countries and donors in projects aiming at the elimination or reduction of trans-boundary pollution from the priority hot spots, as well as in environmental protection projects and comprehensive integrated management plans in the selected pollution-sensitive areas. GEF funds are being used for the preparation of pre-investment studies in GEF eligible countries.

2.2. THE DEVELOPMENT OF ECONOMIC INSTRUMENTS FOR THE SUSTAINABLE IMPLEMENTATION OF THE SAP

When the costs for the SAP remedial actions are considered (Table 1.), it becomes evident that the success of the SAP will largely depend on the sustainable financing of its individual components at a national level.

Table 1. Estimated costs of priority pollution remedial actions in the region

	Estimated cost (million USD)
Hot Spots	6,453.00
Sensitive Areas	195.25
Cities (air pollution, transport, solid wastes, etc.)	2,800.00
Capacity Building	13.00
National Action Plans	11.2
Clean Production	460.7
Monitoring and Enforcement	37.14
Information and Public Participation	2.98
TOTAL	**9,973.27**

Source: UNEP/MEDU, 1997

As a result, the Secretariat is developing administrative, legal and fiscal mechanisms for the sustainable financing of the SAP and is assisting the governments in implementing these mechanisms by adapting them to meet their national requirements.This involves setting priorities for financing and mobilizing the financial community and international donors. In the first stage pilot projects are being implemented in six Mediterranean GEF eligible countries.

In addition to the identification and testing of financial instruments at the national level, large efforts are being made in the framework of the implementation of the SAP objectives to mobilise other funds in the form of grants or donations. To this end, a "Donors Committee" is being established in the region, composed of large international and regional financial institutions and donor countries, which will examine the progress made by the countries in making specific pollution reduction interventions (for example, in the identified hot spots) and will consider participating in specific projects.

Setting the administrative, legal and technical groundwork for the implementation of the SAP by the countries is steered through regional guidelines and plans, which should be integrated into the National Action Plans. Funds from the Mediterranean GEF project have been allocated for their preparation. The selected guidelines and plans address those processes and activities for which the MED POL Programme identified the need for further assistance, including sewage treatment and disposal, disposal of urban solid wastes, industrial wastewater treatment and disposal, riverine and estuarine pollution monitoring and environmental inspection systems. A number of guidelines and plans are intended to strengthen the technical capacity of the countries to take on board the principles introduced by the amended LBS Protocol such as the application of clean technology and best environmental practice.

In addition to adequate financial and technical resources, combating pollution from land based activities also requires specific skills in areas such as environmental policy formulation and enforcement, scientific capability in the assessment of pollution, for example river pollution monitoring, as well as technical and managerial capabilities for the implementation of clean production techniques and environmentally sound technologies, for example the proper operation and maintenance of wastewater treatment facilities. The SAP makes provision for a series of regional training for trainers that aims to enhance the capacities of Mediterranean countries in the above fields as well as to assist countries in overcoming existing inadequacies. Funds from the Mediterranean GEF project support a number of training courses earmarked for 2001-2003. Modern training techniques are deployed and the training package delivered to the trainers at the end of regional courses includes transparencies and explanatory notes in hard copy and software form, prepared in a way that facilitates translation and desk top publishing in any of the Mediterranean languages.

MAP is actively involved in the distribution of information material on all its projects and carrying out public information campaigns and special activities involving the public in environmental protection. MAP has recently set up an ambitious international information and public awareness strategy, which will reach out to broader audiences such as consumers, the private sector and youth, using multilingual literature and modern dissemination methods such as the Internet. Funds from the Mediterranean GEF project will be used to organise a workshop in 2003 on the role the public and non-governmental organisations (NGOs) in particular can play in the implementation of the SAP.

2.3. APPLY CLEANER PRODUCTION

The amended LBS Protocol and the SAP state as one of the general commitments of the countries the obligation to take into account the best available techniques and the best environmental practice including clean technologies when adopting action plans and taking measures to control land-based pollution.

Through the cooperation of MED POL with the Clean Production Regional Activity Centre (CP/RAC) based in Barcelona, Spain, MAP is currently assisting businesses in applying cleaner production by giving priority to pollution prevention at source and the minimisation of waste flows. Businesses are encouraged to adopt these alternatives as a preferable strategy to end-of-pipe treatment. Over 20 tailor-made case studies on pollution prevention have been prepared and implemented by small Mediterranean industrial enterprises in Spain, Malta, Croatia, Turkey, Israel, Lebanon, Egypt, Tunisia and Morocco.

At the same time businesses are being informed of useful tools for the assessment of their industrial activity to detect potential opportunities for preventing and reducing pollution at source and for providing them with sufficient data to orientate their policy towards cleaner practices and technology that are technically and economically viable. One such tool is known as the Minimisation Opportunities Environmental Diagnosis (MOED). The CP/RAC has prepared a manual on the MOED containing a specific work methodology and guidance for experts who work in the environmental sector and intend to carry out the diagnosis for businesses in Mediterranean countries.

2.4. TRACKING THE POLLUTANTS

A priority ongoing MED POL initiative that stems from the commitments in the Strategic Action Programme is to encourage the Mediterranean countries to introduce public tracking and reporting systems of pollutants, known as Pollutant Release and Transfer Registers (PRTR). A PRTR is an environmental database or inventory of potentially harmful releases or transfers to air, water and soil as well as wastes transported off site for treatment and disposal. In addition to collecting data for PRTRs from stationary sources, industries in particular, PRTRs are also designed to include estimates of releases from diffuse sources such as agricultural and transport activities.

A first pilot MED POL PRTR project has been launched in Egypt, in cooperation with the Alexandria branch of the Egyptian Environmental Affairs Agency (EEAA) and ICS/UNIDO in Italy. Six industries in the Alexandria region have been reporting on specific industrial activities and chemical releases and making this information available to interested parties. The successful outcome of this project is expected to accelerate the launching of similar activities throughout the Mediterranean. A workshop will be held in 2003 to present the first results of the Alexandria project and to agree on the methodology to be used throughout the region.

In addition to continuing providing assistance to countries with the preparation of compliance monitoring programmes, the MED POL Programme is introducing inspection and legislation enforcement initiatives to complement the actions for pollution control in the Mediterranean. MED POL has set-up an informal regional network aimed at creating contacts and the exchange of information with regional environmental protection professionals and other networks involved in compliance. The International Network for Environmental Compliance and Enforcement (INECE) is cooperating with MED POL in the preparation of guidelines for environmental inspectorates in the Mediterranean to assist countries in checking compliance with nationally adopted authorisations and regulations.

To optimise the campaign against land-based pollution and to thoroughly integrate environmental concerns with as many aspects of social and economic development in the Mediterranean as possible, MED POL is working closely with the Mediterranean Commission for Sustainable Development (MSCD). Set up as an advisory body to MAP in 1996, the MSCD is a think-tank for the promotion of sustainable development policy in the Mediterranean, making recommendations to the Contracting Parties on future actions for the promotion of sustainable development in areas such as agriculture, industry, energy, transport, tourism and urban management. "Agriculture and rural development", "Desertification and soil erosion", Urban waste management", Energy and transport", "Natural risks" and "Regional cooperation" are examples of some recent themes on the agenda of the MSCD.

2.5. THE IMPLEMENTATION OF THE SAP AS A CONTRIBUTION TO ACHIEVING SUSTAINABLE DEVELOPMENT

The SAP is an ambitious undertaking, spanning a lengthy period of 25 years and addressed to countries with different degrees of socio-economic development, technical, scientific and administrative skills, different cultural values and environmental priorities. To increase the prospects for the success of the SAP, a mechanism has been set up from the onset which covers issues such as the nature of the body that will coordinate the SAP activities at the national level, the effective conveyance and assimilation of all the "support structures" of the SAP at the national administrative level, the successful dissemination and assimilation of the tasks by the local stakeholders in each country, as well as the nature of the body which will monitor the progress of all the activities that are being carried out in each country and of the body which will evaluate the outputs. In addition, the SAP operational plan tackles important practical issues for achieving the financial sustainability of the SAP in the long term. Equally importantly, the operational plan provides detailed instructions for the countries on how to address the technical concerns that are being raised in the deliberations leading to the updating of the SAP. As an example, after providing detailed instructions to the countries on how to calculate the national budgets for individual targeted pollutants and the actual pollutant reductions required on a national basis in order to reach the targets stated in the SAP, the process of estimating the total load of pollutants being discharged into the sea has actually started in many countries. The operational plan also provides directions for the identification of the baseline values against which the pollutant reductions should be estimated. These are just some of the crucial issues that are being tackled by a SAP implementation operational plan that has been developed by MED POL and adopted by the Contracting Parties in November 2001 in Monaco [3].

2.6. THE REDUCTION OF INDUSTRIAL POLLUTION: THE PROCESS HAS STARTED

When elaborating the SAP operational details, the Secretariat and the Mediterranean experts were faced with a real challenge to find the appropriate strategy to translate the regional binding commitment to reduce the releases of pollution from industrial sites into a package of realistic actions for the Mediterranean countries.

It is in fact worth stressing that the SAP is being implemented in a region where social and cultural differences are prevailing between countries, a region where 75% of the countries belong to the developing world with heterogeneous environmental policies and priorities, a region that considers economic development as the first priority to improve the quality of life.

In view of the above, in proposing a methodology for the reduction of pollutant releases, should the Mediterranean countries adopt a "flat rate' or "differentiated" approach as the basis for the implementation of the SAP and the compliance with their commitments?

At first glance, the concept of "differentiation" in its different forms of application based on volumes of releases, volumes of reduction and on the cost of reduction, would seem to be preferable. This would partially allow the national situations of every Mediterranean country regarding their releases of pollutants into the Mediterranean and their relative shares in the degradation of the marine environment, to be taken into account.

However, as a result of a complex calculation of the different criteria of differentiation based on releases/inhabitant, GDP/inhabitant and releases/GDP, it was concluded that at present this approach cannot be adopted in the actual Mediterranean context, mostly because it is not quantifiable or traceable either at national or regional levels. In the present situation, with data and information still scarce, it would be almost impossible to reasonably estimate the share of each Mediterranean country in the degradation of the Mediterranean region and its environment. In addition, one of the results of this exercise was that the costs associated with the reduction of releases were not in favour of the developing countries in the region.

The "flat rate' approach, which was finally agreed upon, consists of adopting for all the Mediterranean countries the same rate of reduction of releases that is indicated under the SAP provisions. Obviously, at the national level, an "internal flexibility" instrument was introduced and, as a result, a different rate of reduction could be imposed among stakeholders according to the individual records (e.g. to favour industries that have already reduced their releases to acceptable levels). In other words, any country may transfer internally release targets between different activities generating the same targeted pollutants. In this context, the estimation exercise carried out by the Secretariat shows that the adoption of the "flat rate" approach would enable the Mediterranean countries to fulfil their SAP commitments in a much more reliable and equitable way.

The adoption of the "flat rate" approach requires the definition of the basis on which the Mediterranean countries would achieve and track the reductions in order to comply with the SAP binding commitments. It was decided to consider the year 2003 as the base year for the establishment of the national "Baseline Budget of emissions/releases" for each substance targeted in SAP. The methodology was prepared and was based, whenever data was not available, on emission factors commonly used. The Secretariat has organized a number of missions and meetings to assist the countries in this challenging initiative that will represent a turning point in the history of pollution control in the region.

The adoption of the SAP operational strategy by the Contracting Parties in Monaco in November 2001 and the implementation of its first actions are indeed creating new and promising opportunities in the region; for the national environmental administration, to take in hand the effective control of pollution, for the industries, to improve the quality of production and sustainability, for the local population, to track the trends of the quality of their environment and finally for the MAP, to participate actively in the regional process of the environmental management of the Mediterranean region. If, as expected, the process is successful, the region will witness a gradual reduction of pollution as a result of more concrete investments from national budgets, more environmental awareness from the industry and more interest from donor countries and institutions.

The prospects for a healthier Mediterranean are therefore positive: during 2003 the pollutants discharged into the sea will be qualified and quantified and during 2004 the countries will prepare National Action Plans for the reduction of releases; the plans are expected to contain all the necessary elements to ensure long-term implementation of the Strategic Action Programme (SAP), i.e. technical, scientific, socio-economic and financial. They will commit the countries to comply with the objectives and targets of the SAP and will describe specific action and the expected outcomes. A comprehensive process, therefore, that is expected to result in a gradual but steady reduction of pollution inputs into the sea and that takes into account the long-term aim of the SAP to stop polluting inputs by the year 2025. This process, which will be supported by the Mediterranean Action Plan through a comprehensive capacity building programme, will substantially contribute to the achievement of sustainable development in the region.

References

1. Meybeck M., and Ragu, A. (1997) *River discharges to the oceans: An assessment of suspended solids, major ions and nutrients*, UNEP, Environmental Information and Assessment, April 1997.
2. UNEP (1998) The Strategic Action Programme to Address Pollution from Land-based Activities. *MAP Technical Report Series 119,* UNEP, Athens.
3. UNEP (2001) Operational Document for the implementation of the Strategic Action Programme to Address Pollution from Land-based Activities *(UNEP(DEC)MED IG.13/4)*.
4. UNEP/MAP (1997) Draft Transboundary Diagnostic Analysis for the Mediterranean Sea *(TDA/MED) (UNEP(OCA)MED IG.11/inf.7)*
5. UNEP/WHO (2000) Municipal Waste Water Treatment Plants in Mediterranean Coastal Cities. *MAP Technical Report Series 128,* UNEP, Athens

Editor

Antonio Marquina: PhD., Chair for International Security and Cooperation and Professor of international relations at Complutense University in Madrid (CUM) since 1985, founder and director of the Research Unit on Security and International Cooperation (UNISCI) at CUM. He has been the director of 14 research projects on European, Mediterranean, North African, Turkey and Caucasus security and on Spanish foreign and security policy. Since 2002 he is a member of the french Conseil Economique de la Défense. From 1996 to 1997 he was the first president of STRADEMED, a network of research institutes working on security, defence and development issues in the Mediterranean. Publications: He is author, editor and co-author of many books and reports in English and Spanish, among them several on CBMs in the Mediterranean. Books on Mediterranean Security issues include: (Ed.): *El Flanco Sur de la OTAN*, (1993); (Ed.): *El Magreb. Cooperación, Concertación y Desafíos*, (1993); (Coed.): *Nuclear Non-proliferation and the Mediterranean*, 1994; (Coed.): *Confidence Building and Partnership in the Western Mediterranean. Tasks for Preventive Diplomacy and Conflict Avoidance*, (1994); (Ed.): *Confidence Building and Partnership in the Western Mediterranean. Issues and Policies for the 1995 Conference*, (1995); (Ed.): *Elites and Change in the Mediterranean*, (1997); (Ed.): *Flujos Migratorios Norteafricanos hacia la Unión Europea. Asociación y Diplomacia Preventiva*, (1997); (Ed.): *Mutual Perceptions in the Mediterranean*, (1998); (Coed): *Political Stability and Energy Co-operation*, (2000); (Coed): *The definition of the Mediterranean Space*, (2001); *España e Irán: Globalización, Cuestiones Regionales y Diálogo Cultural*, (2002).

Authors

Shaden Abdel-Gawad is Vice chairperson of the National Water Research Center of Egypt (NWRC), affiliated to the Ministry of Water Resources and Irrigation. She obtained her M.Sc in 1980 from Memorial University of Newfoundland (Canada) and her Ph.D. in 1985 from University of Windsor (Canada) in the area of environmental engineering. She has a wide experience in water quality management and environmental protection with more than 100 technical published papers. She has supervised and managed several foreign funded projects and she is also an active member of national and international water organization such as ENCID, ICID, IWRA, IPTRID.

Mohamed Ait Kadi is a Professor at the Institut of Agronomy and Veterinary Medecine Hassan II in Rabat. He is also visiting Professor at the Institut of Mediterranean Agronomy in Bary (Italy) Prof. Ait Kadi is author of numerous publications in the field of rural development, irrigation and water management. He is President of the General Council of Agricultural Development belonging to the Ministry of Agriculture Rural Development and Forestry (Morocco) where he has held several positions.

A. Marquina (ed.), Environmental Challenges in the Mediterranean 2000-2050, 377–381.
© 2004 *Kluwer Academic Publishers. Printed in the Netherlands.*

Hassan Abdelhamid is an Assistant Professor at the University of Ain Chams in Cairo (Egypt) where he obtained his M.A. in Law. He holds a Ph.D in Law. He belongs to several scientific organizations such as Réseau Droits Fondamentaux (Paris X), The Committee for the Mediterranean Studies on Social Science (Sassari University, Italy). He has several publications in French, Arab and Italian.

Nicolas Alexandratos is a D.Phil (Economics,University of Sussex, UK) and former Chief of Global Perspective Studies Unit of FAO. Currently he works as a consultant of FAO. Hi main areas or work are developing countries, food security, trade, natural resources and sustainability and national and international policies. He has several publication the most prestigious scientific journals such as, *Food Policy* or *Options Méditerranéennes.*

Sabah Benjelloun is a Professor at the Department of Food Science and Nutrition of the the Institut Hassan II (Morocco) She obtained her B.A from the Iowa State University (USA) and her Ph.d from Tufts University School of Nutrition (Medford, USA) She has several articles in the most important scientific journals, such as *Public Health Nutrition, Cahiers Agriculture, Ecology of Food and Nutrition* or *Options Méditerranéennes*

Manuel de Castro studied Atmospheric Sciences at the University Complutense of Madrid where he obtained his Ph.D in 1981. He is Professor of Meteorology at the Faculty of Environmental Sciences at the University of Castilla la Mancha (Toledo, Spain) He has published more than 20 peer-review papers and several book chapter on radiative transfer, air pollution and meteorological and climate regional modelling. He is also head of MOMAC group and the Spanish Scientific representative for the Woorld Climate Research Program (WCRP)

Francesco Saverio Civili has been working for the UNEP since 1977. After a few years of experience at the University of Rome, Dr. Civili, was instrumental for the launching of the UNEP Mediterranean Programme and in particular for the formulation and implementation of the Marine Pollution Assessment and Control Programme (MED POL) He has written many articles and technical and scientific papers published by international journals, and the internal UNEP Reports Series.

Juan Díez Nicolás is a Professor of Sociology at the Complutense University of Madrid. He has published more than 25 books, *inter alia El Libro Blanco sobre la Enfermedad de Alzheimer y Trastornos Afines, La Inmigración en España: Una Década de Investigaciones* or *La Voz de los Inmigrantes.* He held political responsibilities during the full period of the political transition in the governments of Adolfo Suarez. He is President of ASEP and member of the European Academy of Sciences and Letters. The European Commission appointed him as a member of the High-Level Advisory Group on Dialogue between Peoples and Cultures of the Mediterranean.

Shlomi Dinar is a doctoral student at Johns Hopkins-SAIS working on the topic of conflict and negotiation over transboundary water issues. His research is primarily concerned with the effects of geographical, political and economic variables on the make-up of international water treaties. Mr. Dinar co-edited a special number of International Negotiation: A journal of theory and Practice (vol.5, number 2, 2000) titled "Negotiating in International Watercourses: Water Diplomacy, Conflict and Cooperation" and, most recently, authored an article titled "Water, Security, Conflict and Cooperation" in SAIS Review (Vol. 22, number 2, 2002)

Petra Döll obtained her PhD in 1996, (Technical University of Berlin) is a senior researcher and lecturer at the Center for Environment System Research at the University of Kassel. She has 15 years of experience in the mathematical modelling of water issues from local to global scales, mainly focusing on the sustainable use of water under conditions of global change. Petra Döll has authored approximately 25 peer-review papers in journals like Water Resources Research, Journal of Hydrology and Regional Environmental Change, and approximately 35 conference papers. She is member of the American Geophysical Union, the International Association of Hydrological Sciences as well a of the German board UNESCO's International Hydrological Program.

José María García Álvarez-Coque is a Professor of Economics and Agricultural Policy at the Technological University of Valencia (UPV) where he is also Head of the Department of Economics and Social Science. He has conducted a number of research projects on international trade of agricultural products. Chairman and/or consultant/ researcher in different European and International groups and/or institutions. Since 2001, Chairman of the Spanish Association of Agricultural Economists.

Koray Haktanir obtained his PhD at the Faculty of Agriculture of the Ankara University in 1973. He is a senior lecturer of the Soil Science Department at the Ankara University. Author and co-author of a number of books and articles on Soil, Environmental Issues, Sustainable Land Management and Desertification. He is President of the Soil Science Society of Turkey (SSST).

Rusen Keles is a Professor of Urban Studies and Local Government at Eastern Mediterranean University and Ankara University. He has degrees in Public Administration, Political Science and Law. He served as Dean of the Faculty of Political Science, Ankara University, and Director of the Centre for Urban Studies. His major and more recent publications are *Policies of Environment* (Imge, 2002) and *Introduction to Environmental Law* (Imge, 2002). He served as an advisor to the UNCHS (UNEP) and the Council of Europe.

Hesham Makhlouf, Professor of Demography at the Institute of Statistical Studies and Research of Cairo University (I.S.S.R.), obtained his PhD in the City University, London, 1977. He is the Director of the Cairo Demographic Center and Chairman of the Egyptian Demographers Association and the Egyptian Volunteers Association for Environment & Population. He has been also Chief of the Population Section in the Scientific Research Academy of the Ministry of Higher Education.

Jean Margat graduated in Science and Geology. He is vice-chairman of the association "Blue Plan for the Mediterranean Sea" and the Mediterranean Institute of Water. He has worked at the Office National des Irrigations in Morocco and the Bureau de Recherches Géologiques et Minières in France. Consultant for several international organizations (UNDP, UNESCO, FAO, OECD, CCE, World Bank, Observatoire du Sahara et du Sahel). Author of a number of publications dealing with the assessment and the management of water resources, cartography, terminology and economy of water.

Teresa Mendizabal, PhD in Physics, is a Research Professor at the Spanish Council for Scientific Research and Scientific Adviser of its President. She is also Vice-President of the Club of Rome's Spanish Chapter. She is co-author of *Perspectives on desertification: Western Mediterranean* (1998) and *Population and Land-use Changes: Impacts on Desertification in the Southern Europe and in the Maghreb* (2003). She is co-editor of the book: *Desertification and Migrations* (1995).

Mamdouh Nasr is a Professor of Agricultural Economics at the faculty of Ain Shams University, deputy director of the Centre for Economic and Development Studies and Senior Fellow at the Centre for Development Research at Bonn University. He published several works on development and environmental economics in English and Arabic, being the last work: *Assessing Desertification in the Middle East and North Africa: Policy Implications*, in Brauch, H., Liottta, P., Marquina, A. and others (2003) Security and Environment in the Mediterranean, Springer.

Jean Palutikof is a Professor in the School of Environmental Sciences, and co-Director of the Climatic Research Unit at the University of East Anglia, Norwich, UK. She worked on the MEDALUS project (Mediterranean Desertification and Land Use) between 1991 and 1999 and currently coordinates the MICE project (Modelling the Impacts of Climate Extremes). She is author and co-author of a number books and articles dealing with climate change and development of regional scenarios of climate change and associated impacts. A recent relevant reference would be *Analysis of Mediterranean climate data: measured and modelled* in Bolle, H.J. (2003) Mediterranean Climate: Variability and Trends, Springer

Boris A. Portnov is an Associate Professor at the Department of Natural Resources & Environmental Management of the University of Haifa, Israel. He has authored or edited five books and more than a hundred articles on various aspects of urban and regional development. The most recent books are *Regional Inequalities in Small Countries* (co-edited with Daniel Felsenstein and still under preparation) and *Urban Clustering: The Benefits and Drawbacks of Location* (authored with Evyataar Erell and published by Ashgate Publishers)

Juan Puigdefábregas, PhD in Biological Sciences, is a Senior Research Scientist at the Spanish Council for Scientific Research (CSIC) where heads the Desertification and Geo-Ecology Research Group at the Estación Experimental de Zonas Áridas. Scientific adviser on Environment to Spanish Departments of Research, Environment and Foreign Affairs. He has published a hundred of articles in scientific journal and books on the anthropogenic effects on land degradation from a global change and desertification perspective.

Halima Slimani is a lecturer at the University of Algiers, Faculté des Sciences Biologiques de l'Université des Sciences et de la Technologie Houari Boumédiène. She obtained her Master in Environment in 1990 and holds a PhD since 1998. She has some publications on environment, soil and desertification such as *Effects of Grazing on Soil and Desertification: a View from the Southern Mediterranean Rim*, in Papanastasis, V.P., (1998) Ecological Basis of Livestock Grazing in the Mediterranean, Ecosystems.

Manuel Vázquez has worked at the Instituto Astrofísico de Canarias since 1970. He obtained his Ph.D at La Laguna University. His research areas are solar convection, sunspots, solar-terrestrial and astrobiology. He has published several articles in the most prestigious scientific journals such as *Astrophysical Journal, Astronomy and Astrophysics* or *Solar Physics*. He is also editor of the proceedings and member of the respective SOC and LOC in seven international conferences. He has held several international charges like: Spanish representative at the Council of the LEST Foundation (1986-1995), Spanish National Commission for Astronomy (1993-2000) or Head of the Research Department from (1986-1991).

Subject Index

-A-

Abstraction, 166, 167, 178, 182, 188, 190, 209, 210
Adaptation, 5, 11, 18-24, 71, 75, 88, 101, 158, 168, 259, 356, 369
ACP, 295
Adana, 346, 347
Adige, 367
Aerosols, 30, 41, 77, 79, 88
Afforestation, 127, 128
Agglomerations, 331, 338, 339, 342, 363
Agricultural, 49, 50, 52, 59, 118, 123, 125, 127, 128, 130-133, 135, 137, 139, 140, 142, 144, 145, 147-151, 153-158, 160, 162, 164, 165, 167, 169-171, 176, 177, 180, 181, 184, 189, 190, 192, 195, 196, 199, 202, 203, 206, 207, 209, 211, 213-215, 220, 221, 226, 227, 239, 256, 275, 283, 284, 287, 292, 294-296, 300-302, 307, 308, 315-317, 319-328, 339, 344, 365, 367, 373
Agriculture, 49, 61, 62, 95, 111-113, 123, 124, 126-129, 131, 132, 136, 137, 144, 146, 147-149, 151, 153, 155, 157, 158, 161-163, 165-168, 170, 171, 175, 177, 180, 181, 183, 184, 188-190, 193, 197, 198, 200, 208-210, 214, 215, 220, 226, 242, 274, 275, 277, 283, 284, 288, 294-297, 299, 300, 317, 319, 323-327, 365-367, 369, 374
Ain, 109, 241
Albania, 235, 237, 266, 267, 269, 270
Alexandria, 188, 191, 333, 334, 373
Algeria, 93-96, 102-106, 113, 175, 266-270, 278, 283, 284, 293, 298, 301, 303-307, 309, 311-313, 315, 319-321, 323
Aliakmon, 366, 367
Alkalinisation, 111
Anatolia, 140, 142, 146, 149, 150, 346, 347, 358
Ankara, 139, 154, 346, 347, 349, 353, 361
Anthropogenic, 29, 30, 41, 42, 62, 64, 65, 87, 90, 93, 142, 156, 167, 363
Aqaba, 209, 210, 223, 227, 228
Aquifer, 171, 188, 202, 207, 209, 210, 213, 223, 225
Atmosphere, 30, 35, 39, 43, 64, 69, 76, 79, 83, 143, 167, 354
Ashdod, 221, 222
Athens, 234, 243, 244, 297, 363, 376, 293, 297

-B-

Attica, 234
Austria, 250-252, 254, 255, 257, 266, 267, 269, 270, 274, 275, 278, 293, 297
Axios, 366, 367
Azores, 38, 63

Balkans, 57, 240
Barcelona, 234, 319, 321, 323, 326-328, 363, 368, 369, 373
BCM, 187
Belarus, 249
Belgium, 266-270, 274, 278
Biodiversity, 123, 139, 143, 152, 366, 370
Biosphere, 6, 18, 76
Birth, 248, 249, 251-253, 258, 261, 269, 271-277, 280, 312, 331, 333, 337, 338, 346
Black Sea, 140, 141, 347, 348, 349, 358
Brackish water, 132, 188, 209, 211, 215, 218, 220-222, 369
Brenta, 367
Bursa, 346, 347

-C-

Cairo, 109, 122, 188, 191, 203, 265, 331, 333-344
Camargue, 241
Camp David, 217, 225, 230
Canals, 191
CAP, 295, 319, 323-328
Catalonia, 234
Case study, 93-96, 108, 175
CBR, 268, 270-273, 275, 276, 279
CDR, 268, 270-273
Cereal, 94-96, 103, 120, 125, 153, 158, 283, 284, 287-289, 292, 295, 301, 302, 306, 313, 314, 316, 321, 325
Challenge, 8-12, 14, 16, 17, 52, 156, 175, 177, 180-182, 187, 193, 196, 202, 253, 261, 262, 342, 375
China, 157, 247, 289, 296, 297
City, 130, 134-136, 138, 158, 177, 180, 183, 191, 223, 331, 332, 334, 338-346, 349, 350, 353-360, 364, 365
Climate change, 7, 8, 10, 19-21, 24, 30, 33, 34, 36, 37, 40, 43, 44, 47, 49, 52-62, 64, 65, 69, 75, 76, 78, 79, 83, 85, 87, 88, 106, 107, 113, 139, 142, 143, 151-153, 155, 156, 165-167, 171, 172, 239, 241-244
Climate model, 19, 43-46, 48, 55, 56, 59, 61, 63-68, 71, 75, 76, 78, 88-90
Coastal aquifer, 241
Common Agricultural Policy, 288, 295, 319, 328

A. Marquina (ed.), Environmental Challenges in the Mediterranean 2000-2050, 383–390.
© 2004 *Kluwer Academic Publishers. Printed in the Netherlands.*

-D-

-E-

-T-